Astrophysics and Space Science Proceedings

Volume 44

More information about this series at http://www.springer.com/series/7395

Gerard Gómez • Josep J. Masdemont
Editors

Astrodynamics Network AstroNet-II

The Final Conference

Springer

Editors
Gerard Gómez
Institut d'Estudis Espacials de Catalunya
and Universitat de Barcelona
Barcelona, Spain

Josep J. Masdemont
Institut d'Estudis Espacials de Catalunya
and Universitat Politècnica de Catalunya
Barcelona, Spain

ISSN 1570-6591 ISSN 1570-6605 (electronic)
Astrophysics and Space Science Proceedings
ISBN 978-3-319-23984-2 ISBN 978-3-319-23986-6 (eBook)
DOI 10.1007/978-3-319-23986-6

Library of Congress Control Number: 2016945295

Printed on acid-free paper

This Springer imprint is published by Springer Nature
The registered company is Springer International Publishing AG Switzerland

Preface

From June 15 to June 19, 2015, the *AstroNet-II International Final Conference* was held in Tossa de Mar. This conference was one of the last milestones of the Marie-Curie Research Training Network on Astrodynamics "AstroNet-II" that was funded by the European Commission under the Seventh Framework Programme. The network put together mathematicians, engineers and astronomers from European universities, governmental agencies and industry with the objective of training through research activities.

The AstroNet-II training programme focused on projects for early-stage researchers (ESRs) and experienced researchers (ERs) that crossed the traditional boundaries between mathematics, engineering and industry, ensuring that they obtained an interdisciplinary and multi-sectorial overview of the field. This was supported by an extensive programme of schools, workshops, tutorials and internships and by a close collaboration between academia and space companies. In this framework, the main goals to be achieved were:

- To introduce the fellows of the network to a range of astrodynamic concepts and problems, as well as to the relevant new mathematical theories and techniques
- To develop their technical expertise and to train them to conduct research, to collaborate and to communicate their results
- To deepen and broaden the knowledge and skills of the fellows working in the areas of astrodynamics, dynamical systems, control theory and numerical methods
- To offer in the fellows' project opportunities for doing research in private companies, international organisations and academia
- To provide them also with the complementary communications and project management skills that are needed for a successful career

Nowadays space missions are being required to fulfil many different types of functions and, as a consequence, are becoming increasingly more complex. Furthermore, a number of different spacecraft architectures, such as constellations, formations, tethered spacecraft or solar sails, have been proposed for some specific objectives. To achieve their functions and goals, future missions sometimes require new, faster and unusual kinds of trajectories. They are very often also merged with precise attitude control dynamics whose determination, in many cases, raises new major dynamics and control questions.

A broad range of mathematical objects, theories and techniques are needed for these new concepts and applications. The research topics of the network included innovative new methods for designing spacecraft trajectories and controlling their dynamics. The scientific programme, centred on a number of key astrodynamical objectives of current interest to space agencies and industry but also of innovative mathematical interest, was mainly divided in three sections: Trajectory Design and Control, Attitude Control and Structural Flexibility of Spacecraft and Formation Flying. Particular emphasis was placed on optimising trajectories and control to minimise fuel usage to extend mission ranges. Many times this is achieved by maximising the use of "natural dynamics" and employing ideas and techniques from dynamical systems theory.

Since January 2012, 18 young researchers were involved in the network, working on the above-mentioned astrodynamic problems and the interface between mathematics and astrodynamics which provided an exciting research arena from the point of view of both applications and mathematics. When the network came to its end, it was time to communicate to an international specialised audience some of the work carried out during those four years. Together with the presentations of the fellows, we had also the opportunity to hear about the work of other young scientists in the field, as well as a number of invited talks delivered by relevant people in astrodynamics. This present book of proceedings contains some of the contributions in the conference, and it is intended to be a summary for young scientists and researchers interested in the field.

We would like to thank all the people who made the network, the conference and this book possible, starting from the ESR and ER fellows: Albert Caubet, Marta Ceccaroni, Luca de Filipis, Mohammad M. Gomroki, Elisabetta Iorfida, Pawel Kicman, Junquan Li, Pedro J. Llanos, Zubin P. Olikara, Rocío I. Páez, Fabrizio Paita, Claudiu-Lucian Prioroc, Leon Simpson, Stefania Soldini, Andrea Turconi, Willem van der Weg, Alexander Wittig and Mattia Zamaro. Next are their scientist in charge and associated partners: Franco Bernelli (Politecnico di Milano), James Biggs (Strathclyde University), Juan Luís Cano (Elecnor Deimos), Alessandra Celetti (Universita di Roma Tor Vergatta), Jordi Fondecaba (Thales-Alenia Space), Jesús Gil (GMV), Mariela Graziano (GMV), Steve Greenland (Clyde Space), Steve Kemble (Astrium Limited Satellites), Ugo Locatelli (Universita di Roma Tor Vergatta), Andrzej Maciejewski (University of Zielona Gora), Vincent Martinot (Thales-Alenia Space), Seppo Mikola (University of Turku), Phil Palmer (University of Surrey), Maria Przybylska (University of Zielona Gora), Mark Roberts (University of Surrey), Mariano Sánchez (Elecnor Deimos), Johannes Shoenmakers (ESA/ESOC) and Ozan Tekinalp (Middle East Technical University). Last but not the least are the other invited speakers who also participated in this final conference: Miguel Belló (Elecnor Deimos), Fabrizio Bernardi (SpaceDyS), Franco Boldrini (Selex ES), Kathleen Howell and Natasha Bosanac (Purdue University), Chris Brunskill (Catapult Satellite Applications), Mateo Ceriotti (University of

Glasgow), Pierluigi di Lizia (Politecnico di Milano), Elena Fantino (Universitat Politècnica de Catalunya), Ariadna Farrés (Universitat de Barcelona), Elisabet Herrera-Sucarrat (Mathworks), Tomasz Kwiatkowski (University of Zielona Gora), Martin Lara (Universidad de La Rioja), Thomas Peters (GMV), Marco Sansottera (University of Milan), Hanspeter Schaub (University of Colorado) and Francesco Topputo (Politecnico di Milano). Finally we want also to thank Pilar Montes, Anna Bertolin and Eva Notario (Institut d'Estudis Espacials de Catalunya) for their administrative support and the care they had managing all the events.

During the conference, we also wanted to pay tribute to the astrodynamicist José Rodríguez-Canabal who died aged 68 in 2013. For their contribution to this special session, we want to thank his family, Eugenia, Marta, Pablo and Adrián, and also friends who approached us to his scientific and personal career: Walter Flury (ESA/ESOC), Miguel Belló (Elecnor Deimos), Vicente Companys (ESA/ESOC), Guy Janin (ESA/ESOC), Javier Jiméenez (Univ. Politécnica de Madrid), Jaume Pagès (Universia) and Carles Simó (University of Barcelona) who chaired the session. The reader will also find a summary of their presentations to the special session edited in the header of the present book.

Barcelona, Spain — Gerard Gómez
Barcelona, Spain — Josep J. Masdemont

Contents

José Rodríguez-Canabal Memorial (1945–2013)

Gerard Gómez and Josep J. Masdemont

1 Introduction

Dr. José Rodíguez-Canabal was an aerospace engineer that graduated at E.T.S.I. Aeronáuticos of the Universidad Politécnica de Madrid. In 1972, he finished his Ph.D. in control theory at the University of Southern California under the supervision of Prof. R.S. Bucy. After completing his PhD, he moved to Toulouse, where he worked at the Laboratoirte d'Automatique et ses Applications Spatiales of the CNRS during a short period of time; in 1973 he got a position at ESOC, from which he retired in 2010.

During almost 37 years of his professional career at ESOC, Dr. Rodríguez-Canabal worked in the mission analysis of many of the most relevant missions launched or participated by ESA, including Exosat, Cluster II, Double Star, Giotto, Lisa, Rosetta, SOHO and Venus Express.

The AstroNet-II International Final Conference devoted a special session to his memory, where there took part, together with his closest family and some of his best friends and colleagues at ESOC. All of them recalled not only his high professionalism but also his kindness and friendly personality. The memorial included the following lectures:

- Walter Flury (ESA/ESOC): "35 years of mission analysis at ESOC"
- Miguel Belló (Elecnor Deimos): "Lessons learned"
- Vicente Companys (ESA/ESOC): "The Rosetta mission: Flight operations"
- Guy Janin (ESA/ESOC): "The office no. 411"
- Javier Jiménez (Univ. Politécnica Madrid): "José Rodríguez-Canabal: The early years"
- Jaume Pagès (Universia): "José Rodríguez-Canabal: A personal approach"
- Johannes Schoenmaekers (ESA/ESOC): "Interplanetary trajectory design for Rosetta and Solar Orbiter"

This paper is a short summary on the contents of the above lectures in the memorial and has been prepared by the editors of the present book. Although it is difficult to separate the human aspects from the professional side, especially in a such rich personality like the one of José, for this note we have preferred to do so, in order to highlight some of his many qualities.

2 The Human Approach

Pepe, which is how all his friends called him, was born in 1945 in Plasencia, a beautiful city in western Spain. It was in the midst of the post-civil war era, during a period of autocracy with very few foreign relations. He was the fourth son of a family of five (four boys and a girl).

He was an intelligent man with great intellectual curiosity, especially in the areas of technology where he had truly unique skills. He enjoyed a brilliant academic career; according to one of his professors, Dr. A. Liñan, and his Ph.D. supervisor, Dr. Richard S. Bucy, this was due to his dedication, curiosity, open mind and desire to learn, which all of them are required qualities to succeed in science.

Exploring new possibilities was something Pepe did his whole life, both alone and in company, to face amazing challenges and intergalactic dreams. He was musically gifted; when he was young, he used to play the harmonica by ear and as an adult he took clarinet lectures. He also explored gardening, art, cooking, organising parties and trips, librarianship, document management and sailing, although after a trip from Barcelona to Mallorca by boat the captain diploma ended up in a drawer (Fig. 1).

Fig. 1 Exploring new possibilities was something Pepe did his whole life

Fig. 2 His relationship with Eugenia was the most important thing in his life

In 1963, he moved to Madrid to study aeronautical engineering. This was a very significant event for Pepe with consequences throughout his life: he found Eugenia there. They met, fell in love, got married and left to the USA. His relationship with Eugenia was the most important thing in his life. As one would expect, the initial partnership of Eugenia and Pepe grew. Their daughter Marta was born in California and their son Pablo in Germany. His grandchildren, Adrián, Nicolás and Hernán, undoubtedly gave him the greatest pleasure over the last few years (Fig. 2).

Pepe wanted to share his life and interests with many friends, and especially with ESOC fellows and young people, to which he delivered help, support and guidance. He always listened to their concerns and he did everything in his hands to be there for them and to help in key moments. He was always there to share their worries, either at the office or at home. Together with Eugenia, they worked for many years to improve schooling and to help the children of immigrants to go to a gymnasium. Every year, he organised a group day out with the ESOC students, which ended with a barbecue at his house, all with the aim of socialising together and gaining their trust. He did this with his utmost and absolute conviction and devotion (Fig. 3).

And as a last personal trait, Pepe was a committed pro-European. The idea won him over and he was convinced that today's generations have a huge responsibility to build Europe as a place to share and where the errors of the not-so-distant past shall not be repeated. He was proud to have done his bit working with a European transnational organisation. He rejected offers to move to the USA on a number of occasions, and one of the underlying reasons for these decisions was the personal connection he had built with the idea of Europe.

Fig. 3 Pepe shared his life and interests with many friends, ESOC students and young people

3 The Professional Approach

As it has already been mentioned, during his professional career, José took part in the mission analysis of many of the most relevant missions launched or participated by ESA. Here we just remember some of his contributions to SOHO, LISA, Cluster and Rosetta and what we all learned from his approach (Figs. 4 and 5).

Launched in December 1995, the SOHO mission is a joint ESA/NASA project to monitor the solar activity from a halo orbit in the L_1 libration point of the Earth-Sun system. The main activities of José for SOHO were the mission analysis, the definition and computation of the nominal trajectory and the definition and design of the orbit maintenance. José developed a very clever way to numerically generate an accurate nominal halo orbit. He used Richardson theory to obtain a first approximation together with very small velocity manoeuvres at the ecliptic plane crossings. He also implemented a cheap and efficient station-keeping procedure for this new kind of unstable orbits. As Guy Janin recalled during the memorial, all these novel questions could only be solved in a so imaginative and brilliant way by the occupant of the office no. 411 of ESOC.

The Laser Interferometer Space Antenna (LISA) mission is a joint ESA/NASA project to develop and operate a space-based gravitational wave detector by laser interferometry. For this purpose, a fleet of three probes are located at the vertex of a triangle in an Earth-like orbit with 20° phase delay and 60° inclined with respect to the ecliptic. The transfer of the three satellites to a trailing orbit, 20° behind the Earth and targeting the LISA operational configuration (with a distance between satellites of 5×10^6 km), is not an easy problem. It was solved as an optimal control problem, after iteratively computing several optimal solutions in a sequential process and augmenting at each step the complexity of the problem (Fig. 6).

Fig. 4 José with Guy Janin and Walter Flury in the farewell of Dr. Lin (CNSA) in 1999

Fig. 5 Artist's impression of the Solar and Heliospheric Observatory (SOHO). Credit: ESA

From the approach used by José together with M. Bello for the LISA mission, one learns that in mission analysis, in general, there does not exist a simple optimal solution for a given problem. For most of them, a systematic search is needed to find all possible optimal candidates, and the solution of complex problems is usually obtained after the computation of a sequence of solutions of simpler problems with incremental level of difficulty.

The Cluster ESA mission is currently investigating the Earth's magnetic environment and its interaction with the solar wind in three dimensions. For this purpose, four satellites follow a high eccentric polar orbit. The four Cluster satellites must configure a tetrahedron with different prescribed distances, according to the region of the magnetosphere (North and South cusp, bow shock, magnetopause, etc.) to be analysed (Fig. 7).

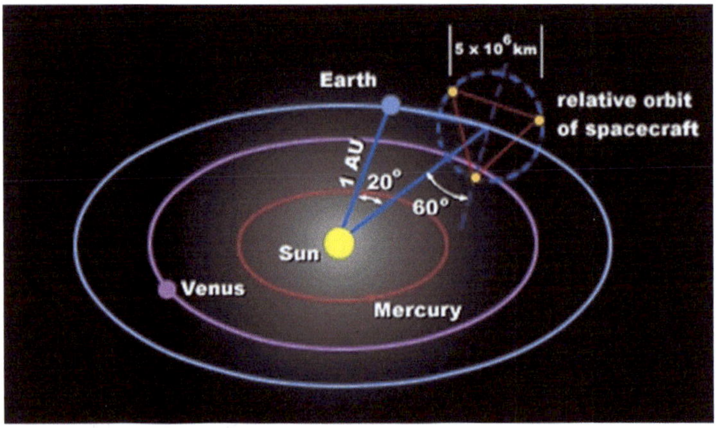

Fig. 6 A schematic diagram of the LISA spacecraft in formation as they orbit around the Sun. Credit: ESA

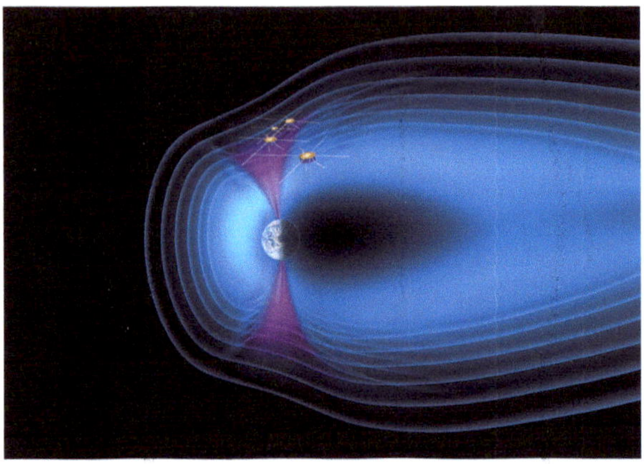

Fig. 7 Artist's impression of the Cluster spacecraft crossing the northern cusp of Earth's magnetosphere. Credit: ESA

Several problems were studied by José and M. Bello related to this mission. The first was the optimisation of the Cluster orbit injection strategy from its launcher delivery state. The second was the optimisation of the orbital manoeuvre sequence, in such a way that the required spacecraft configuration, with the required satellite distances, was achieved at each phase of the mission with a minimum cost in terms of Delta-v.

Miguel explained us that one of the most important things that he learned from José during the realisation of this project was how to behave in a very difficult and changing team environment of scientists, principal investigators, engineering and managers: always with respect for the work performed by others.

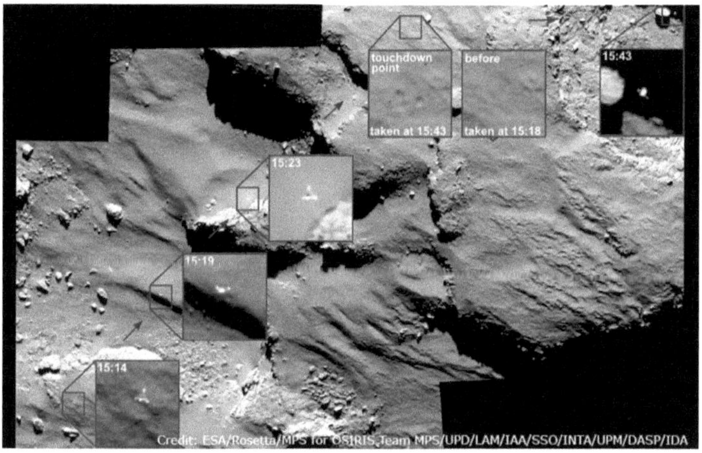

Fig. 8 The European Space Agency's Rosetta spacecraft captured these photos of the Philae lander descending towards, and then bouncing off, the surface of Comet 67P/Churyumov–Gerasimenko during its historic touchdown on Nov. 12, 2014. Credit: ESA/Rosetta/MPS for OSIRIS Team MPS/UPD/LAM/IAA/SSO/INTA/UPM/DASP/IDA

Rosetta was one of the last missions that José took part. In this iconic ESA mission, the in situ analysis of a comet nucleus is done for the first time. For this mission, José participated in the mission analysis, the interplanetary trajectory generation, the interplanetary navigation, the near-comet operations, the comet outgassing model and the lander release strategy (Fig. 8).

As what V. Companys explained during his lecture, in 1998 a Rosetta meeting took place at ESOC involving members of the project group and of the ground segment. José was at the meeting in his function of mission analyst. The topic of the meeting was to explore potential measures for mass reduction: the spacecraft development had consumed mass margins and the fear was that the maximum Ariane V lift capability would be exceeded. Several options were explored, some of them involving spacecraft hardware modification. Also, the possibility to reduce propellant reserves (and hence available Delta-v) was considered. In particular, it was questioned whether the allocation of 185 m/s for launcher dispersion correction was adequate. Also, an allocation of 140 m/s for adjusting the time of asteroid Otawara and Siwa fly-bys was challenged. José fiercely defended the propellant budget and refused any reduction.

Due to the failed Ariane 157, the original Rosetta mission to comet 46P/Wirtanen could not be flown. The backup mission to 67P/Churyumov–Gerasimenko required 1770 m/s nominal cruise Delta-v instead of the 1550 m/s of the Wirtanen mission. The propulsion contingency in 2006 further degraded the propellant situation. People responsible of operations always take extreme care to preserve propellant resources. But if in 1998 José would not have managed to defend the propellant budget, the Rosetta spacecraft would not be today flying around Churyumov–Gerasimenko. The flight dynamics department of ESOC is very aware of this, and with these lines, it wanted to give herewith credit to José for maybe one of his most remarkable contributions to ESA.

4 Summary

Several keywords, related to the personality and professionalism of José, often appeared during the memorial lectures. We think that they are a good summary of all it was said. So, we want to include them here just to remember what we learned from him:

- *Tolerance:* All ideas are respectable.
- *Justice:* Hate all type of injustices and help all affected groups.
- *Sense of humour:* Be always in a good mood.
- *Support:* Be always supportive for whenever need of a person.
- *Creativity:* In science and engineering, not all is invented; there is a lot of room for innovation.
- *Persistence:* Success is a 1 % inspiration and a 99 % transpiration.
- *Confidence:* Be confident in the work that we perform, not to be confused with arrogance.
- *Patience:* Results come if we put all means.

We all, friends and colleagues, are indebted to José that has gone but not forgotten

Leveraging Discrete Variational Mechanics to Explore the Effect of an Autonomous Three-Body Interaction Added to the Restricted Problem

Natasha Bosanac, Kathleen C. Howell, and Ephraim Fischbach

Abstract With recent improvements in ground and space-based telescopes, a large number of binary systems have been observed both within the solar system and beyond. These systems can take the form of asteroid pairs or even binary stars, with each component possessing a similar mass. In this investigation, periodic motions near large mass ratio binaries are explored using the circular restricted three-body problem, which is modified to include an additional three-body interaction. Discrete variational mechanics is leveraged to obtain periodic orbits that exhibit interesting shape characteristics, as well as the corresponding natural parameters. Shape characteristics and structural changes are explained using the stability and existence of equilibrium points, enabling exploration of the effect of an additional three-body interaction and conditions for reproducibility in a natural gravitational environment.

1 Introduction

With recent improvements in ground and space-based telescopes, a large number of binary systems have been observed both within the solar system and beyond. These systems can take the form of asteroid pairs or even binary stars (Margot et al. 2002; Raghavan et al. 2006). In each of these examples, however, the system is typically described by larger values of the mass ratio than the Sun-planet or planet-moon combinations commonly examined within the solar system. Furthermore, limitations on the accuracy and capability of various observational techniques may

N. Bosanac (✉) • K.C. Howell
School of Aeronautics and Astronautics, Purdue University, 701 W. Stadium Ave., West Lafayette, IN 47907, USA
e-mail: nbosanac@purdue.edu; howell@purdue.edu

E. Fischbach
Department of Physics and Astronomy, Purdue University, 525 Northwestern Ave., West Lafayette, IN 47907, USA
e-mail: ephraim@purdue.edu

G. Gómez, J.J. Masdemont (eds.), *Astrodynamics Network AstroNet-II*,
Astrophysics and Space Science Proceedings 44,
DOI 10.1007/978-3-319-23986-6_1

result in an uncertainty in the mass ratio of the binary, significantly impacting the potential motions of a nearby small body orbiting within a chaotic system.

In this investigation, the motion of a small body, such as an exoplanet or moonlet, is examined using the Circular Restricted Three-Body Problem (CR3BP) as a foundation. For a binary star system, the CR3BP may provide a simple, autonomous approximation to a dynamical environment that is otherwise influenced by additional forces such as gravitational radiation. In addition, near a binary asteroid, the restricted problem can provide an approximation to complex higher-order gravitational effects for preliminary analysis of the motion of a spacecraft or a moonlet (Chappaz 2015). When using the CR3BP to represent the dynamics in either form of binary system, the motion of a comparatively small body is governed by the existence of periodic orbits, which form an underlying dynamical structure. For instance, stable orbits may aid in identifying potential motions of an exoplanet that persist for a long time interval near a binary star system. Alternatively, unstable orbits may be valuable in studying natural transport of matter near a binary asteroid. In either case, periodic orbits offer useful information in identifying the possible motions near large mass ratio binary systems.

An alternative dynamical model for the binary system is also derived based on the CR3BP, but extended to incorporate an additional autonomous term in the potential function, producing a Modified Circular Restricted Three-Body Problem (MCR3BP) (Bosanac 2012). Given the absence of experimental data gathered within the vicinity of a binary star, for example, it is possible that the gravitational field within this system might not be accurately modeled solely using pairwise gravitational forces. Thus, in this investigation, the impact of an additional three-body interaction is examined. Many-body forces are well-established in nuclear physics as critical contributions to the accurate modeling of a force field on the atomic scale (Fischbach 1996). The motion of a small body orbiting a binary object, however, serves as an interesting larger scale application for studying the impact of a three-body interaction in gravitational systems. This three-body interaction is added to the potential function of the CR3BP and is assumed to possess a form that is inversely proportional to the product of the distances between all three bodies in the system: when the three bodies are in a tighter configuration, the three-body interaction is stronger. A constant k is used to scale the strength of this additional contribution relative to the gravitational field (Bosanac 2012). In this modified dynamical environment, families of orbits are inherited from the CR3BP, but may undergo structural changes for a sufficiently disturbing three-body interaction, causing interesting and potentially new periodic motions to emerge.

Interesting behaviors, identified in previous work via dynamical systems theory techniques, are explored using discrete variational mechanics (Bosanac et al. 2015). Although this technique has typically been used in astrodynamics to identify optimal paths under a control force, the underlying formulation is applied to the computation of natural periodic motions that resemble a given reference path, as well as the corresponding values of the system parameters, μ and k (Moore 2011). The constrained optimization problem formulated using discrete variational mechanics potentially overcomes some of the sensitivity associated with using continuous shooting techniques to compute motions in a chaotic system where the natural parameters are not accurately known. This methodology is leveraged to compute periodic orbits at various values of the system natural parameters, μ and k, that exhibit desired shape characteristics (Bosanac et al. 2015). By reproducing a motion of interest in the MCR3BP, discrete variational mechanics enables the exploration of the effect of a three-body interaction, as well as the identification of any unique solutions that may emerge. Although the additional autonomous term examined in this investigation is assumed to take the form of a three-body interaction, a similar analysis can be performed for autonomous forces that are derivable from a potential function, such as a time-averaged quantity or a higher-order gravitational term for a body fixed in a given coordinate frame.

2 Dynamical Models

The dynamical environment in the vicinity of a binary system is modeled using the CR3BP as a foundation. This dynamical model reflects the motion of a massless particle under the influence of the point-mass gravitational attractions of two primaries. To study the effect of a three-body interaction, an autonomous term is added to the pairwise gravitational interactions in the potential function in the CR3BP (Bosanac 2012). The resulting model, denoted the MCR3BP, has been introduced and explored by Bosanac et al. (2013) The resulting autonomous potential in the MCR3BP influences the equations of motion and still possesses an integral of motion. Accordingly, families of periodic orbits still exist throughout various regions of the configuration space (Bosanac et al. 2015).

2.1 Circular Restricted Three-Body Problem

When modeling the dynamics in the vicinity of a binary system using the CR3BP, a body of interest, P_3, is examined as it orbits near the larger and smaller primaries, P_1 and P_2. Each of these bodies, P_i, possesses a mass m_i and remains fixed in a right-handed coordinate frame, $\hat{x}\hat{y}\hat{z}$, that rotates with the primaries and is centered at the system barycenter. This rotating frame is oriented such that the $\hat{z}$ axis is oriented parallel to the angular momentum of the primaries as they orbit their mutual barycenter. Furthermore, nondimensional quantities are used to enable comparisons between different systems of similar mass ratios. In particular, length quantities are normalized such that the distance between the two primaries is equal to unity, while time quantities are nondimensioned to ensure that the mean motion of the primaries is also equal to unity. A characteristic mass quantity is also introduced and set equal to the sum of the masses of the primaries. Accordingly, P_1 and P_2 possess nondimensional mass values equal to $(1-\mu)$ and μ, respectively (Szebehely 1967). Using this normalization, the two primaries, P_1 and P_2, remain fixed along the $\hat{x}$-axis of the rotating frame at the locations $(-\mu, 0, 0)$ and $(1-\mu, 0, 0)$, respectively. When modeled using the CR3BP, the equations of motion of P_3, located at the coordinate (x, y, z) relative to the barycenter, are compactly written in the rotating frame as:

$$\ddot{x} - 2\dot{y} = \frac{\partial U^*}{\partial x}, \qquad \ddot{y} + 2\dot{x} = \frac{\partial U^*}{\partial y}, \qquad \ddot{z} = \frac{\partial U^*}{\partial z} \tag{1}$$

where the pseudo-potential function, $U^* = \frac{1}{2}(x^2+y^2) + \frac{1-\mu}{r_1} + \frac{\mu}{r_2}$, and the distances between P_3 and each of the two primaries are $r_1 = \sqrt{(x+\mu)^2 + y^2 + z^2}$ and $r_2 = \sqrt{(x-1+\mu)^2 + y^2 + z^2}$. The autonomous psuedo-potential function, U^*, admits an energy integral, labeled the Jacobi constant, equal to $C_J = 2U^* - \dot{x}^2 - \dot{y}^2 - \dot{z}^2$ (Szebehely 1967).

2.2 *Modified Circular Restricted Three-Body Problem*

Using the CR3BP as a foundation, the equations of motion of P_3 in the MCR3BP are derived using the derivatives of a potential function. In particular, the scalar pseudo-potential governing the motion of P_3 is assumed to consist of the following terms when formulated in the $\hat{x}\hat{y}\hat{z}$ frame that rotates with the primaries:

$$U_k^* = \underbrace{\frac{1}{2}(x^2+y^2) + \frac{1-\mu}{r_1} + \frac{\mu}{r_2}}_{\text{pairwise potential}} + \underbrace{\frac{k}{r_1 r_2}}_{\text{three-body potential}} \tag{2}$$

where k is a constant that scales the potential term corresponding to the three-body interaction. Since the distance between the two primaries is equal to a constant value of unity, only the variables r_1 and r_2 appear in the denominator of the three-body potential term. Furthermore, the magnitude and sign of k can be selected as either positive, negative or zero. A positive value of k corresponds to an attracting three-body interaction, while a negative value of k produces a repulsive interaction. A value of k equal to zero, however, causes the dynamics in the MCR3BP to reduce to those of the CR3BP. The particular values of k are constrained within the range $k = [-0.20, 0.70]$ for numerical and dynamical reasons. The lower limit of $k = -0.20$ produces a repulsive three-body interaction that is approximately half as strong as the natural gravitational environment, while an upper limit of $k = 0.70$ reduces numerical sensitivities during integration. Using derivatives of the potential function as defined in (2), the equations of motion for P_3 are compactly written as:

$$\ddot{x} - 2\dot{y} = \frac{\partial U_k^*}{\partial x}, \qquad \ddot{y} + 2\dot{x} = \frac{\partial U_k^*}{\partial y}, \qquad \ddot{z} = \frac{\partial U_k^*}{\partial z} \tag{3}$$

Since the three-body potential term is formulated to be autonomous, a constant energy integral still exists and is equal to $C_k = 2U_k^* - \dot{x}^2 - \dot{y}^2 - \dot{z}^2$. As the value of k approaches zero, this energy constant reduces to the Jacobi constant in the CR3BP.

2.3 Equilibrium Points

Particular solutions in the form of equilibrium points offer insight into the nonlinear dynamical systems modeled in the CR3BP and the MCR3BP. When the equations of motion of the CR3BP are formulated in the rotating frame, five equilibrium points exist, as depicted via green dots in Fig. 1 (top) for a system with a mass ratio of $\mu = 0.30$, and are labeled L_i, for $i = 1, 2, 3, 4, 5$. The presence of a three-body interaction, indicated by nonzero values of k, impacts the location, stability and even existence of these equilibrium points (Bosanac et al. 2015). At the sample mass ratio of $\mu = 0.30$, the planar equilibrium points are numerically computed for values of k within the range $k = [-0.20, 0.70]$ and plotted in Fig. 1 (top) with blue dots corresponding to positive values of k, and red dots indicating the location of the L_i for negative values of k. For an increasingly attractive three-body interaction, the collinear equilibrium points are displaced farther from the primaries. The converse is true for an increasingly repulsive three-body interaction. In either case, L_4 and L_5 are no longer located at the vertices of equilateral triangles formed with the two primaries. At a critical negative value of k equal to -0.1985, however, Fig. 1 (top)

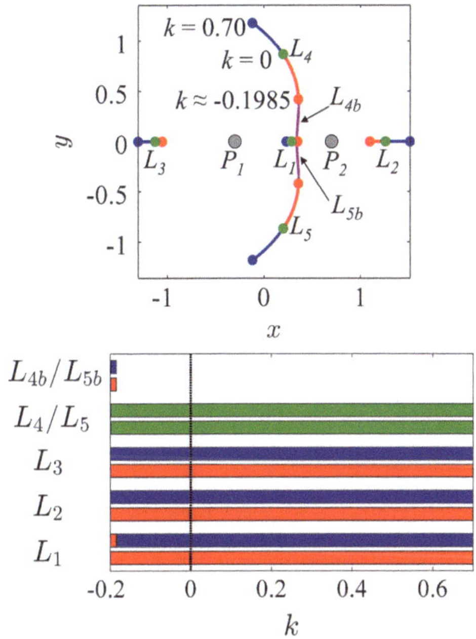

Fig. 1 Equilibrium points in the MCR3BP for $\mu = 0.30$. *Top figure*: Location of planar equilibrium points in the CR3BP (*green*) and the MCR3BP for $\mu = 0.30$: $k > 0$ (*blue dots*), $k < 0$ (*red dots*), and $-0.1985 < k < -0.1839$ (*purple*). *Bottom figure*: Stability of the two planar modes for each equilibrium point as a function of k for $\mu = 0.30$, with purely oscillatory modes (*blue*), real eigenvalues (*red*), or complex conjugate eigenvalues (*green*)

reveals that L_4 and L_5 disappear. Furthermore, purple dots in Fig. 1 (top) depict the existence of two additional equilibrium points, labeled L_{4b} and L_{5b}, that exist for large negative values of k within the range $k = [-0.1985, -0.1839]$ as identified by Douskos (2014).

To explain the formation of L_{4b} and L_{5b}, an overview of the stability of each equilibrium point is useful. In Fig. 1 (bottom), a qualitative measure of the linear stability of each equilibrium point is displayed for a mass ratio equal to $\mu = 0.30$. In this figure, the two horizontal bars associated with each equilibrium point are colored such that blue portions indicate an oscillatory mode, red sections correspond to the presence of a stable and unstable pair of real modes, and green depicts complex conjugate eigenvalues. While the qualitative stability of the planar modes of L_2 and L_3 remain unchanged across the examined range of values of k, L_1 undergoes a stability change at $k = -0.1839$. In particular, the oscillatory mode that corresponds to the existence of the planar Lyapunov family near L_1 becomes real. As explored in previous work by the authors, this stability change in L_1 is accompanied by a pitchfork bifurcation that produces the equilibrium points L_{4b} and L_{5b}. An increasingly repulsive three-body interaction causes these equilibrium points to evolve away from the x-axis in the rotating frame, as depicted in Fig. 1 (top) via purple dots. As k approaches the critical value of $k = -0.1985$, L_4 and L_{4b} meet and disappear.

Prior to examination of the reproducibility of periodic orbits that exhibit a given behavior, a summary of the in-plane stability and existence of each equilibrium point over various large values of μ is useful. Figure 2 portrays the qualitative planar stability of L_1 and L_2 for negative values of k, represented on the horizontal axis, and values of μ in the range $\mu = [0.10, 0.30]$ on the vertical axis. Blue points in this two-dimensional space indicate that the associated equilibrium point possesses one pair of real modes and one pair of oscillatory modes, as inherited from the CR3BP. Red

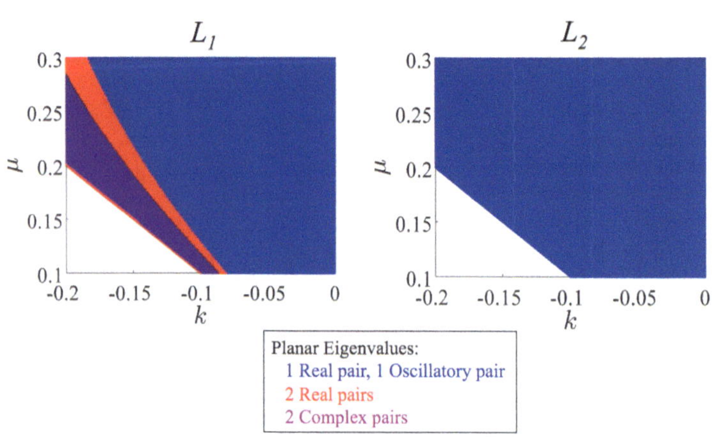

Fig. 2 In-plane stability of L_1 and L_2 for combinations of (k, μ) within the range $\mu = [0.10, 0.30]$ and $k = [-0.20, 0.0]$

points correspond to values of the system natural parameters at which an equilibrium point possesses two real pairs of modes, indicating that motion in its vicinity may not remain nearby indefinitely. Purple regions identify values of μ and k where the planar modes are comprised of two pairs of complex conjugate eigenvalues. Any white regions of Fig. 2 locate combinations of μ and k where the associated equilibrium point does not exist. Analysis of the left plot in Fig. 2 reveals that at each value of μ between 0.10 and 0.30, L_1 undergoes a stability change. As μ is decreased towards 0.10, this stability change occurs with a less negative value of k. Beyond a sufficiently large negative value of k, L_1 disappears, as indicated by the white triangular region. In comparison to the white region on the rightmost plot of Fig. 2, L_2 also apparently disappears at the same critical value of k as L_1. However, the blue region that spans each value of the natural parameters for which L_2 exists reveals that no stability change occurs prior to its disappearance. The qualitative stability of these equilibrium points can influence the underlying structure of the entire system, potentially affecting additional types of steady-state solutions, including periodic and quasi-periodic motions. Such an impact on the dynamical environment in a binary system under the influence of a three-body interaction may be examined using both dynamical systems theory and discrete variational mechanics (Bosanac et al. 2015).

2.4 *Periodic Orbits*

Since the MCR3BP produces an autonomous dynamical fields when formulated in the rotating frame, families of planar, periodic orbits exist within the configuration space. These continuous families of orbits, characterized by motion that repeats after a minimal period, T, form an underlying structure within the phase space. In particular, stable orbits attract or bound nearby motion, while unstable orbits cause nearby trajectories to depart (Contopoulos 2002). Stable periodic orbits are invaluable in both trajectory design and astronomy applications, as they are typically surrounded by families of quasi-periodic orbits. Unstable orbits, however, are associated with stable and unstable manifold structures that may describe natural transport within the vicinity of a binary system. Regardless of their stability, periodic orbits can also be classified by their direction of motion relative to one or both primaries. In particular, a trajectory that is prograde at some instant of time possesses a state with an angular momentum vector, relative to either primary, that is oriented parallel to the $\hat{z}$-axis (Bosanac 2012). Conversely, a state that is retrograde possesses an angular momentum vector that is oriented anti-parallel to the $\hat{z}$-axis. By examining the periodic orbits within various regions of configuration space, the effect of an additional autonomous force contribution on the underlying dynamical structures near a binary system can be investigated.

3 Constrained Optimization Using Discrete Variational Mechanics

To explore the effect of an additional autonomous force contribution, discrete variational mechanics is used to compute the natural parameters of the system for which periodic orbits exhibit a desired shape characteristic. By determining the range of values of μ and k for which a desired motion is possible, the reproducibility of the effect of a three-body interaction in the natural gravitational environment is examined. Such analysis reveals whether the effect of this additional term is unique, and enables identification of periodic motions that otherwise do not exist in the CR3BP. Unlike collocation or multiple-shooting, which require that the dynamics of a continuous-time system be exactly satisfied at a discrete set of nodes or along multiple trajectory arcs, discrete variational mechanics begins with a discretization of the action integral (Ober-Blobaum et al. 2011). Using the variation of the discrete action at a finite set of nodes, a discrete version of Hamilton's principle is used to constrain the motion along a path (Moore 2011). This methodology does not require sequential integration, and may mitigate the difficulties associated with a poor initial guess. To demonstrate this process, some background on variational mechanics is presented, followed by a discussion of its numerical implementation in computing natural paths in the MCR3BP.

3.1 *Variational Principles for Continuous Time Systems*

Variational principles in Lagrangian mechanics are first summarized for continuous time systems. Consider an autonomous mechanical system described by the generalized coordinates q and generalized velocities $\dot{q}$. The associated continuous Lagrangian, $L(q(t), \dot{q}(t))$, is then integrated along a path from a time $t_0 = 0$ to a subsequent time t to construct the action integral, defined as follows (Greenwood 1988):

$$A = \int_{t_0}^{t} L(q(t), \dot{q}(t))dt \tag{4}$$

In a holonomic system, a natural path, $q(t)$, results in a stationary action integral with respect to path variations, given fixed endpoints (Lanczos 2012). This statement, known as Hamilton's principle, can be mathematically expressed as:

$$\delta A = \delta \int_{t_0}^{t} L(q(t), \dot{q}(t))dt = \int_{t_0}^{t} \left(\frac{\partial L}{\partial q} - \frac{d}{dt}\left(\frac{\partial L}{\partial \dot{q}}\right)\right) \delta q dt = 0 \tag{5}$$

The solution to this equation is written as follows:

$$\frac{\partial L}{\partial q} - \frac{d}{dt}\left(\frac{\partial L}{\partial \dot{q}}\right) = 0 \tag{6}$$

Thus, a natural path in a continuous time autonomous and holonomic system must satisfy the well-known Euler-Lagrange equations (Greenwood 1988).

3.2 *Discrete Variational Mechanics*

The continuous variational concepts in Lagrangian mechanics are straightforwardly modified to accommodate discrete time systems, as summarized by Marsden and West (2001). Since numerical integration or observation of the motion of an object inherently results in a discretely-sampled path, a discrete time Lagrangian formulation is valuable. Consider, for instance, a continuous path $q(t)$ discretely sampled N times at constant time intervals of length h to produce a discrete path $\tilde{q}$ (Marsden and West 2001). While the generalized coordinates along this discrete path are denoted $q_i = \tilde{q}(ih)$, velocities can be replaced by finite difference approximations such as $\dot{q}_i \approx \frac{(q_{i+1}-q_i)}{h}$. A discrete time Lagrangian can also be defined for autonomous systems as $L_d(q_i, q_{i+1}, h)$, approximating the integral of the continuous time Lagrangian, $L(q(t), \dot{q}(t))$, over the i-th time interval of length h as the system travels from q_i to q_{i+1} (Moore 2011). Once this integral is approximated numerically using a quadrature rule, a discrete action is constructed as the summation of $L_d(q_i, q_{i+1}, h)$ over $N-1$ time intervals, such that:

$$A_d = \sum_{i=0}^{N-2} L_d(q_i, q_{i+1}, h) \approx \int_{t_0}^{t} L(q(t), \dot{q}(t))dt \tag{7}$$

By taking the variation with respect to path variables, assuming fixed points and leveraging summation by parts, a discrete time version of Hamilton's principle can be mathematically written as:

$$\delta A_d = \delta \sum_{i=0}^{N-2} L_d(q_i, q_{i+1}, h) = \sum_{i=1}^{N-2}\left[\left(\frac{\partial L_d(q_i, q_{i+1}, h)}{\partial q_i} + \frac{\partial L_d(q_{i-1}, q_i, h)}{\partial q_i}\right)\delta q_i\right] = 0 \tag{8}$$

The solution to this expression supplies the discrete Euler-Lagrange equations, which must be satisfied across each time interval $[t_i, t_{i+1}]$:

$$\frac{\partial L_d(q_i, q_{i+1}, h)}{\partial q_i} + \frac{\partial L_d(q_{i-1}, q_i, h)}{\partial q_i} = 0 \tag{9}$$

for $i = [1, N-2]$. Although this solution is only true for holonomic autonomous systems with no external forcing, modifications to the discrete Euler-Lagrange equations have been presented and used by numerous authors in previous works to enable extension to forced and nonautonomous systems (Moore 2011; Marsden and West 2001).

3.3 Formulation of Constrained Optimization Problem

The discrete variational principles can be used in a constrained optimization problem to find a natural periodic orbit that resembles a given reference path. In a continuous time system, the objective function to be minimized can be written as the integral of a cost function, C, from time $t_0 = 0$ to t along the path such that $J(q, \dot{q}) = \int_{t_0}^{t} C(q(t), \dot{q}(t))dt$. This functional is transformed into a finite-dimensional objective function for a discretely-sampled path by replacing the integral with a summation of discrete cost functions evaluated along the $N-1$ time intervals:

$$J_d(\tilde{q}) = \sum_{i=0}^{N-2} C_d(q_i, q_{i+1}, h) \approx \int_{t_0}^{t} C(q(t), \dot{q}(t))dt \tag{10}$$

where the discrete cost function, $C_d(q_i, q_{i+1}, h)$, is constructed using an appropriate quadrature rule to compute the numerical integral of the continuous cost function C over the i-th time interval (Moore 2011). Since the optimization problem is used to compute a periodic orbit that resembles a given reference path, a continuous cost function is defined using the distance between a point located at $q = (x, y)$ and the corresponding point along the reference path, $q_{ref} = (x_{ref}, y_{ref})$, i.e., $C(x, y) = (x - x_{ref})^2 + (y - y_{ref})^2$. The discrete cost function is then calculated via application of the midpoint rule, and summed over each time interval to produce the objective function in Eq. (10).

Prior to implementing the constrained optimization problem, an expression for the discrete Lagrangian must be defined. This approximation, $L_d(q_i, q_{i+1})$, to the integral of the exact Lagrangian over the i-th time interval is calculated using the well-known midpoint rule such that:

$$L_d(q_i, q_{i+1}, h) = hL\left(\frac{q_i + q_{i+1}}{2}, \frac{q_{i+1} - q_i}{h}\right) \tag{11}$$

The continuous Lagrangian used to evaluate this expression for the MCR3BP is written as:

$$L(x, y, \dot{x}, \dot{y}) = \frac{1}{2}(\dot{x} - y)^2 + \frac{1}{2}(\dot{y} + x)^2 + \frac{1-\mu}{r_1} + \frac{\mu}{r_2} + \frac{k}{r_1 r_2} \tag{12}$$

As the value of k approaches zero, this function reduces to the Lagrangian of the CR3BP. Once this Lagrangian is used to compute $L_d(q_i, q_{i+1})$ via the midpoint rule, it can be differentiated and used in Eq. (9) to form the discrete Euler-Lagrange equations, which must be satisfied at each interior node along the discrete path.

A constrained optimization problem is used to compute a discretely-sampled path that minimizes the objective function, while also satisfying the dynamics of the system (Bosanac et al. 2015). Accordingly, the discrete Euler-Lagrange equations must be satisfied at each interior node. These constraints are met by modifying the planar position variables $q_i = (x_i, y_i)$ at each node, as well as the time interval h and the natural parameters, μ and k. Since there are N discrete nodes along each path, the resulting minimization problem can affect $2N + 3$ design variables and is summarized as:

$$\min \qquad J_d(\tilde{q}) = \sum_{i=0}^{N-2} C_d(q_i, q_{i+1}, h)$$

subject to the constraints from Eq. (9), reflecting the system dynamics:

$$\frac{\partial L_d(q_i, q_{i+1}, \mu, k, h)}{\partial q_i} + \frac{\partial L_d(q_{i-1}, q_i, \mu, k, h)}{\partial q_i} = 0 \qquad i = 1, \ldots, N-2$$

Additional constraints are also applied to enforce periodicity along the discrete path via position and momentum continuity. Furthermore, linear equality constraints can also be applied to the values of h, μ, or k to enforce the orbital period or properties of the dynamical system. This optimization problem, constructed using discrete variational mechanics, is used in the sequential quadratic programming algorithm available in MATLAB's *fmincon* routine to compute a natural path that exhibits a desired motion as well as the corresponding system natural parameters.

4 Exploring the Effect of an Additional Autonomous Force Contribution

The use of discrete variational mechanics enables analysis of the effect of an additional three-body interaction via the computation of orbits that exhibit desired shape characteristics as well as the corresponding values of the natural parameters, μ and k. Previous work by the authors has identified structural changes in the evolution of various families of periodic orbits, resulting in identification of interesting behaviors that exist under the influence of a three-body interaction (Bosanac et al. 2015). These motions of interest have been isolated and are explored in this investigation. Using discrete variational mechanics, such behaviors are reproduced for various values of μ and k. The existence of a desired shape characteristic is compared to the qualitative stability of the equilibrium points, potentially aiding the identification of unique motions under the influence of a three-body interaction. The example explored in this investigation involves interesting shape characteristics that emerge along a family of large retrograde circumbinary orbits.

4.1 Retrograde Circumbinary Orbits

If observational data suggests that an exoplanet, for example, follows a circumbinary orbit in a binary star system, a simply-periodic family of retrograde orbits that exists in the exterior region may be of interest. For this family of orbits, the effect of a three-body interaction has been explored using composite stability representations in previous work by the authors (Bosanac et al. 2015). In the CR3BP, orbits in the retrograde circumbinary family evolve from large and circular paths to those existing closer to the primaries and exhibiting loops in the vicinity of L_4 and L_5 as displayed in Fig. 3a (Bosanac et al. 2014). Several members of this family are plotted in black in the rotating frame, with the arrow indicating direction. Gray filled circles locate the primaries, while red diamonds indicate the equilibrium points. As the orbital period is increased further, additional loops form near these equilibrium points in a fractal manner. For nonzero values of k in the MCR3BP, L_4 and L_5 are shifted in configuration space and no longer exist at the vertices of equilateral triangles. Accordingly, the loops that form along orbits in this family will also be shifted in configuration space. Beyond a critical value of k, however, the loops along the retrograde exterior orbits exhibit 'pointed tips' that evolve towards L_1, as depicted in Fig. 3b. To investigate the significant changes in the physical configurations of the orbits in the retrograde exterior family, retrograde exterior orbits with 'pointed tips' are reproduced at various values of the natural parameters, μ and k, using discrete variational mechanics. These results are then visualized in a (μ, k) two-dimensional space to determine the correlation between the stability of L_1 and the existence of this interesting behavior. By examining the values of μ and

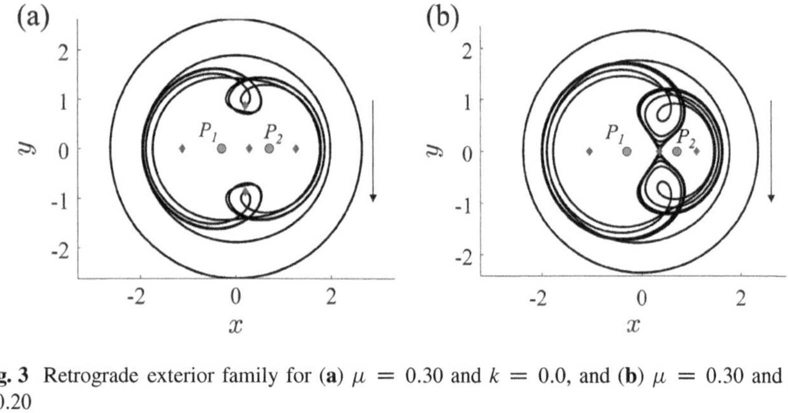

Fig. 3 Retrograde exterior family for (**a**) $\mu = 0.30$ and $k = 0.0$, and (**b**) $\mu = 0.30$ and $k = -0.20$

k at which an orbit resembling a given reference path exists, requirements on the dynamical environment for reproducibility can be determined. These bounds on the values of μ and k are valuable in determining if such motion exists only under the influence of an additional three-body interaction.

Retrograde circumbinary orbits that exhibit 'pointed tips' are recreated for various values of the system natural parameters using discrete variational mechanics to explore the effect of a three-body interaction. Consider the retrograde exterior orbits depicted in Fig. 4, existing in the MCR3BP for $\mu = 0.30$ and $k = -0.20$, and an orbital period of $T = 26.26$ nondimensional time units. Each primary is identified via a gray filled circle, with equilibrium points located using red diamonds. This orbit, which exhibits the motion of interest, is discretized using 600 nodes and used as a reference path in the constrained optimization problem formulated using discrete variational mechanics (Bosanac et al. 2015). Periodic orbits exhibiting a similar behavior are then computed for various values of μ sampled within the range $\mu = [0.10, 0.30]$. For each value of the mass ratio, sampled within the given range, the minimization problem is solved using sequential quadratic programming in Matlab's *fmincon* for a single, unconstrained value of k. Then, the value of k is constrained and successively increased (or decreased) away from this initial value of k until the minimization problem either cannot be solved or produces an orbit that exhibits significantly different behavior, identified via an uncharacteristically large value of the cost function. The resulting process, applied to the reproduction of retrograde circumbinary orbits with the desired characteristics, recovers orbits within a specific region of the two-dimensional space of the natural parameters, μ and k, as displayed in Fig. 5. In this plot, each black point in (k, μ) space indicates the existence of an orbit which exhibits loops with 'pointed tips' that approach L_1, with two sample orbits at distinct values of μ and k displayed in the margins to confirm the recovery of the desired motion. Using Fig. 5 as a reference, the desired motion only appears to exist for negative values of k, i.e., when the three-body interaction is repulsive. However, to explore the specific nonzero values of k bounding this region, additional analysis, using concepts from dynamical systems theory, is required.

To explain the existence of 'pointed tips' that occur along the retrograde exterior family for a repulsive three-body interaction, a comparison to the stability of L_1 is

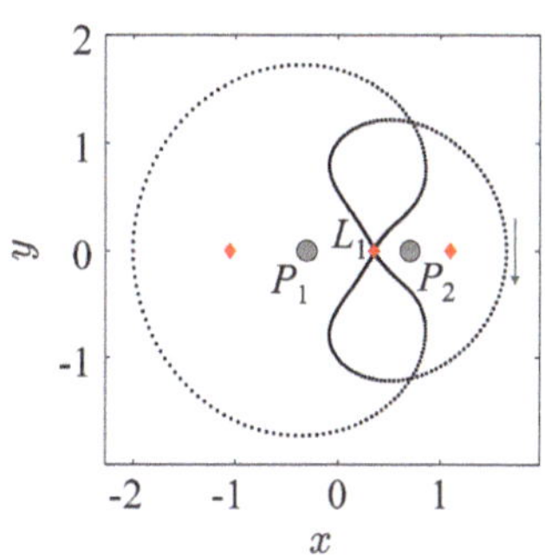

Fig. 4 Reference retrograde exterior orbit for $\mu = 0.30$ and $k = -0.20$, exhibiting 'pointed tips'

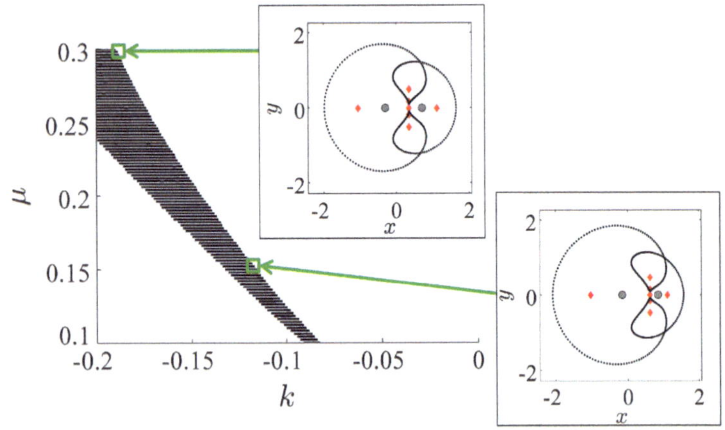

Fig. 5 Existence of retrograde circumbinary orbits with loops that exhibit 'pointed tips' for combinations of k and μ that are indicated via *black points*. Sample orbits are displayed in the margins

valuable. Recall from Fig. 2 that a straightforward stability analysis for L_1 reveals a change in stability for all values of μ within the examined range of $\mu = [0.10, 0.30]$. In fact, for sufficiently negative values of k, the pair of eigenvalues corresponding to the planar oscillatory mode inherited from the CR3BP undergoes a stability change, resulting in L_1 possessing only stable and unstable modes, and causing the disappearance of the well-known L_1 Lyapunov family that exists at all values of μ in the CR3BP. These changes to the equilibrium points, their manifolds and, therefore, the underlying structure of the MCR3BP, influence the foundational dynamical environment near a binary system. For instance, this change in the stability of L_1 may cause the observed structural change along the retrograde exterior family of orbits. The combinations of μ and k at which this family of orbits exhibits 'pointed tips' that evolve towards L_1, as depicted in Fig. 5, are overlaid on the qualitative stability of L_1 using the same color scheme as in Fig. 2: blue points indicate the existence of a planar oscillatory mode, while red regions indicate only real stable/unstable modes, and purple shading corresponds to two pairs of complex conjugate eigenvalues indicating spiraling motion towards and away from the equilibrium point. The resulting plot, displayed in Fig. 6, features black points for each combination of μ and k at which the observed motion exists. Using this figure as a reference, this motion of interest appears to exist only when L_1 possess no oscillatory modes within the plane of motion of the primaries. Such stability properties are not characteristic of L_1 at any value of the mass ratio in the CR3BP. Furthermore, since this black region does not encompass $k = 0$ for any large values of mass ratio, the effect of the three-body interaction on these retrograde exterior orbits at large negative values of k may not be exactly reproduced within the CR3BP, but, rather, mimicked. Accordingly, such motion may be a unique effect of the three-body interaction examined within this investigation.

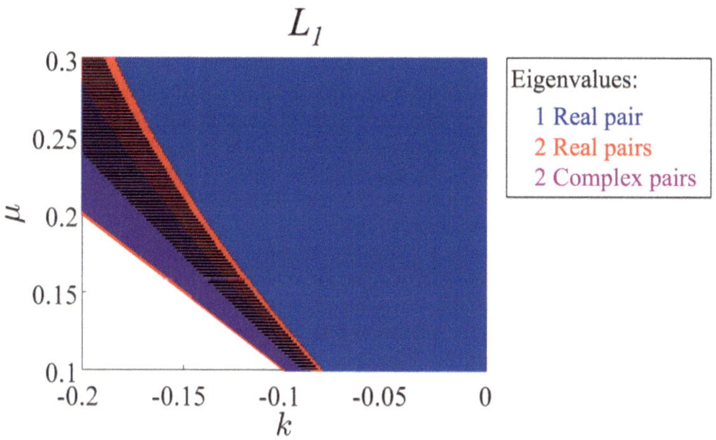

Fig. 6 Existence of retrograde circumbinary orbits with loops, indicated via *black points*, overlaid on a plot of the qualitative stability of L_1

5 Summary

The influence of an additional autonomous force contribution, in the form of a three-body interaction, on the natural gravitational environment of the circular restricted three-body problem is examined using discrete variational mechanics. Previous works by the authors have used stability analysis and dynamical systems theory to study the evolution of families of orbits and identify interesting motions that occur under the influence of a three-body interaction. In this investigation, the existence of these motions of interest is examined in the MCR3BP, as well as their potential for reproducibility in a natural gravitational environment. Discrete variational mechanics is used to overcome the inherent sensitivity in using a shooting method to compute periodic orbits and to enforce the reproduction of desired shape characteristics. The resulting constrained optimization problem enables the computation of periodic orbits exhibiting a particular behavior and the associated bounds on the values of μ and k. A straightforward comparison to the qualitative stability of the equilibrium points in the MCR3BP facilitates explanation of the existence of these interesting orbits and a determination of their uniqueness. Accordingly, dynamical systems theory and discrete variational mechanics are leveraged to explore the effect of a three-body interaction on the dynamics in the restricted problem.

Acknowledgements The authors wish to acknowledge support from the Zonta International Amelia Earhart Fellowship during this work, as well as from the School of Aeronautics and Astronautics at Purdue University.

References

Bosanac, N.: Exploring the influence of a three-body interaction added to the gravitational potential function in the circular restricted three-body problem: a numerical frequency analysis. M.S. Thesis, School of Aeronautics and Astronautics, Purdue University, West Lafayette (2012)

Bosanac, N., Howell, K.C., Fischbach, E.: Exploring the impact of a three-body interaction added to the gravitational potential function in the restricted three-body problem. In: 23rd AAS/AIAA Space Flight Mechanics Meeting, Hawaii (2013)

Bosanac, N., Howell, K.C., Fischbach, E.: Stability of orbits near large mass ratio binary systems. Celest. Mech. Dyn. Astron. **122**(1), 27–52 (2014)

Bosanac, N., Howell, K.C., Fischbach, E.: A natural autonomous force added in the restricted problem and explored via stability analysis and discrete variational mechanics. In: AAS/AIAA Space Flight Mechanics Meeting, Williamsburg (2015)

Chappaz, L.: Exploration of bounded motion near binary systems comprised of small irregular bodies. Celest. Mech. Dyn. Astron. (2015). doi:10.1007/s10569-015-9632-5

Contopoulos, G.: Order and Chaos in Dynamical Astronomy. Springer, Berlin (2002)

Douskos, C.N.: Effect of three-body interaction on the number and location of equilibrium points of the restricted three-body problem. Astrophys. Space Sci. **356**(2), 251–268 (2014)

Fischbach, E.: Long-range forces and neutrino mass. Ann. Phys. **247**, 213–291 (1996)

Greenwood, D.T.: Principles of Dynamics, 2nd edn. Prentice-Hall, Englewood Cliffs (1988)

Lanczos, C.: The Variational Principles of Mechanics. Courier Dover, Toronto (2012)

Marsden, J.E., West, M.: Discrete mechanics and variational integrators. Acta Numer. **10**, 357–514 (2001)

Margot, J.L., Nolan, M.C., Benner, L.A.M., Ostro, S.J., Jurgens, R.F., Giorgini, J.D., Slade, M.A., Campbell, D.B.: Binary asteroids in the Near-Earth object population. Science **296**(5572), 1445–1448 (2002)

Moore, A.: Discrete mechanics and optimal control for space trajectory design. Ph.D Thesis, Control and Dynamical Systems, California Institute of Technology, Pasadena (2011)

Ober-Blobaum, S., Junge, O., Marsden, J.E.: Discrete mechanics and optimal control: an analysis. ESAIM Control Optim. Calc. Var. **17**(2), 322–352 (2011)

Raghavan, D., Henry, T.J., Masion, B.D., Subasavage, J.P., Jao, W., Beaulieu, T.H., Hambly, N.C.: Two suns in the sky: stellar multiplicity in exoplanet systems. Astrophys. J. **646**, 523–542 (2006)

Szebehely, V.: Theory of Orbits: The Restricted Problem of Three Bodies. Academic, London (1967)

An Efficient Sub-optimal Motion Planning Method for Attitude Manoeuvres

Albert Caubet and James D. Biggs

Abstract A motion planning technique for efficiently generating smooth spacecraft attitude slew manoeuvres is presented. The attitude trajectory (using quaternions) is shaped by a polynomial, determined by matching prescribed boundary conditions and the manoeuvre time. This method allows constraints such as limits on velocity, acceleration, jerk, and torque to be evaluated via inverse dynamics. Pointing constraints are also considered. A spin-to-spin case is presented whereby an axis-azimuth parameterisation is used. The problem of time minimization (within the set of trajectories defined by the given polynomials) is addressed, and a method for analytically estimating the minimum time of a manoeuvre is proposed. The method requires low computational capacity, and a comparison with optimal control solutions shows its relative performance.

1 Introduction

Motion planning refers to the problem of defining a feasible trajectory subject to differential constraints, both kinematic and dynamic (LaValle 2006), and path constraints. Motion planning methods are able to generate a trajectory that satisfies the constraints, while optimising a certain performance index. In this paper polynomials are used to obtain the attitude profile of attitude slew maneuvers of spacecraft. Polynomial motion planning is widely used in robotics (Froissart and Mechler 1993) and computer graphics (Hou and Andrews 1978), for their efficiency and ease of manipulation. Cubic splines are a popular method for connecting path points, but they fail to meet acceleration smoothness (Guan et al. 2005). Additionally,

A. Caubet (✉)
Mechanical and Aerospace Engineering, University of Strathclyde, 75 Montrose Street, G1 1XJ Glasgow, UK
e-mail: albert.caubet@strath.ac.uk

J.D. Biggs
Aerospace Sciences and Technology, Politecnico di Milano, B12, Via Privata Giuseppe la Masa 34, 20156 Milano, Italy
e-mail: jamesdouglas.biggs@polimi.it

G. Gómez, J.J. Masdemont (eds.), *Astrodynamics Network AstroNet-II*,
Astrophysics and Space Science Proceedings 44,
DOI 10.1007/978-3-319-23986-6_2

obstacle avoidance usually requires the computation of path points, which has been applied to spacecraft maneuvers in Kim et al. (2004); Cheng et al. (2010); Kjellberg and Lightsey (2014) with various degrees of computational expense. In previous work on attitude motion planning, McInnes (1998) shows the potential of applying inverse control to attitude maneuvers using polynomials, while Biggs and Horri (2012) solves an optimal control problem to plan the reorientation of a spinning satellite, and Zhang and Zhang (2013) uses a fifth degree polynomial to obtain a smooth eigenaxis rotation on a flexible spacecraft. Ventura et al. (2015) compares the performance (computational time and optimization cost) of different attitude parameterisations using splines for motion planning, and shows that the exponential functions in the quaternion parameterization require a larger computational expense.

In this paper, quaternions are used in polynomial motion planning for attitude maneuvers with arbitrary boundary conditions. Common rest-to-rest manoeuvres are possible, but also move-to-move ones such as in Earth observation satellites required to image different ground areas. Apart from the focus on differential constraints such as torque and reaction wheels' speed, obstacle avoidance is also addressed. Essentially, this is a guidance approach to attitude manoeuvres, as opposed to using classic control (where the commanded torque depends on the state error). However, the trajectory obtained needs to be tracked by a controller, although in this case the state error is small—avoiding stability issues.

As to the spin-to-spin case, the method presented allows spinning spacecraft to repoint without altering significantly its spin rate, as opposed to 3-axis stabilized spacecraft having to spin-down, re-point, and spin-up (Bonnamy et al. 2006). This application of the method might prove useful for solar sails (which usually deploy and maintain their extended state through centrifugal forces), spacecraft rotating in Sun acquisition mode, and other spin-stabilized spacecraft.

The work presented here focuses on developing a motion planning method with a very low computational overhead, while featuring the benefits of planning. The degree of the polynomial is the minimum required to match the boundary conditions (and any selected waypoint). Therefore, there are no "free coefficients" in the polynomial that would need to be chosen by an optimiser. Extra coefficients can improve the performance, but this requires the implementation of an on-board constrained optimisation algorithm, which is difficult to validate and reduces the overall robustness of the system. The motion planning framework presented here does not require any gradient-based optimisation. The general aim of this method is to allow an increase in the autonomy of spacecraft bypassing the usually significant increase in computational cost.

The smoothness in the trajectories generated by this method (also at the endpoints), and the ability to monitor the acceleration and jerk, make this method particularly suited to spacecraft with flexible appendages. Discontinuities in the torque profile such as in bang-bang maneuvers and non-zero endpoints torque result in infinite jerk leading to the excitation of flexible modes and spillover (post-maneuver vibrations) (Singh et al. 1989; Skaar and Tang 1986). Kim and Agrawal (2006) and Byers et al. (1990) propose solutions to limit jerk by smoothing the torque switches with a variety of functions. However, the time-optimality of a bang-bang manoeuvre is lost by smoothing the switches.

This paper first outlines the motion planning method, then focuses on the spin-to-spin case, followed by a discussion on inverse dynamics, presents the strategies for time minimisation and obstacle avoidance, and finally compares its performance with an optimal control solver.

2 Motion Planning Method Outline

The proposed method represents the attitude of the body with a prescribed analytically defined function of time. In this paper polynomial functions were chosen since they are smooth, and easy to derive and manipulate. Polynomials are parameterised to match prescribed boundary conditions on attitude, velocity, and higher order derivatives. Once the desired attitude trajectory has been obtained, the torque profile can be obtained with inverse dynamics. Quaternions are used in this paper to parameterise attitude, as they are non-singular and computationally efficient.

The trajectory of each quaternion in the S^3 unit sphere is shaped by the rational polynomial function

$$q_i(t) = \frac{q_i^*(t)}{\|\mathbf{q}^*(t)\|} \tag{1}$$

for $i = 1, \ldots, 4$, where $q_i^*(t)$ is a polynomial:

$$q_i^*(t) = a_{i0} + a_{i1}t + a_{i2}t^2 + \ldots + a_{in}t^n = \sum_{j=0}^{n} a_{ij}t^j \tag{2}$$

Since these quaternions (depicted by the * superscript) are individually shaped, they form a vector whose norm is not constant, thus each component i is normalised in Eq. (1) using the quaternion unit norm

$$\|\mathbf{q}^*(t)\| = \sqrt{q_1^*(t)^2 + q_2^*(t)^2 + q_3^*(t)^2 + q_4^*(t)^2} \tag{3}$$

Arbitrary boundary conditions can be matched for any manoeuvre time t_f. This results in solving a simple linear equation to find the value of the polynomial coefficients in Eq. (2). The $m = n + 1$ boundary conditions of the manoeuvre determine the degree of the polynomial and provide a system of linear equations from which the coefficients a_{ij} can be obtained, given a final maneuver time. The minimum number of boundary conditions to define a slew manoeuvre are the initial and final attitude and velocity (requiring a third degree polynomial). Additionally, acceleration boundary conditions can be introduced, with the purpose of having zero torque at the trajectory endpoints. This may be required for flexible spacecraft, to avoid vibration-inducing discontinuities in angular acceleration. For the same reason, the boundary jerk (time derivative of acceleration) can be forced to zero so that torque is smooth also at the endpoints.

If a scenario with $m = 8$ boundary conditions is considered, matching attitude, velocity, acceleration, and jerk, the coefficients that make the trajectory match the boundary conditions are determined by the following linear system of equations:

$$\begin{bmatrix} 1 & 0 & 0 & 0 & 0 & 0 & 0 & 0 \\ 0 & 1 & 0 & 0 & 0 & 0 & 0 & 0 \\ 0 & 0 & 2 & 0 & 0 & 0 & 0 & 0 \\ 0 & 0 & 0 & 6 & 0 & 0 & 0 & 0 \\ 1 & t_f & t_f^2 & t_f^3 & t_f^4 & t_f^5 & t_f^6 & t_f^7 \\ 0 & 1 & 2t_f & 3t_f^2 & 4t_f^3 & 5t_f^4 & 6t_f^5 & 7t_f^6 \\ 0 & 0 & 2 & 6t_f & 12t_f^2 & 20t_f^3 & 30t_f^4 & 42t_f^5 \\ 0 & 0 & 0 & 6 & 24t_f & 60t_f^2 & 120t_f^3 & 210t_f^4 \end{bmatrix} \begin{bmatrix} a_{i0} \\ a_{i1} \\ a_{i2} \\ a_{i3} \\ a_{i4} \\ a_{i5} \\ a_{i6} \\ a_{i7} \end{bmatrix} = \begin{bmatrix} q_i(0) \\ \dot{q}_i(0) \\ \ddot{q}_i(0) \\ \dddot{q}_i(0) \\ q_i(t_f) \\ \dot{q}_i(t_f) \\ \ddot{q}_i(t_f) \\ \dddot{q}_i(t_f) \end{bmatrix} \tag{4}$$

where the vector on the right-hand side contains the selected boundary conditions. The coefficients a_{ij} of the ith quaternion are then a function of the manoeuvre final time t_f and the corresponding boundary conditions.

Additionally, k trajectory waypoints can be selected if the degree of the polynomial is increased beyond the matching of boundary conditions, and whose coefficients are found by adding k equations to the system. In this case, the trajectory will pass through the specified attitudes at the specified times. At the waypoints, given the use of a single polynomial between endpoints, the curve is smooth (i.e. of differentiability class C^∞).

The system depicted in Eq. (4) can be ill-conditioned for polynomials of high degree and large values of t_f. In order to reduce the condition number of the system's matrix, the time domain can be scaled so that the final time is 1, to prevent some elements in A from being too large. For instance, with the new variable $\tau \in [0, 1]$, where $\tau = t/t_f$, the condition number is reduced from 10^{18} (for $m = 8$ and $t_f = 300$ sec) to 10^4. The scaled coefficients can be calculated by solving the corresponding linear system. Note that the boundary values must be re-calculated accordingly due to the differentiation with respect to a scaled variable.

The initial and final values of the quaternions' time derivatives in Eq. (4) can be obtained, given the boundary angular velocities, via the kinematics equation:

$$\begin{bmatrix} \dot{q}_1 \\ \dot{q}_2 \\ \dot{q}_3 \\ \dot{q}_4 \end{bmatrix} = \frac{1}{2} \begin{bmatrix} 0 & \omega_3 & -\omega_2 & \omega_1 \\ -\omega_3 & 0 & \omega_1 & \omega_2 \\ \omega_2 & -\omega_1 & 0 & \omega_3 \\ -\omega_1 & -\omega_2 & -\omega_3 & 0 \end{bmatrix} \begin{bmatrix} q_1 \\ q_2 \\ q_3 \\ q_4 \end{bmatrix} \tag{5}$$

3 Spin-to-Spin Manoeuvres

In the special case of spin-to-spin manoeuvres, only the final pointing of a certain body axis is relevant. Since quaternions define the full attitude (i.e. the three body axes), the final spin phase angle must be determined, which adds another degree of freedom to the problem that must be chosen. In order to avoid this, the direction of the pointing axis is parameterised with two spherical coordinates (s_1 and s_2), which are expressed independently as time polynomials. A third parameter (s_3), describing the rotation angle about the pointing axis, completes the full attitude in what is known as axis-azimuth parameterisation (Shuster 1993). The s_3 is also expressed as a polynomial, however, its final value s_{f3} is not included in the boundary conditions set since it is irrelevant. This strategy is particularly convenient for spin-to-spin manoeuvres, i.e. transferring the spacecraft from one pointing direction and spinning state $\dot{\mathbf{s}}_0 = [0\,0\,\dot{s}_{03}]^T$ to another one with final $\dot{\mathbf{s}}_f = [0\,0\,\dot{s}_{f3}]^T$. In fact, the resulting attitude rotation formalism formed by s_1, s_2, and s_3 is a particular combination of intrinsic Euler angles.

For instance, assume a spacecraft with an instrument aligned with the body axis y, which is required to point in different directions. In this case, the attitude can be described as a $z - x' - y''$ Euler rotation, where the rotation matrix is $R = R_z(s_1)R_x(s_2)R_y(s_3)$. The order of the x and z rotations is not relevant, but the third rotation must be about the pointing axis (y in this case). With eight boundary conditions, the polynomials are of degree 7:

$$s_i(t) = a_{i0} + a_{i1}t + a_{i2}t^2 + \ldots + a_{i7}t^7 \tag{6}$$

For $i = 1, 2$. However, for $i = 3$, there are only seven boundary conditions, thus $a_{37} = 0$. The coefficients are obtained by solving a linear system analogous to Eq. (4).

Figure 1 shows the body axes paths of a spin-to-spin manoeuvre in an inertial frame, where the pointing axis y_b moves towards the target direction (depicted by a point at the end of the path line) while the other two orthogonal axes keep rotating about it. In Fig. 2, the two coordinates defining the direction of the pointing axis (s_1 and s_2) are driven to their final desired values, while s_3 follows a constant rate trajectory (since the prescribed initial and final spin rates are the same) where the final value of the angle is not relevant.

The singularity associated with Euler angles occurs when calculating their time derivatives with the kinematics equation at angles of 90°. However, in this case the kinematic equation is not used, since the time derivatives of the angles are obtained by differentiating the polynomial. Also, no singularities arise when evaluating the angular velocities and accelerations (needed to calculate the torque). The trajectory can be converted to quaternions if required by the attitude control system of the spacecraft, at the cost of having to use trigonometric functions.

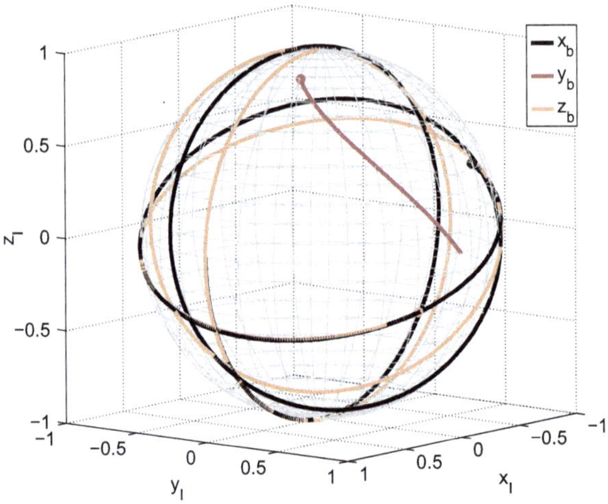

Fig. 1 Path of the body axes on the unit sphere, in a spin-to-spin manoeuvre

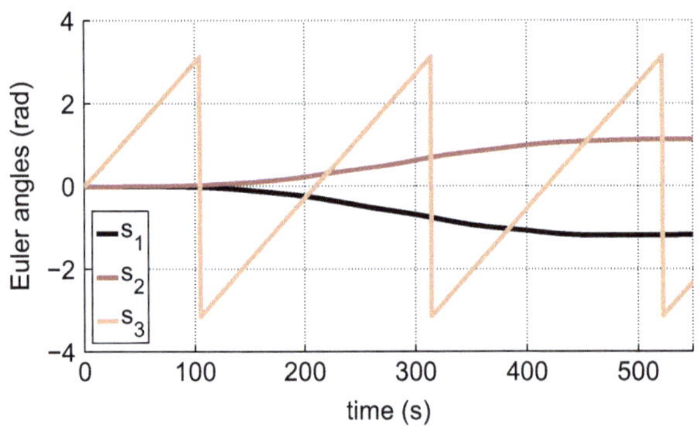

Fig. 2 Trajectory of the attitude coordinates, in a spin-to-spin manoeuvre

4 Inverse Dynamics

A simple model of a fully actuated rigid body has been used to obtain the torque profile. The Euler's equation of rigid-body dynamics relates the torque u_i (along the body ith axis) to the angular velocity ω_i and acceleration $\dot{\omega}_i$ and principal moments of inertia I_i, as

$$
\begin{aligned}
u_1 &= I_1\dot{\omega}_1 - (I_2 - I_3)\omega_2\omega_3 \\
u_2 &= I_2\dot{\omega}_2 - (I_3 - I_1)\omega_1\omega_3 \\
u_3 &= I_3\dot{\omega}_3 - (I_1 - I_2)\omega_1\omega_2
\end{aligned}
\tag{7}
$$

The angular velocities and accelerations are related to quaternions and their time derivatives through the rotational kinematics (Wie 2008), as

$$\begin{aligned} \omega_1 &= 2(\dot{q}_1 q_4 + \dot{q}_2 q_3 - \dot{q}_3 q_2 - \dot{q}_4 q_1) \\ \omega_2 &= 2(\dot{q}_2 q_4 + \dot{q}_3 q_1 - \dot{q}_1 q_3 - \dot{q}_4 q_2) \\ \omega_3 &= 2(\dot{q}_3 q_4 + \dot{q}_1 q_2 - \dot{q}_2 q_1 - \dot{q}_4 q_3) \end{aligned} \tag{8}$$

$$\begin{aligned} \dot{\omega}_1 &= 2(\ddot{q}_1 q_4 + \ddot{q}_2 q_3 - \ddot{q}_3 q_2 - \ddot{q}_4 q_1) \\ \dot{\omega}_2 &= 2(\ddot{q}_2 q_4 + \ddot{q}_3 q_1 - \ddot{q}_1 q_3 - \ddot{q}_4 q_2) \\ \dot{\omega}_3 &= 2(\ddot{q}_3 q_4 + \ddot{q}_1 q_2 - \ddot{q}_2 q_1 - \ddot{q}_4 q_3) \end{aligned} \tag{9}$$

Analytical expressions for the quaternion derivatives are obtained by differentiating Eq. (1) with respect to time:

$$\dot{q}_i(t) = \frac{\dot{q}_i^*(t)}{\|\mathbf{q}^*(t)\|} - \frac{q_i^*(t)}{\|\mathbf{q}^*(t)\|^3}\left(\sum_{i=1}^{4} q_i^*(t)\dot{q}_i^*(t)\right) \tag{10a}$$

$$\ddot{q}_i(t) = \frac{\ddot{q}_i^*(t)}{\|\mathbf{q}^*(t)\|} - \frac{\dot{q}_i^*(t)}{\|\mathbf{q}^*(t)\|^3}\left(\sum_{i=1}^{4} q_i^*(t)\dot{q}_i^*(t)\right) + q_i^*(t)\left[\frac{3}{\|\mathbf{q}^*(t)\|^5}\left(\sum_{i=1}^{4} q_i^*(t)\dot{q}_i^*(t)\right)^2 - \frac{1}{\|\mathbf{q}^*(t)\|^3}\left(\sum_{i=1}^{4}(\dot{q}_i^*(t)^2 + q_i^*(t)\ddot{q}_i^*(t))\right)\right] \tag{10b}$$

Since the quaternions in Eqs. (8) and (9) can be replaced by their corresponding time-dependent polynomials (Eq. (1) and their derivatives), whose coefficients are a function of the manoeuvre time t_f, ultimately the torque is a function of t and t_f (given a set of boundary conditions for a particular manoeuvre). Evaluating the torque along the trajectory is essential to ensure that the actuators always remain within their operative limits. Similarly, an additional time differentiation of Eq. (9) allows for the evaluation of jerk along the manoeuvre.

If the actuators are reaction wheels, it can be useful to assess the speed buildup during the manoeuvre, to ensure that they will not become saturated. Assuming that the wheels are aligned with the body axes, the planned torque can be related to the derivative of their angular momentum. The body angular velocity is considered negligible compared to the magnitude of typical wheels' speeds. Therefore the wheel's acceleration and moment of inertia can be related to the torque provided along its axis by

$$\mathbf{u} \approx -\mathbf{I}_W \dot{\boldsymbol{\omega}}_W \tag{11}$$

Where the vector $\dot{\boldsymbol{\omega}}_W$ contains the wheels' angular acceleration and $\mathbf{I}_W = \text{diag}(I_{W1}, I_{W2}, I_{W3})$ contains their moments of inertia. The wheel speeds are obtained by replacing Eq. (7) into Eq. (11) and integrating:

$$\boldsymbol{\omega}_W(t) \approx -\frac{1}{\mathbf{I}_W}\left(\mathbf{I}(\boldsymbol{\omega}(t) - \boldsymbol{\omega}(0)) + \int_0^t \boldsymbol{\omega} \times (\mathbf{I}\boldsymbol{\omega})dt\right) + \boldsymbol{\omega}_W(0) \tag{12}$$

where the angular velocity of the body $\boldsymbol{\omega}$ can in turn be replaced by Eq. (8). While the integral in Eq. (12) has a closed-form solution as a function of time and the polynomial coefficients, it is so complex that it is computationally more efficient to evaluate it numerically.

5 Manoeuvre Time Minimization

Once the boundary conditions of a particular manoeuvre have been selected, the remaining variable necessary to determine the polynomial coefficients [Eq. (4)] is the manoeuvre final time t_f. In some applications, such as Earth observation or waypoint tracking, a higher layer in the planning hierarchy (either on-board or ground-based) provides all the manoeuvre requirements including final time. In this section the goal of finding a minimum t_f is addressed, which requires the evaluation of the torque. The manoeuvre duration affects the torque profile, with a shorter final time resulting in higher torques. A criterion for choosing t_f is to find its minimum value such that the maximum estimated torque in the manoeuvre (of any axis, in absolute value) is equal to the actuator's torque limit u_{lim}. In an analogous way, a limit on jerk can be considered when operating a flexible spacecraft. Also, if reaction wheels are used, their speed can be evaluated to ensure they don't saturate.

It is possible to obtain an analytic expression of the planned torque as a function of time and t_f by combining the kinematics and dynamics equations and the polynomials representing the attitude parameters. The minimum t_f could be obtained by deriving the expression with respect to time and evaluating its roots to find the maximum torque (within the manoeuvre time). However, due to the high non-linearity of the expression obtained this way, the calculation of the roots cannot be done analytically and the numerical computation is complex. A more efficient strategy consists in discretizing the manoeuvre in time, and evaluating the torque at each node (u_{ik} of ith axis and kth node). The difference with the torque limit u_{lim} is calculated for each node and the maximum value of the set is obtained [Eq. (13)].

$$J_i = \max_k(|u_i(t_k)| - u_{lim}) \tag{13}$$

Where J_i is the largest difference among all nodes of the ith torque profile. The three axes can be combined in $J = \max\{J_1, J_2, J_3\}$.

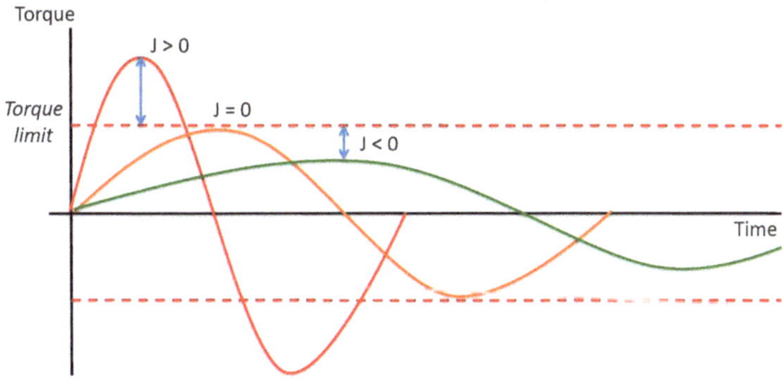

Fig. 3 Illustration of torque profiles with different final times, with J being a metric of the peak height

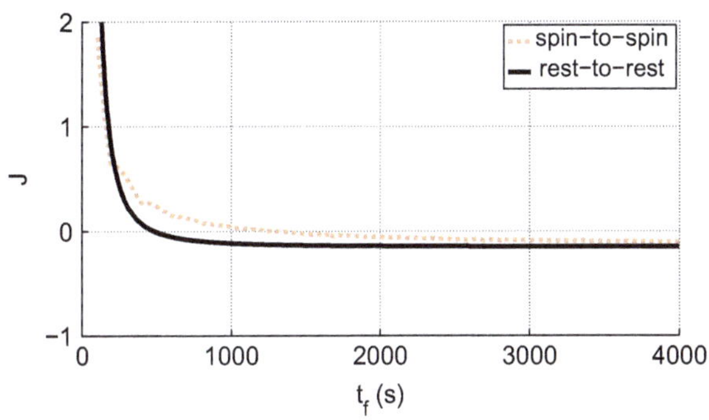

Fig. 4 Evolution of J (maximum torque over the limit) with manoeuvre time

As shown in Fig. 3, the optimum point corresponds to $J = 0$, whereas if $J > 0$, the maximum torque is above the limit. While gradient-based optimisation algorithms would use J^2 to find the optima, this performance index allows to (a) know if a trajectory is feasible (in terms of torque) simply by checking the sign of J, and (b) use a root-finding algorithm, which are more efficient than numerical gradient-based ones. For general manoeuvres with arbitrary boundary angular velocities, accelerations, and jerks, the shape of J as a function of t_f may feature multiple local optima—some of which could be unfeasible. Thus, an algorithm capable of finding the global optimum would be required to compute the minimum feasible t_f. However, in the cases of rest-to-rest and spin-to-spin manoeuvres, the evolution of J with t_f is monotonically decreasing, as shown in the example of Fig. 4. In this case, finding the root of this curve, corresponding to the minimum t_f of that particular path, is performed with very few iterations.

Note that the planned torque profile obtained is a prediction based on the internal model, thus its accuracy will determine how large the difference is between the planned and real torque. Considering an additional safety margin on the torque limit or on t_f is advised to account for model inaccuracies and other disturbances (such as solar radiation pressure, atmospheric drag, etc).

In an effort to keep the computational overhead as low as possible, an analytic function for calculating the minimum final time was developed—the so-called *t_f-function*. This strategy is based on the assumption that the minimum time of a manoeuvre mainly depends on a number of selected parameters, such as the traveled distance, and that there are no obstacles.

5.1 Rest-to-Rest t_f-Function

In the rest-to-rest case using quaternions, the rotation angle θ_f, representing the length of the slew, is one of the drivers of the manoeuvre duration. In an asymmetric spacecraft, the actuators require more power if the manoeuvre is performed along high inertia axes. Therefore, another main driver of t_f is the moment of inertia along the rotation axis (I_e), calculated using

$$I_e = \hat{\mathbf{e}}_B^T \mathbf{I} \hat{\mathbf{e}}_B \tag{14}$$

where $\mathbf{I}$ is the inertia matrix and $\hat{\mathbf{e}}_B$ the rotation axis as previously defined.

When a large enough Monte-Carlo simulation is performed with random manoeuvres, by plotting t_f with θ_f and I_e a cloud of points is obtained which has a surface pattern. With a data-fitting tool a parametric cubic surface can be well approximated (see Fig. 5).

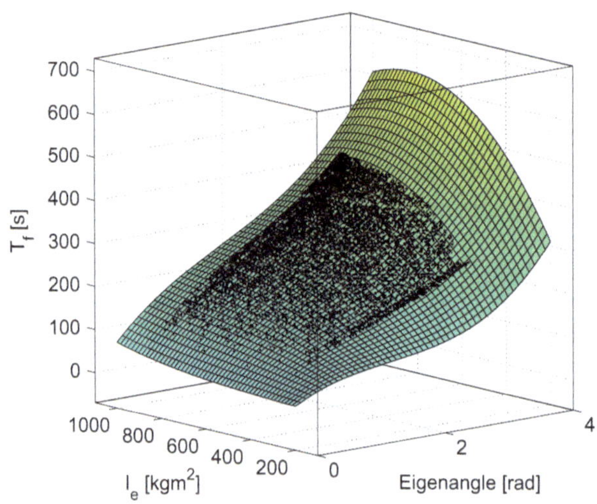

Fig. 5 Cubic surface fitting a cloud of points corresponding to random manoeuvres

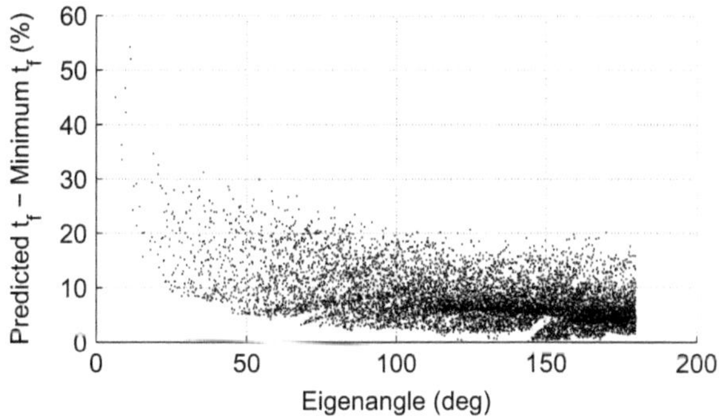

Fig. 6 Difference between the predicted and minimum final times, in % (rest-to-rest)

However, the surface $t_f = f(\theta_f, I_e)$, as obtained by data-fitting, is essentially an average, thus a margin on the estimated t_f must be added so that the torque limit is respected in all possible manoeuvres. Despite the advantage of quickly estimating t_f, some performance is lost since the calculated manoeuvre duration will not be the minimum possible. As an example, with the model studied in this work, with the rest-to-rest t_f-function most of the predicted times are under 30 % larger than the actual minimum final time (Fig. 6). The larger errors correspond to small displacements, which have short associated manoeuvre times. The R-squared value for this example is 0.986.

5.2 Spin-to-Spin t_f-Function

The manoeuvre final time can also be analytically estimated in the spin-to-spin scenario. However, in this case there is no fixed rotation axis, so the moment of inertia metric is no longer valid. The angle between the initial and final pointing vectors (α), analogue to θ_f, can still be used as a geometric parameter. An alternative parameter can be selected to form a surface pattern in the Monter-Carlo analysis based on the difference between endpoints of s_1 and s_2 [Eq. (15)].

$$L_s = \sqrt{|s_{10} - s_{1f}|^2 + |s_{20} - s_{2f}|^2} \tag{15}$$

A parametric surface is obtained as $t_f = f(\alpha, L_s)$ via data-fitting. In the case under study, a fourth degree surface is considered, giving an R-squared value of 0.995. The difference between estimated and optimum t_f can be as high as 400 %. However, Fig. 7 shows that, as in the previous case, high differences correspond to short duration manoeuvres, thus the error in absolute terms is not large.

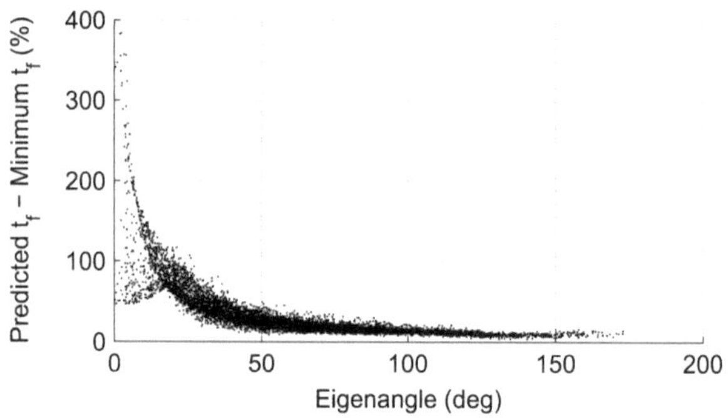

Fig. 7 Difference between the predicted and minimum final times, in % (spin-to-spin)

A single t_f-function in the spin-to-spin case is, again, model-specific, and valid only for fixed endpoint spin rates (in this work equal spins of 0.03 $\frac{rad}{s}$ were considered).

6 Obstacle Avoidance

The path of the body axes on the unit sphere can be diverted in order to avoid pointing constraints. The static path constraints (Kim et al. 2004) or keep-out areas are represented by cones intersecting the unit sphere. The resulting circle should not be trespassed by the path of the corresponding body axis i, in other words, the angle between the body axis $\mathbf{v}_i^I(t)$ (resolved in the inertial frame) and the cone axis $\mathbf{w}_c$ should not be lower than the cone angle γ_c:

$$\mathbf{v}_i^I(t) \cdot \mathbf{w}_c \leq \cos(\gamma_c) \tag{16}$$

The pointing body axis can be drawn from quaternions with

$$\mathbf{v}_i^I = (q_4^2 - \|\mathbf{q}\|^2)\mathbf{v}_i^B + 2(\mathbf{q}^T\mathbf{v}_i^B)\mathbf{q} + 2q_4(\mathbf{q} \times \mathbf{v}_i^B) \tag{17}$$

where $\mathbf{q} = [q_1, q_2, q_3]^T$ and $\mathbf{v}_i^B$ is the pointing axis resolved in the body frame.

Obstacle avoidance can be achieved in a deterministic way using a path point [thus adding an additional equation to the linear system in Eq. (4)]. The point is selected by calculating the nearest point of the nominal path to the cone center and moving it to the closest point on the circle (Fig. 8). This implies rotating the attitude about the axis $\mathbf{w}_c \times \mathbf{v}_i^I(\tau_k)$ by an angle $\gamma_c - \gamma$. Since time is adimensional, the τ_k associated with the new point is the same as the original one. The resulting path of the pointing axis is tangent to the keep-out cone. This approach is suboptimal, but as it is deterministic it avoids the optimisation process and a feasible motion is obtained.

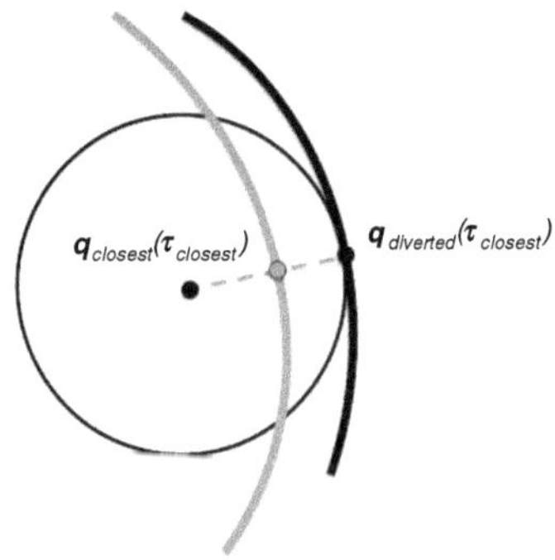

Fig. 8 A path crossing the keep-out area is diverted by forcing it to pass through a determined waypoint

The limitation of this approach is on manoeuvres with arbitrary endpoint velocities. In this case the variable t_f affects the path, thus the time-minimisation and the obstacle avoidance problems are coupled. However, in the case of spin-to-spin using the previously described parameterisation, the path of the pointing axis is not dependent on t_f, and an analogous approach to the rest-to-rest scenario can be used. Also, note that the manoeuvre time needs to be re-computed for the diverted trajectory (but the t_f-*function* cannot be applied).

7 Performance Comparison with Optimal Control

Using polynomials to shape the trajectories yields suboptimal results, in terms of torque and time. Thus, it is worth comparing the results of this method with those of a pseudospectral optimal control solver (PSOPT in this case). Using a direct transcription method for optimal control, close-to-optimal solutions can be obtained. Firstly, simulations have been realised with a fixed final time and the cost function:

$$J = \int_0^{t_f} \mathbf{u}^T \mathbf{u} \, dt \tag{18}$$

which is a measure of the energy spent on the manoeuvre.

Figure 9 shows the attitude and velocity profile of both methods, fixing the manoeuvre time. The attitude trajectory, or the path of the body axes, is similar. The angular velocity profile obtained with the polynomial method is symmetric, accelerating the first half of the trajectory (almost linearly) and decelerating the second half. The optimal control solution, however, has a less symmetric profile since it accounts for the moments of inertia of the spacecraft. The torque profiles are compared in Fig. 10.

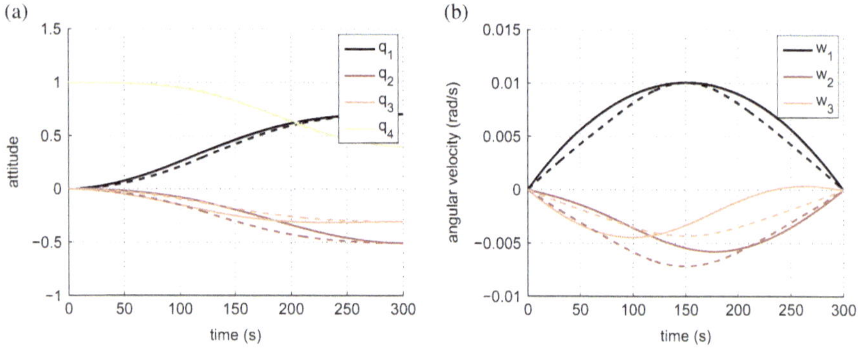

Fig. 9 Attitude (quaternions) and velocity profiles of fixed-time, minimum energy trajectory. *Solid lines*: PSOPT, *dashed lines*: polynomial

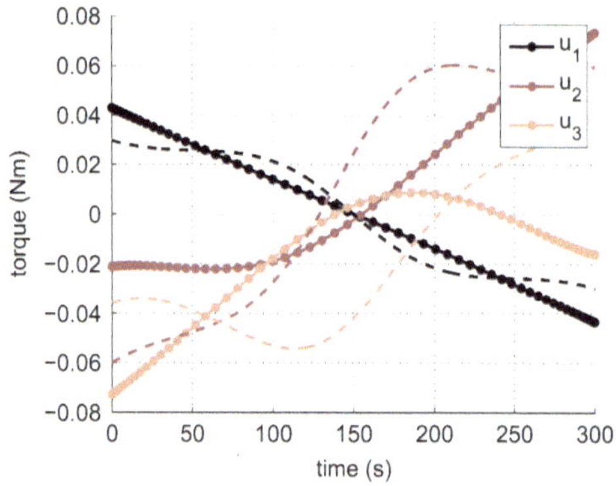

Fig. 10 Torque profile of fixed-time, minimum energy trajectory. *Marked lines*: PSOPT, *dashed lines*: polynomial

While the time of the trajectory is fixed, the cost as in Eq. (18) is 60 % higher in the polynomial method than in PSOPT. Other random manoeuvres fall within this order of magnitude.

An interesting result arises when simulating a rotational manoeuvre about one of the spacecraft's axes. In this case, as shown in Fig. 11, the trajectory is very similar, and the cost of the polynomial trajectory in terms of energy is higher only by 13 % (Fig. 12).

These results suggest that the class of polynomial trajectories provide solutions with a relatively low energy cost, considering that no optimisation of this cost function has been undertaken.

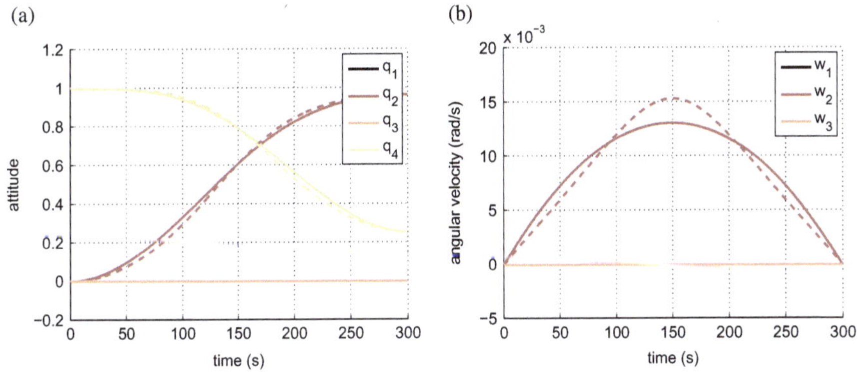

Fig. 11 Attitude (quaternions) and velocity profiles of fixed-time, minimum energy trajectory about one axis. *Solid lines*: PSOPT, *dashed lines*: polynomial

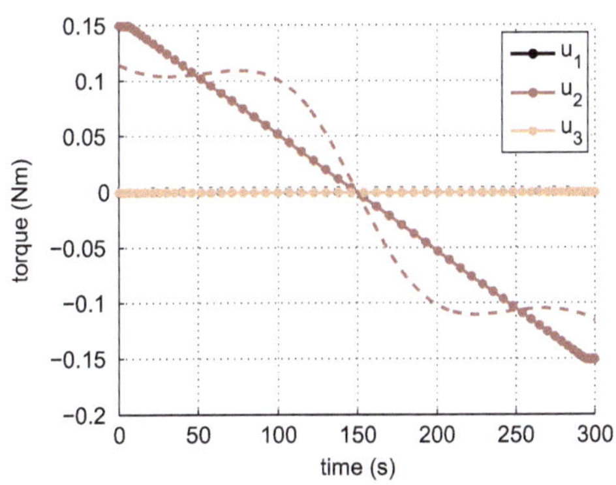

Fig. 12 Torque profile of fixed-time, minimum energy trajectory about one axis. *Marked lines*: PSOPT, *dashed lines*: polynomial

Figures 13 and 14 depict a manoeuvre where time is minimised. As expected, the optimal control solution is of bang-bang type. The shape of the polynomial trajectory is essentially the same as in the fixed-time case, but the maximum torque is allowed to be equal to the actuators' limit—as mentioned in Sect. 5. In this case, PSOPT improves the manoeuvre time by 19 %, but the energy cost is increased by a factor of two with respect to the polynomial solution. Therefore, the polynomial method yields trajectories with a final time relatively close to the optimal solution, while maintaining the energy efficiency.

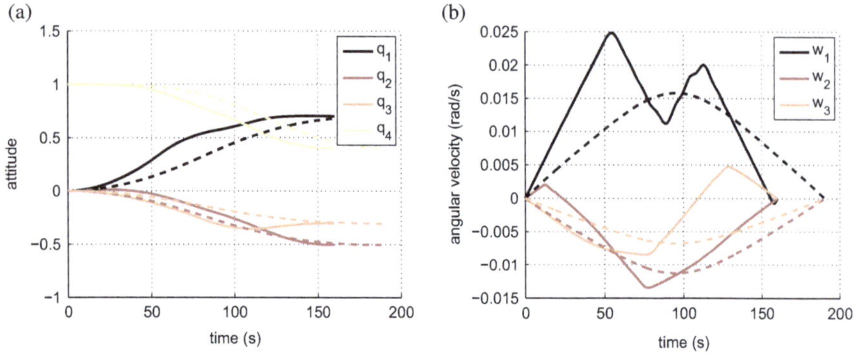

Fig. 13 Attitude (quaternions) and velocity profiles of minimum-time trajectory. *Solid lines*: PSOPT, *dashed lines*: polynomial

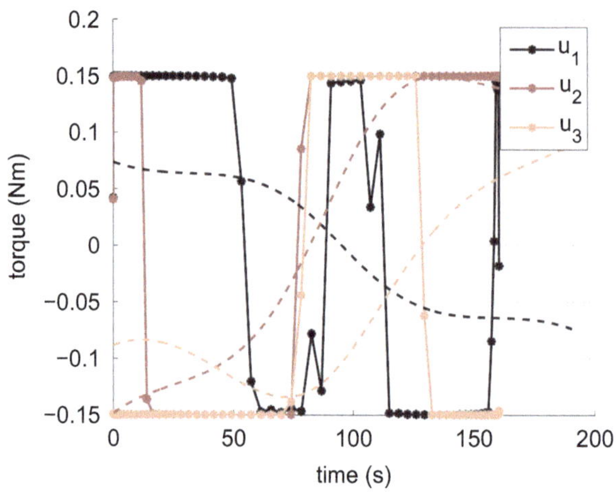

Fig. 14 Torque profile of minimum-time trajectory. *Marked lines*: PSOPT, *dashed lines*: polynomial

8 Conclusions

The general method presented can match arbitrary boundary conditions on attitude, velocity, and higher derivatives (acceleration, jerk) if required. The torque profile of the actuators can be obtained via inverse dynamics, thus torque constraints can be enforced (by adjusting the manoeuvre time t_f). Similarly, other constraints such as maximum velocity, acceleration, or reaction wheel speed can be checked. By further increasing the degree of the polynomial, selected waypoints can be included in the path, as a way to address pointing constraints. The minimum final time can be efficiently obtained through root-finding. However, for arbitrary boundary velocities, this function can have multiple local minima. For rest-to-rest and spin-to-spin manoeuvres free of obstacles, a method has been proposed to obtain the final time using an analytical function. The optimality of the trajectories obtained by this method has been assessed by comparing them to optimal control solutions. Trajectories shaped by polynomials are naturally energy efficient without needing further optimisation, while the loss of optimality in terms of manoeuvre time is not large.

The method presented combines the benefits of inverse dynamics motion planning with computational efficiency, and has potential for on-board implementation to enhance the capabilities of autonomous spacecraft. Moreover, it is able to provide trajectories with a smooth torque profile (also at the endpoints), which is relevant for spacecraft with flexible appendages and continuous control actuators such as reaction wheels. However, for highly constrained spaces, this method might be unable to provide a feasible trajectory.

References

Biggs, J.D., Horri, N.: Optimal geometric motion planning for a spin-stabilized spacecraft. Syst. Control Lett. **4**(61), 609–616 (2012)

Bonnamy, O., Bonneau, C., Ecale, E., Denis, M.: Mars express AOCS in-flight behaviour. In: Proceedings of the 6th International ESA Conference on Guidance, Navigation and Control Systems, vol. 33, pp. 17–20 (2006)

Byers, R.M., Vadali, S.R., Junkins, J.L.: Near-minimum time, closed-loop slewing of flexible spacecraft. J. Guid. Control Dyn. (1990). doi:10.2514/3.20517

Cheng, X., Cui, H., Cui, P., Xu, R.: Large angular autonomous attitude maneuver of deep spacecraft using pseudospectral method. In: 3rd International Symposium on Systems and Control in Aeronautics and Astronautics (2010). doi:10.1109/ISSCAA.2010.5632498

Froissart, C., Mechler, P.: On-line polynomial path planning in Cartesian space for robot manipulators. Robotica (1993). doi:10.1017/S0263574700016118

Guan, Y., Yokoi, K., Stasse, O.: On robotic trajectory planning using polynomial interpolations. In: IEEE International Conference on Robotics and Biomimetic, pp. 111–116 (2005)

Hou, H.H.H., Andrews, H.: Cubic splines for image interpolation and digital filtering. IEEE Trans. Acoust. Speech Signal Process. (1978). doi:10.1109/TASSP.1978.1163154

Kim, Y., Mesbahi, M., Singh, G., Hadaegh, F.Y.: On the constrained attitude control problem. In: AIAA Guidance, Navigation, and Control Conference and Exhibit, pp. 10.2514/6.2004–5129 (2004)

Kim, J., Agrawal, B.N.: Experiments on jerk-limited slew maneuvers of a flexible spacecraft. In: AIAA Guidance, Navigation, and Control Conference and Exhibit, pp. 1–20 (2006)

Kjellberg, H.C., Lightsey, E.G.: A constrained attitude control module for small satellites. In: Proceedings of the AIAA/USU Conference on Small Satellites, Series SSC12-XII-1 (2014)

LaValle, S.M.: Planning Algorithms, pp. 3–5. Cambridge University Press, Cambridge (2006)

McInnes, C.R.: Satellite attitude slew manoeuvres using inverse control. Aeronaut. J. **102**, 259–265 (1998)

Shuster, M.D.: A survey of attitude representations. J. Astronaut. Sci. (1993). doi:10.2514/6.2012-4422

Singh, G., Kabamba, P.T., Mcclamrochj, N.H.: Bang-bang control of flexible spacecraft slewing maneuvers: guaranteed terminal pointing accuracy. J. Guid. Control Dyn. (1989). doi:10.2514/3.56512

Skaar, S.B., Tang, L.: On-Off attitude control of flexible satellites. J. Guid. Control Dyn. (1986). doi:10.2514/3.20140
Ventura, J., Romano, M., Walter, U.: Performance evaluation of the inverse dynamics method for optimal spacecraft reorientation. Acta Astronaut. (2015). doi:10.1016/j.actaastro.2014.11.041
Wie, B.: Space Vehicle Dynamics and Control, pp. 344–345, 2nd edn. AIAA, Washington (2008)
Zhang, Y., Zhang, J.-R.: Combined control of fast attitude maneuver and stabilization for large complex spacecraft. Acta Mech. Sin. (2013). doi:10.1007/s10409-013-0080-8

Bifurcations Thresholds of Halo Orbits

Marta Ceccaroni, Alessandra Celletti, and Giuseppe Pucacco

Abstract In this work an analytical study of the bifurcation of the halo orbits around the collinear points L_1 and L_2 for the circular, spatial, restricted three–body problem is presented. The energy level at which the bifurcation takes place, for arbitrary values of the mass ratio, is found by reducing the Hamiltonian of the problem into a synchronous resonant normal form by means of Lie Transformations. This naturally provides an integrable approximation the system, which yields to the reduction of the system to the center manifold. The bifurcation thresholds of the 1 : 1 resonant periodic orbit families are estimated, among which the 'halo' orbits. Analytical results are compared with the numerical ones existing in the literature. Initial conditions for generating halos are found inverting this analytical process.

1 Introduction

This work investigates the motion of a small body with negligible mass in the gravitational field of two primaries which move on circular trajectories around their barycenter, namely the *spatial, circular, restricted* three–body problem (SCR3BP). This model admits five equilibrium positions (see, e.g., Szebehely 1967) in the synodic reference frame, the barycentric frame rotating with the angular velocity of the primaries. Three of these equilibria are located on the line joining the primaries, thus called *collinear* , and shown to be unstable.

The dynamics around the collinear points has gained an increasing interest in the space era. Since then, several space missions have fully exploited the capabilities of such equilibrium positions. Nowadays collinear equilibria and the respective *halo orbits*, are widely used, especially in low–energy space missions. Halos are

M. Ceccaroni (✉) • A. Celletti
Department of Mathematics, University of Roma Tor Vergata, Via della Ricerca Scientifica 1, 00133 Roma, Italy
e-mail: ceccaron@mat.uniroma2.it; celletti@mat.uniroma2.it

G. Pucacco
Department of Physics, University of Roma Tor Vergata, Via della Ricerca Scientifica 1, 00133 Roma, Italy
e-mail: pucacco@roma2.infn.it

G. Gómez, J.J. Masdemont (eds.), *Astrodynamics Network AstroNet-II*,
Astrophysics and Space Science Proceedings 44,
DOI 10.1007/978-3-319-23986-6_3

three-dimensional periodic trajectories around the collinear points resulting from the interaction between the gravitational pull of two planetary bodies, and the Coriolis and centrifugal accelerations acting on the spacecraft.

It was suggested to use the Earth–Moon L_2 halo orbit as a communication relay station for an Apollo mission to the far side of the Moon, as it would enable continuous views of both the Earth and the hidden Moon. Yet, the establishment of a bridge for radio communication is a significant problem for future space missions, planning to use the outer side of the Moon as a launch site for space explorations or as an observation point.

Moreover, a number of missions have used the Sun-Earth L_1 halo orbits, like the International Sun-Earth Explorer (ISEE-3 1978), the Solar and Heliospheric Observatory (SOHO 1996) and Genesis (2001). All these space missions have a strategic importance for solar-wind physics, cosmic-ray physics, and astrophysics. Remarkably, the Next Generation Space Telescope (NGST) and Lisa Pathfinder will also use halo orbits.

There are extensive results in the literature about the determination of accurate approximations of such equilibrium orbits. Just to quote some results, in 1973 Hénon (1973) studied the stability of the planar Lyapunov orbits with respect to vertical perturbations, see also Richardson (1980). A center manifold reduction was used by Howell et al. (1997), Jorba and Masdemont (1999) and Gómez and Mondelo (2001) in combination with the Lindstedt-Poincaré method, which enabled them to develop a semi-analytical technique to describe and compute solutions in the extended neighborhood of an equilibrium point. A method for the analytic evaluation of the bifurcation thresholds in terms of the energy in the rotating frame has been progressively illustrated in Celletti et al. (2015), Bucciarelli et al. (2015), and Ceccaroni et al. (2016a,b).

2 The Model

Consider a synodic reference frame centered in the barycenter of the primaries P_1, P_2, and rotating with the angular velocity of the primaries. The X axis is set along the line joining P_1 and P_2, the Z axis along the angular momentum and the Y axis in such a way to have a positively oriented frame. We normalize the units of measure so that the gravitational constant as well as the sum of the masses and the mutual distance of the primaries are unitary. The period of rotation of the primaries is thus equal to 2π. Let μ be the (scaled) mass of the smaller primary: the larger primary results to be located at $(\mu, 0, 0)$, while the smaller is at $(-1 + \mu, 0, 0)$. Define the kinetic moments as $P_X = \dot{X} - Y\,, \qquad P_Y = \dot{Y} + X\,, \qquad P_Z = \dot{Z}\,.$

The Hamiltonian describing the motion of a third body P_3 of negligible mass is given by

$$H^{(IN)}(P_X, P_Y, P_Z, X, Y, Z) = \frac{1}{2}(P_X^2 + P_Y^2 + P_Z^2) + YP_X - XP_Y - \frac{1-\mu}{r_1} - \frac{\mu}{r_2}\,, \tag{1}$$

where r_1, r_2 denote the distances from the primaries:

$$r_1 = \sqrt{(X-\mu)^2 + Y^2 + Z^2}\,, \qquad r_2 = \sqrt{(X-\mu+1)^2 + Y^2 + Z^2}\,.$$

Introduce the *pseudo-potential* (scalar function) (compare with Murray and Dermott 1999),

$$\Omega(X,Y,Z) \equiv \frac{1}{2}(X^2+Y^2) + \frac{1-\mu}{r_1} + \frac{\mu}{r_2}\,;$$

then, the equations of motion can be written in compact form as

$$\begin{aligned} \ddot{X} - 2\dot{Y} &= \frac{\partial \Omega}{\partial X}\,, \\ \ddot{Y} + 2\dot{X} &= \frac{\partial \Omega}{\partial Y}\,, \\ \ddot{Z} &= \frac{\partial \Omega}{\partial Z}\,. \end{aligned} \tag{2}$$

Calling $\gamma_j, j = 1,2$, the distance of the collinear equilibria, L_1 or L_2 respectively, from the closest primary, solution of

$$\gamma_j^5 \mp (3-\mu)\gamma_j^4 + (3-2\mu)\gamma_j^3 - \mu\gamma_j^2 \pm 2\mu\gamma_j - \mu = 0 \tag{3}$$

(the uppers sign for L_1 and the lower for L_2) we shift and scale the system of reference to the equilibria, such that the primary is now at distance 1 from the origin, namely:

$$X = -\gamma_j x + \mu + a\,, \qquad Y = -\gamma_j y\,, \qquad Z = \gamma_j z\,.$$

Moreover, setting $a = -1 + \gamma_1$ for L_1 and $a = -1 - \gamma_2$ for L_2 and denoting by $P_n = P_n(\chi)$ the Legendre polynomial of order n and argument χ, the equations of motion in the new variables can be written as:

$$\begin{aligned} \ddot{x} - 2\dot{y} - (1+2c_2)x &= \frac{\partial}{\partial x}\sum_{n\geq 3} c_n(\mu)\rho^n P_n\left(\frac{x}{\rho}\right) \\ \ddot{y} + 2\dot{x} + (c_2-1)y &= \frac{\partial}{\partial y}\sum_{n\geq 3} c_n(\mu)\rho^n P_n\left(\frac{x}{\rho}\right) \\ \ddot{z} + c_2 z &= \frac{\partial}{\partial z}\sum_{n\geq 3} c_n(\mu)\rho^n P_n\left(\frac{x}{\rho}\right)\,, \end{aligned} \tag{4}$$

where $\rho = \sqrt{x^2 + y^2 + z^2}$ and where the coefficients c_n, $n \geq 2$, are given by the following expressions:

$$c_n(\mu) = \frac{1}{\gamma_1^3}\left(\mu + (-1)^n \frac{(1-\mu)\gamma_1^{n+1}}{(1-\gamma_1)^{n+1}}\right), \qquad \text{for } L_1 ,$$

$$c_n(\mu) = \frac{(-1)^n}{\gamma_2^3}\left(\mu + \frac{(1-\mu)\gamma_2^{n+1}}{(1+\gamma_2)^{n+1}}\right), \qquad \text{for } L_2 . \tag{5}$$

The Hamiltonian becomes

$$H^{(in)}(p_x, p_y, p_z, x, y, z) = \frac{1}{2}\left(p_x^2 + p_y^2 + p_z^2\right) + yp_x - xp_y - \sum_{n\geq 2} c_n(\mu)\rho^n P_n\left(\frac{x}{\rho}\right), \tag{6}$$

with $p_x = \dot{x} - y$, $p_y = \dot{y} + x$, $p_z = \dot{z}$.

The relation between $H^{(IN)}$ in (1) and $H^{(in)}$ in (6) is given by (see Gómez et al. 1991)

$$H^{(IN)} = H^{(in)}\gamma_j^2 - \frac{1}{2}(1 \mp \gamma_j - \mu)^2 - \frac{\mu}{\gamma_j} - \frac{1-\mu}{1 \mp \gamma_1}, \tag{7}$$

where the upper sign and $j = 1$ hold for L_1 and the lower sign for L_2, with $j = 2$.

3 Normalization and Center Manifold Reduction

In the present section we describe the procedures to construct the integrable approximation of the resonant dynamics around the collinear points, performing a resonant Birkhoff normalization of (6) by means of Lie transformations and then reduce the system to the center manifold.

It can be easily shown that the motion in the z-direction decouples with $\omega_z = \sqrt{c_2}$ frequency of the motion in such direction (see Jorba and Masdemont 1999; Celletti et al. 2015 for full details). Also, the eigenvalues associated with the planar system are

$$\eta_1 = \frac{c_2 - 2 - \sqrt{9c_2^2 - 8c_2}}{2}, \qquad \eta_2 = \frac{c_2 - 2 + \sqrt{9c_2^2 - 8c_2}}{2}. \tag{8}$$

As $c_2 > 1$, it results that $\eta_1 < 0$ and $\eta_2 > 0$, thus the equilibria is a saddle $\times$ center $\times$ center point. Let be $\omega_y \equiv \sqrt{-\eta_1}$, $\lambda_x = \sqrt{\eta_2}$; according to Celletti et al. (2015); Jorba and Masdemont (1999), there exists a symplectic change of variables, such that the quadratic part of the Hamiltonian is diagonalised as

$$H_2^{(d)}(p_1,p_2,p_3,q_1,q_2,q_3) = \lambda_x q_1 p_1 + i\omega_y q_2 p_2 + i\omega_z q_3 p_3 \,, \tag{9}$$

where $(p_1,p_2,p_3,q_1,q_2,q_3)$ are the diagonalising variables.

In such variables the Hamiltonian becomes

$$H^{(d)}(p_1,p_2,p_3,q_1,q_2,q_3) = \sum_{n\geq 2} H_n^{(d)}(p,q) \,, \tag{10}$$

where $H_n^{(d)}$ are homogeneous polynomials of degree n.

A resonant perturbation theory is thus performed in the neighborhood of the synchronous resonance $\omega_y = \omega_z$ (see Ferraz-Mello 2007) by constructing a canonical transformation, $(p,q) \longrightarrow (P,Q)$, which conjugates (10) to the form:

$$\begin{aligned} K^{(NF)}(P_1,P_2,P_3,Q_1,Q_2,Q_3) &= \lambda_x Q_1 P_1 + i\omega_y Q_2 P_2 + i\omega_z Q_3 P_3 \\ &+ \sum_{n=3}^{N} K_n^{(NF)}(Q_1P_1,P_2,P_3,Q_2,Q_3) + R_{N+1}(P,Q) \,, \end{aligned} \tag{11}$$

where the homogeneous polynomials $K_n^{(NF)}$, $n = 3,\ldots,N$ are in *normal form* with respect to the resonant quadratic part $K_2^{(NF)} = H_2^{(r)}$ with $H_2^{(r)}$ given by

$$H_2^{(r)}(P_1,P_2,P_3,Q_1,Q_2,Q_3) \equiv \lambda_x Q_1 P_1 + i\omega_z (Q_2 P_2 + Q_3 P_3) \tag{12}$$

and $R_{N+1}(P,Q)$ is a remainder function of degree $N+1$. By *normal form* we mean that each term up to order N in the series (11) satisfies the condition $\{H_2^{(r)}, K_n^{(NF)}\} = 0$, where $\{\cdot,\cdot\}$ denotes the Poisson brackets. The resonant quadratic Hamiltonian in (12), $H_2^{(r)}(P_1,P_2,P_3,Q_1,Q_2,Q_3)$ is obtained by the original quadratic part in (9) expressed in the new variables and modified in order to be resonant in the elliptic components. We can justify this assumption by observing that, for any $\mu \in (0,1/2]$ the two elliptic frequencies are such that the quantity $\delta = \omega_z\tilde{\delta} \equiv \omega_y - \omega_z$ to which we refer to as the *detuning*, is always a small quantity (of the order of 10^{-2}). The detuning provides a measure of the distance in the frequency from the synchronous resonance. In this way, even if the unperturbed system is strictly not resonant, we are able to describe the resonant dynamics of the perturbed system determined by the nonlinear coupling. The detuning parameter will be used to obtain series expansions of indicators, such as the bifurcation thresholds to halo orbits.

Since the normalization involving the hyperbolic components is a standard Birkhoff normalization, the normal form only depends on Q_1, P_1 through their product, while the remainder $R_{N+1}(P,Q)$ might depend on Q_1, P_1 separately.

Another change of variables is performed:

$$\begin{cases} Q_1 = \sqrt{I_x} e^{\theta_x} \\ Q_2 = \sqrt{I_y}(\sin\theta_y - i\cos\theta_y) = -i\sqrt{I_y} e^{i\theta_y} \\ Q_3 = \sqrt{I_z}(\sin\theta_z - i\cos\theta_z) = -i\sqrt{I_z} e^{i\theta_z} \\ P_1 = \sqrt{I_x} e^{-\theta_x} \\ P_2 = \sqrt{I_y}(\cos\theta_y - i\sin\theta_y) = \sqrt{I_y} e^{-i\theta_y} \\ P_3 = \sqrt{I_z}(\cos\theta_z - i\sin\theta_z) = \sqrt{I_z} e^{-i\theta_z} . \end{cases} \tag{13}$$

From the structure of the normal form Hamiltonian (11) we see that the 'action' variable $I_x = Q_1 P_1$ is a constant of motion, whenever the remainder is neglected. Therefore, given an initial condition $I_x(0) = 0$ and neglecting R_{N+1}, we obtain an integrable Hamiltonian in two degrees of freedom (hereafter DOF), which provides the dynamics in the center manifold within an approximation to order N. Within the center manifold, we describe the motion by the following 2-DOF Hamiltonian in action–angle variables:

$$K^{(CM)}(I_y, I_z, \theta_y, \theta_z) = K_0(I_y, I_z) + K_r(I_y, I_z, \theta_y - \theta_z) + R^{(r)}(I_y, I_z, \theta_y, \theta_z) , \tag{14}$$

where K_0 depends only on the actions; K_r is the resonant part depending on the actions as well as on the angles, but just through the combination $\theta_y - \theta_z$, which corresponds to the synchronous resonance; $R^{(r)}$ represents the reduced remainder function. This procedure will lead to have, by construction, that $\dot{I}_y + \dot{I}_z = 0$ up to the remainder.

4 Analytical Estimates of the Bifurcation Values

In this section we provide a method to give an analytical estimate of the value at which a bifurcation to halo orbits occurs. This method, inverted, provides an analytical formula for approximated initial conditions halo orbits. Truncating the normal form whenever the first resonant terms appear can be considered a *first-order* resonant perturbation approach. From the order of the resonance generated by (12) it is straightforward to check that the odd degree terms in the normal form vanish and that the first non-trivial term is $K_4^{(NF)}$. Therefore, truncating (14) to degree two in the actions leads to the following first-order normal form:

$$K^{(CM1)}(I_y, I_z, \theta_y, \theta_z) = \omega_y I_y + \omega_z I_z + [\alpha I_y^2 + \beta I_z^2 + I_y I_z(\gamma + 2\tilde{\gamma}\cos(2(\theta_y - \theta_z)))] \tag{15}$$

with suitable coefficients α, β, γ, $\tilde{\gamma}$.

By reduced 1DOF dynamics we mean the following. Let us make the change of variables:

$$\begin{aligned} \mathscr{E} &= I_y + I_z \\ \mathscr{R} &= I_y \\ \nu &= \theta_z \\ \psi &= \theta_y - \theta_z \, . \end{aligned} \tag{16}$$

Then, the Hamiltonian (15) is transformed into

$$H_6^{(tr)}(\mathscr{E},\mathscr{R},\nu,\psi) = \mathscr{E}+\tilde{\delta}\mathscr{R}+a\mathscr{R}^2+b\mathscr{E}^2+c\mathscr{E}\mathscr{R}+d(\mathscr{R}^2-\mathscr{E}\mathscr{R})\cos(2\psi) \, , \tag{17}$$

where the constants are defined as follows: $a = (\alpha + \beta - \gamma)/\omega_z$, $b = \beta/\omega_z$, $c = (\gamma - 2\beta)/\omega_z$, $d = -2\tilde{\gamma}/\omega_z$. Finally, Hamilton's equations associated to (17) take the form

$$\begin{aligned} \dot{\mathscr{E}} &= 0 \\ \dot{\mathscr{R}} &= 2d\mathscr{R}(\mathscr{R} - \mathscr{E})\sin(2\psi) \\ \dot{\nu} &= 1 + 2b\mathscr{E} + c\mathscr{R} - d\mathscr{R}\cos(2\psi) \\ \dot{\psi} &= \tilde{\delta} + 2a\mathscr{R} + c\mathrm{e} + d(2\mathscr{R} - \mathrm{e})\cos(2\psi) \, . \end{aligned} \tag{18}$$

Therefore we obtain a 1DOF system, in which e is constant. As$\tilde{\delta} = \frac{\omega_y - \omega_z}{\omega_z}$, the theory developed in Pucacco and Marchesiello (2014); Celletti et al. (2015) translates into the following constraints, which provide the existence of resonant orbits, bifurcating from the normal modes:

$$\mathscr{E} \geq \mathscr{E}_{iy} \doteq \frac{\delta}{2(\alpha - \tilde{\gamma}) - \gamma} \quad \text{or} \quad \mathscr{E} \geq \mathscr{E}_{iz} \doteq \frac{\delta}{2(\beta - \tilde{\gamma}) - \gamma} \tag{19}$$

for the inclined orbits (namely with the fixed phase relation $\theta_y - \theta_z = 0, \pi$) and

$$\mathscr{E} \geq \mathscr{E}_{\ell y} \doteq \frac{\delta}{2(\alpha + \tilde{\gamma}) - \gamma} \quad \text{or} \quad \mathscr{E} \geq \mathscr{E}_{\ell z} \doteq \frac{\delta}{2(\beta + \tilde{\gamma}) - \gamma} \tag{20}$$

for the loop orbits (that is the orbits with the fixed phase relation $\theta_y - \theta_z = \pm\pi/2$). The first of (20) is just the occurrence of the bifurcation of the halo family from the planar Lyapunov orbit, which becomes unstable. A second bifurcation may occur at the value given by the first of (19), when the Lyapunov orbit regains stability. The estimate energy level at which the bifurcation takes place to first order is simply

$$E_1 = \omega_z \mathscr{E}_1 = \omega_z C_1 \delta \, , \tag{21}$$

which, coming back to the original coefficients, gives the bifurcation value

$$E_1 = \frac{\omega_z \delta}{\gamma - 2(\alpha + \tilde{\gamma})} . \tag{22}$$

The procedure could be analogously iterated for the upper orders of the Hamiltonian (Ceccaroni et al. 2016b).

5 Results

A comparison of the results between the analytical estimates at different orders and the numerical values available in the literature (Hénon 1973, 1974; Gómez et al. 1991; Gómez and Mondelo 2001, 2014) is provided in Tables 1 and 2. As far as the analytical estimates are concerned, the first-order predictions have been computed by using formula (22). The results obtained computing the normal form at higher orders, up to the sixth order (which corresponds to degree 7 in the actions) are reported as well (see Ceccaroni et al. 2016b).

The analysis of the data in Tables 1 and 2, respectively for L_1 and L_2, show a very good agreement with the numerical data. In particular, the theoretical results improve as the order of normalization increases, reaching the convergence up to the

Table 1 Results for the analytical bifurcation estimates for L_1 up to a normal form of order 6 and the numerical values obtained in Hénon (1973, 1974); Gómez et al. (1991); Gómez and Mondelo (2001), physical energy; the values of the mass ratios are $\mu_{bS} = 3.0404326 \times 10^{-6}$ and $\mu_{EM} = 0.01215058$

	Hill's case	Barycenter–Sun	Earth–Moon	Equal masses
	$\mu \to 0$	$\mu = \mu_{bS}$	$\mu = \mu_{EM}$	$\mu = 1/2$
First order	−1.500000	−1.500415	−1.587193	−1.961675
Second order	−1.500000	−1.500417	−1.587175	−1.961534
Sixth order	−1.500000	−1.500416	−1.587176	−1.961536
Numerical	−1.50000	−1.50042	−1.58718	−1.96154

Table 2 Results for the analytical bifurcation estimates for L_2 up to a normal form of order 6 and the numerical values obtained in Hénon (1973, 1974); Gómez and Mondelo (2001, 2014), physical energy; the values of the mass ratios are $\mu_{bS} = 3.0404326 \times 10^{-6}$ and $\mu_{EM} = 0.01215058$

	Hill's case	Barycenter–Sun	Earth–Moon	Equal masses
	$\mu \to 0$	$\mu = \mu_{bS}$	$\mu = \mu_{EM}$	$\mu = 1/2$
First order	−1.500000	−1.500412	−1.575838	−1.524509
Second order	−1.500000	−1.500413	−1.576087	−1.548191
Sixth order	−1.500000	−1.500413	−1.576060	−1.544820
Numerical	−1.50000	−1.50041	−1.57606	−1.54476

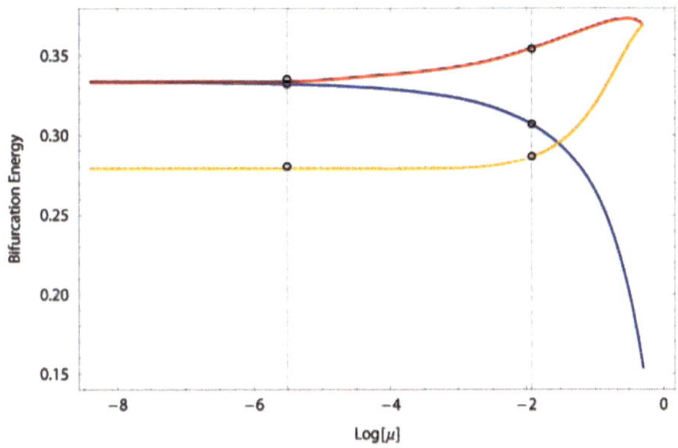

Fig. 1 Bifurcation thresholds as a function of the mass ratio: L_1 (*blue*), L_2 (*red*), L_3 (*yellow*)

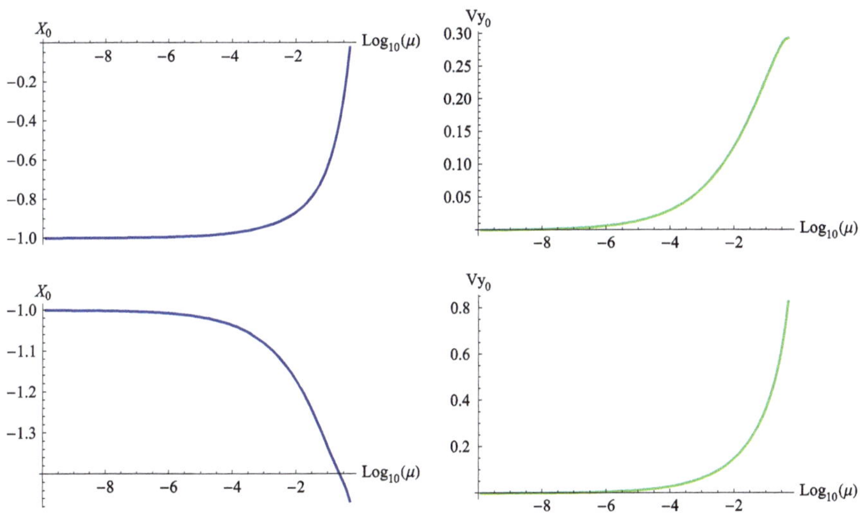

Fig. 2 Initial values for X (*left panels*) and $\dot{Y}$ (*right panels*) of the first halo orbits as a function of the mass ratio for L_1 (*upper panels*) and L_2 (*lower panels*)

6th decimal digit at the fifth order of normalization. Figure 1 shows the results of the table evaluated for 20 values of mass ratio and interpolated.

In view of concrete applications, we find it convenient to give also the initial values of X_0, $\dot{Y}_0$ as a function of μ, which correspond to the bifurcations of the halo orbits. We notice that the plots presented in Fig. 2 are in substantial agreement with the values reported in Hénon (1973, 1974).

References

Bucciarelli, S., Ceccaroni, M., Celletti, A., Pucacco, G.: Qualitative and analytical results of the bifurcation thresholds to halo orbits. Annali di Matematica Pura ed Applicata (2015). http://dx.doi.org/10.1007/s10231-015-0474-2

Ceccaroni, M., Celletti, A., Pucacco, G.: Birth of periodic and artificial halo orbits in the restricted three-body problem. Int. J. Non Linear Mech. **81**, 65–74 (2016a)

Ceccaroni, M., Celletti, A., Pucacco, G.: Halo orbits around the collinear points of the restricted three-body problem. Physica D **317**, 28–42 (2016b)

Celletti, A., Pucacco, G., Stella, D.: Lissajous and Halo orbits in the restricted three-body problem. J. Nonlinear Sci. **25**(2), 343–370 (2015)

Ferraz-Mello, S.: Canonical Perturbation Theories. Springer, New York (2007)

Gómez, G., Mondelo, J.M.: The dynamics around the collinear equilibrium points of the RTBP. Physica D **157**, 283–321 (2001)

Gómez, G., Mondelo, J.M.: Private communication (2014)

Gómez, G., Jorba, À., Masdemont, J., Simó, C.: Study refinement of semi-analytical Halo orbit theory. ESOC Contract 8625/89/D/MD(SC), Final Report (1991)

Hénon, M.: Vertical stability of periodic orbits in the restricted problem. I. Equal masses. Astron. Astrophys. **28**, 415–426 (1973)

Hénon, M.: Vertical stability of periodic orbits in the restricted problem. II. Hill's case. Astron. Astrophys. **30**, 317–321 (1974)

Howell, C., Barden, B.T., Lo, M.: Application of dynamical systems theory to trajectory design for a libration point mission. J. Astronaut. Sci. **45**(2), 161–178 (1997)

Jorba, À., Masdemont, J.: Dynamics in the center manifold of the collinear points of the restricted three body problem. Physica D **132**, 189–213 (1999)

Murray, C.D., Dermott, S.F.: Solar System Dynamics. Cambridge University Press, Cambridge (1999)

Pucacco, G., Marchesiello, A.: An energy-momentum map for the time-reversal symmetric 1 : 1 resonance with $Z_2 \times Z_2$ symmetry. Physica D **271**, 10–18 (2014)

Richardson, D.L.: Analytic construction of periodic orbits about the collinear points. Celest. Mech. **22**, 241–253 (1980)

Szebehely, V.: Theory of Orbits. Academic, New York (1967)

Unconventional Solar Sailing

Matteo Ceriotti

Abstract The idea of exploiting solar radiation pressure for space travel, or solar sailing, is more than a 100 years old, and yet most of the research thus far has considered only a limited number of sail configurations. However solar sails do not have to be inertially-pointing squares, spin-stabilised discs or heliogyros: there is a range of different configurations and concepts that present some advantageous features. This chapter will show and discuss three non-conventional solar sail configurations and their applications. In the first, the sail is complemented by an electric thruster, resulting in a hybrid-propulsion spacecraft which is capable to hover above the Earth's Poles in a stationary position (pole-sitter). The second concept makes use of a variable-geometry pyramidal sail, naturally pointing towards the sun, to increase or decrease the orbit altitude without the need of propellant or attitude manoeuvres. Finally, the third concept shows that the orbit altitude can also be changed, without active manoeuvres or geometry change, if the sail naturally oscillates synchronously with the orbital motion. The main motivation behind these novel configurations is to overcome some of the engineering limitations of solar sailing; the resulting concepts pose some intriguing orbital and attitude dynamics problems, which will be discussed.

1 Introduction

Solar sailing has long been studied and proposed as potential means to accelerate a spacecraft without the use of propellant, hence achieving (at least theoretically) an unlimited Δv. The concept essentially employs a large, thin, reflective surface facing the sun, and exploits the solar radiation pressure (SRP) exerted by the photons to impart an acceleration to the spacecraft. Ideally, the acceleration provided by a

M. Ceriotti (✉)
School of Engineering, University of Glasgow, James Watt South Building, Glasgow G12 8QQ, UK
e-mail: matteo.ceriotti@glasgow.ac.uk

G. Gómez, J.J. Masdemont (eds.), *Astrodynamics Network AstroNet-II*, Astrophysics and Space Science Proceedings 44,
DOI 10.1007/978-3-319-23986-6_4

sail is:

$$\mathbf{a} = \eta\beta\frac{\mu}{r^2}\cos^2\alpha\,\hat{\mathbf{n}} \tag{1}$$

where η is an efficiency coefficient, μ is the sun gravitational constant, α is the angle between the sail normal $\hat{\mathbf{n}}$ (away from the sun) and the sun line, r is the distance of the sun, and $\beta = \sigma^* A/m$ is the lightness number, which essentially depends on the area-to-mass ratio of the spacecraft (σ^* is a constant for the sun). More detailed models include non-ideal effects such as absorption, scattering and re-emission of photons (McInnes 1999).

The literature in the last 100 years has mainly focused on three configurations of solar sails: square, heliogyro and spinning disc. Each of these three configurations has its own advantages and drawbacks. The square sail is the simplest of the three, usually with four booms deployed diagonally, and four triangular sail membranes between each pair. The deployment relies entirely on the booms (which can be motorized or store elastic energy), and the solar radiation pressure force is again transmitted from the membrane to the spacecraft bus through the bending stiffness of the booms. Demonstrators of this technology have flown: NASA NanoSail-D2 (Johnson 2011), The Planetary Society LightSail (Biddy and Svitek 2012). A heliogyro is essentially a spinning propeller, similar to windmills, but powered by photons. It exploits the centrifugal force due to spinning to deploy the "propeller" blades, made of strips of reflective material. It was considered as an option for the (subsequently cancelled) NASA mission to comet Halley in the mid-1970s, mainly due to its sheer size and therefore high cost and risk. The spinning disk type is spinning similarly to the heliogyro, but it uses a flat surface instead of individual blades. JAXA IKAROS (Yamaguchi et al. 2010) was a demonstrator of this type of sail: it was injected into an interplanetary trajectory to Venus in 2010. Apart from demonstrators, the vast majority of solar sail studies envisaged interplanetary missions: this is because solar sailing is particularly advantageous for these types of missions, where the dynamics is slower than in Earth-bound orbits, long time of flights are possible, and the low but continuous acceleration can be exploited. However, this results in high-cost, high-risk missions and spacecraft which are hard to demonstrate and validate on a smaller scale. In other words, solar sailing is still on a relatively low level in the technology readiness level (TRL) scale, and in addition, for the reasons discussed above, its advancement degree of difficulty (AD2) (Macdonald and McInnes 2010) is high. In the last few years, the author and collaborators have been investigating novel ways to use solar sailing, with the aim of being able to use it for low-cost, Earth-based missions, particularly involving small satellites, paving the way for cheaper demonstrators and missions. The research presented in this chapter aims to:

1. Investigate and propose novel configurations and concepts for solar sailing, moving away from the square-heliogyro-spinning disc sail paradigm;
2. Lower the AD2 of solar sailing, by:

a. Adjoining solar sails to other conventional propellant-based propulsion, thus reducing the demand on the sail, and reducing the risk of mission failure should the sail not deploy;
b. Enabling sails to be used for Earth-bound missions, which are less expensive than interplanetary ones;
c. Enabling solar sails to be used on nanosatellite platforms, such as CubeSats, therefore further lowering the cost of the mission.

In doing so, this chapter summarises some results of on-going research on three different mission concepts and corresponding applications: hybridising a solar sail with a solar electric propulsion (SEP) thruster; using variable-geometry to modulate the sail acceleration in Earth orbits; and finally, exploiting SRP-induced oscillatory motion to remove the need of active attitude control.

2 Hybrid Solar Sail Propulsion

Due to the intrinsic limitation of the sail to provide an acceleration away from the Sun and limitations in realistic sail area-to-mass ratios, some of the mission scenarios, in which a continuous Δv is required, simply cannot be implemented with a solar sail.

In order to overcome these limitations, hybrid propulsion has been proposed (Leipold and Götz 2002). In the hybrid system, at the cost of increased spacecraft and mission design complexity, the SEP and the sail complement each other, cancelling their reciprocal disadvantages and limitations. In principle, SEP can provide thrust in almost any direction (as long as the exhaust flow does not interfere with other spacecraft systems), and in particular towards the Sun, where the sail cannot. Similarly, the hybrid spacecraft can be seen as an SEP spacecraft, in which an auxiliary solar sail provides part of the acceleration, enabling a saving of propellant mass, and lowering the demand on the electric thruster, possibly with some intervals in which it could be turned off. In this sense, the hybrid spacecraft can be seen as a way to gradually introduce solar sails for space applications, and hence to reduce the AD2 (Macdonald and McInnes 2010) in the TRL scale. Therefore, hybrid propulsion can, in principle, enable missions that are not feasible using only a solar sail and can extend the mission lifetime with respect to the pure SEP case.

2.1 *Pole-Sitter*

Hybridizing SEP and solar sailing is a comparatively recent idea (Leipold and Götz 2002), nevertheless research is flourishing in this field, investigating novel, interesting applications: artificial equilibria above L_1 in the Sun-Earth system for

Earth observation (Baig and McInnes 2008), optimal interplanetary transfers to Venus and Mars (Mengali and Quarta 2007b,a), displaced periodic orbits in the Earth-Moon system (Simo and McInnes 2009) and displaced non-Keplerian orbits for geostationary coverage (Heiligers et al. 2011).

What we will be discussing here is a satellite which does not orbit, and instead it is stationary above one of the poles, with respect to the Earth. In this position, the footprint of the spacecraft will be constantly at the pole, in the same way as the footprint of a geostationary spacecraft is constantly at some longitude on the equator.

This spacecraft is known in the literature as "pole-sitter" (Ceriotti and McInnes 2011a), and it uses low-thrust propulsion to maintain a stationary position relative to the Earth along its rotation axis. The pole-sitter is the only platform that can offer a truly continuous hemispheric view of one of the poles, enabling real-time imaging over the entire disc. The first study of this concept was apparently made by Driver (1980), although the author notes that the original idea belongs to the mathematician and writer Kurd Lasswitz from 1897. Applications include weather forecasting, polar ice cap monitoring, imagery and observation of the Polar Regions, vessel tracking, and low-bandwidth telecommunications with the high latitudes, where the geostationary spacecraft are not accessible (Ceriotti et al. 2012a,b).

It was shown that a solar sail cannot maintain the spacecraft in this position throughout the year, due to the said limitation in acceleration direction. On the other hand, Driver (1980) showed that the thrust requirements are rather high, and the lifetime relatively limited, if SEP alone is used. It seems therefore natural to investigate the hybrid system.

2.2 *Hybrid Pole-Sitter Orbits*

In the following, we consider that the direction of the polar axis of the Earth is inertially fixed while the Earth is orbiting the Sun. In the synodic ecliptic reference frame (Sun-Earth), the same axis rotates with a motion of apparent precession, due to the obliquity of the ecliptic: its angular velocity is $\omega = -2\pi/\text{year}$. Therefore, the polar axis spans a full conical surface every year, in a clockwise direction (refer to Fig. 1). The cone half angle is the Earth's obliquity on the ecliptic, i.e. 23.5°. The position of the spacecraft is to be constrained to follow the clockwise apparent precession of the polar axis, and hence maintain the pole-sitter condition.

An optimal pole-sitter orbit is defined as the one that minimizes the propellant consumption of the spacecraft, while maintaining the pole-sitter condition at any time, and being 1-year periodic. The implicit approximation that is made here is that the optimal control problem is solved only for the first year (or period) of the mission, and the same trajectory is then used for following years. Details of the numerical optimization process are presented in Ceriotti and McInnes (2011a).

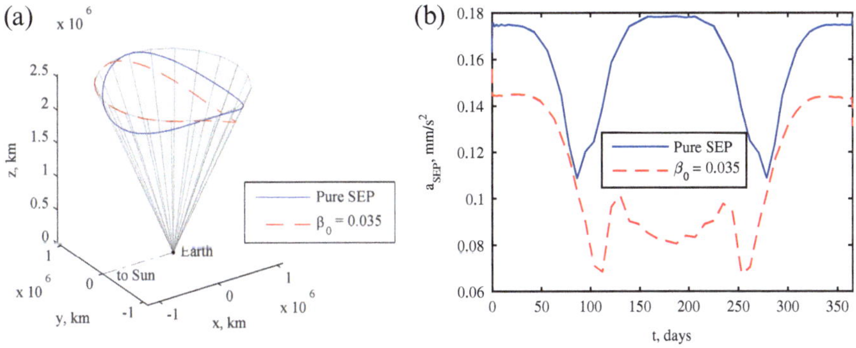

Fig. 1 Fuel-optimal pole-sitter orbits for the pure SEP case and the hybrid case. (**a**) Optimal trajectories in the synodic reference frame. (**b**) SEP acceleration (magnitude) required

The mass of the spacecraft appears in the equations of motion. The initial mass is set arbitrarily to 1000 kg, but in reality it depends on the launcher capability and the transfer phase, which are to be designed yet; however, the solutions found are scalable in mass. The result of the optimization, for the two scenarios, is presented in Fig. 1. The optimal SEP-only path is essentially symmetric around spring and autumn, and the spacecraft is closest to the Earth at the summer and winter solstices. Instead, in the hybrid case, the spacecraft is closest to the Earth in winter and farthest in summer. The SEP spacecraft distance to Earth varies between 2.3457 and 2.7391 million km, while for the hybrid case it varies between 2.0480 and 2.7391 million km. Figure 1(b) plots the modulus of the SEP acceleration as function of time in the two scenarios, which shows that the hybrid case needs less acceleration due to the contribution of the sail.

A preliminary mission design was performed, for pure SEP and hybrid sail-SEP, with the aim of comparing the lifetime for carrying given payload (for a selection of possible payloads for the mission, see Ceriotti et al. 2012a), including the optimisation of the transfer trajectory from launch to pole-sitter operational phase. Two launcher options were considered: Ariane 5 and Soyuz, and their orbit and payload capabilities were extracted from their respective manuals; here the Ariane 5 option is presented. More details on this approach as well as the generation of suitable initial guesses can be found in Heiligers et al. (2012a).

In addition, a transfer phase is introduced that allows the pole-sitter spacecraft to change between pole-sitter orbits above the North and South Poles before the start of the Arctic and Antarctic winters. In fact, due to the tilt of the obliquity of the ecliptic, the North and South Poles are alternately situated in darkness for 6 months per year. For observations performed in the visible part of the spectrum, this significantly constrains the mission scientific return.

The work in Heiligers et al. (2012b), in which more details on the trajectory optimisation can be found, showed that two types of transfers can be considered: a short transfer that takes less than half a year and a long transfer that takes between half a year and 1 year. In order to maximize the observation time during the mission, the short option is chosen. Departure takes place between summer and autumn (June–September), while arrival takes place between autumn and winter (September–December), where this chapter conventionally refers to the seasons in the northern hemisphere. Note that due to the symmetry of the problem, the optimal transfers from north to south can also be used to transfer from south to north.

2.3 *Complete Mission and Spacecraft Design*

Figure 2 represents the full trajectory of the pole-sitter spacecraft, including the launch phase, operational phases and north-to-south transfers, for the hybrid spacecraft launched by Ariane 5. Different trajectories were found for the other options. The plot is again in an Earth-centred synodic reference frame (Earth and Lagrange points L_1 and L_2 are represented with dots).

Now that the trajectory is optimised, a systems design provides the mission lifetime that can be achieved into this orbit, or alternatively the payload mass that can be carried for a given lifetime of the spacecraft. The spacecraft mass of the pure SEP is split into payload, propellant, tank, SEP thrusters, solar arrays and other subsystems; the hybrid spacecraft adds the SEP thruster gimbal, the solar sail and radiators to dissipate excess power. Details are provided in Ceriotti and McInnes (2011b) and Ceriotti et al. (2014c). The resulting payload mass, as a function of the lifetime, is plotted in Fig. 3 for the two launchers and propulsion technologies. It can be noted that the maximum mission lifetime (i.e. no payload carried on-board) does not depend on the injection total mass, but only on the technology that is used to build the spacecraft. The lifetime for the pure SEP system is limited to approximately 4.5 years, while this value extends to about 7 years for the hybrid architecture. This result itself should be sufficient to justify the interest in the hybrid propulsion technology for this type of mission, and in general for all those missions which require a continuous acceleration (Ceriotti and McInnes 2011b). Furthermore, for nearly equal injected masses, the hybrid spacecraft can carry the

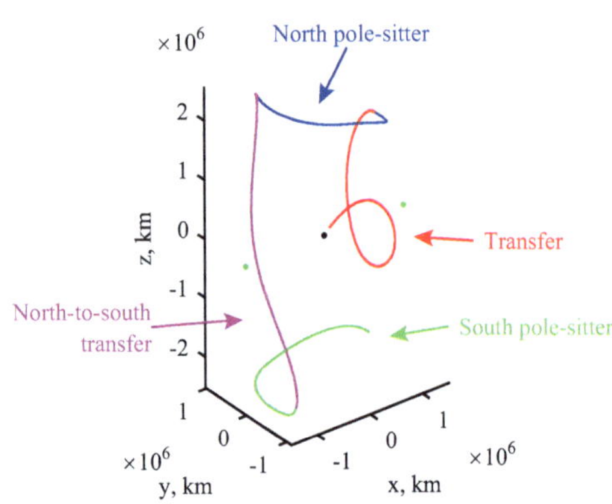

Fig. 2 Mission trajectory (Ariane 5 launch, hybrid propulsion) in synodic reference frame

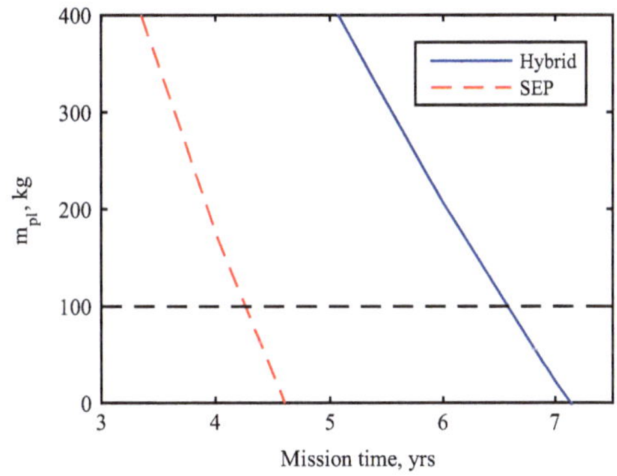

Fig. 3 Payload mass as a function of the mission lifetime, for the pure SEP and hybrid spacecraft, Ariane 5 launch

Table 1 Design points for 100 kg payload mass, Ariane 5 launch. Masses are without margins

Propulsion system	Pure SEP	Hybrid
Lifetime, years	4.2	6.6
Pole-sitter injected mass, kg	4439	4599
SEP mass, kg	348	299
Propellant mass, kg	2192	2242
Other subsystems mass, kg	674	707
Total sail area (reflective + solar cells), m^2	–	324 × 324
Maximum SEP thrust, mN	776	667
Maximum SEP power, kW	17.4	15.0

same payload mass for a longer mission lifetime. Finally, for a particular spacecraft architecture and a given lifetime, the payload mass scales linearly with the injected mass.

By fixing a payload mass, e.g. 100 kg, the design points can be computed, and they are described in Table 1. The design points are computed assuming that the entire capacity of the launcher is used for the pole-sitter spacecraft; however, it is possible to scale down any of the four scenarios (at the cost of a reduction in mission lifetime). The lifetime is 4.2 years if the spacecraft is using pure SEP technology, or 6.6 years if using hybrid propulsion. (3.6 years and 5.6 years respectively for a Soyuz launch). The subsystem design also allows the computation of the mass of the other subsystems, some of which are reported in Table 1. The size of the total sail assembly (reflective surface and thin film solar cells) of the hybrid spacecraft, assuming a square assembly, is $324 \times 324\,m^2$ ($191 \times 191\,m^2$ for the Soyuz launch).

3 Solar Sailing with Variable Geometry

One can argue that having an electric thruster to support a solar sail is no more *solar sailing*: it defeats the point that no fuel, thrusters and other related subsystems need be designed, carried and used on the spacecraft, and most importantly, the lifetime, although extended, is still ultimately limited by availability of propellant.

This objection motivates the research summarised in this section, which will focus on the use of solar radiation pressure only for orbit manoeuvring around the Earth. Studies in the literature have considered planet-centred solar sails, including optimal control laws (Macdonald and McInnes 2005a; Stolbunov et al. 2013), orbit raising (Mengali and Quarta 2005) and escape strategies (Macdonald and McInnes 2005b). In all these studies, however, slew manoeuvres were assumed to change the direction (and magnitude) of the acceleration provided by the sail, and an active attitude control system is required to steer the sail according to the control law.

One of the technological challenges of solar sails is the fact that a large deployed surface, even if thin and lightweight, can increase considerably the moments of inertia of the satellite. It is well known that the inertia is proportional to the mass and the square of the distance from the centre of mass. For a conventional propulsion system that is not mounted on a gimbal, the thrust vector is directed through changing the attitude of the entire spacecraft. For a solar sail, this obviously means a yaw manoeuvre of the entire sail. Attitude manoeuvres do not generally require propellant: gyroscopes and inertia wheels could in principle be used. In addition, the solar sail itself can be exploited, for example with extensible and/or tiltable solar vanes (Wie and Murphy 2007), or moving masses: both these devices displace the centre of mass with respect of the centre of pressure, therefore resulting in a solar radiation pressure torque. Differential reflectivity can also be generated by changing the reflectivity on parts of the membrane (Funase et al. 2011). These techniques can also be used if the sail is of spinning type: IKAROS has demonstrated this concept through the use of photo-chromic coating on the sail surface (Funase et al. 2010). However, the spin adds additional gyroscopic stiffness, making even more demanding to change the spinning axis direction of the sail.

Despite the torque needed for slew attitude manoeuvres can be entirely provided by the SRP, due to the large moments of inertia, these manoeuvres are relatively slow. In fact, research has been done (Borggräfe 2011) on sail trajectory optimisation taking into account the constraint of the maximum slew rate and torque of the spacecraft. However, the small slew rate is generally not a major issue for interplanetary missions, as the trajectory timescale is generally very slow in comparison; however, this becomes an issue on Earth orbits, where orbital periods can be as short as few hours. By observing Gauss' variational equations (Battin 1999), it is apparent that most orbital manoeuvres, which aim at a net change of one or more orbital parameters, require an acceleration vector that changes its direction considerably along one orbital period; for example, a change in inclination requires an out-of-plane acceleration which switches sign every half orbit; a change in semimajor axis requires an acceleration aligned with the velocity vector, which rotates in the orbital plane once per orbit. This issue becomes more important if the spacecraft bus is in the nanosatellite class (such as CubeSats), where the constraints in power and size significantly affect the capabilities of the attitude control system.

3.1 Heliostability and Variable Geometry

We propose the use of a variable-geometry solar sail for spacecraft operating in Earth orbit. Specifically, this novel concept is based on a solar sail deployed in the shape of a quasi-rhombic pyramid (QRP) (Ceriotti et al. 2014a), where the spacecraft bus lies at the apex and deploy booms along the slant edges, with the reflective membranes filling the slant faces. Thanks to the sail geometry and shift between centre of mass and centre of solar pressure (or heliostatic margin), the spacecraft is heliostable, i.e. nominally pointing towards the sun, and when deflected, experiences a restoring moment towards the nominal attitude without need of active control (as shown in Fig. 4). If undamped, the motion is a quasi-harmonic oscillation about the equilibrium position. Additional stability is provided through spin motion, which maintains the nominal attitude even when heliostability is not available, such as during the eclipses (Felicetti et al. 2016).

In addition, motorized hinges allow adjustment of the flare angle of the booms, effectively varying the sail area exposed to the Sun without creating any net torque and without requiring a slew manoeuvre or a change in the sail acceleration direction (as in Fig. 4). The spacecraft under consideration is a 1U, 1 kg CubeSat bus, adjoined by the sail. Data are in Table 2.

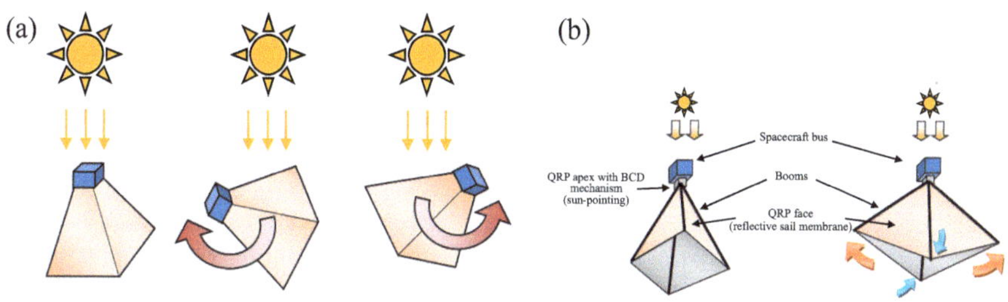

Fig. 4 (**a**) Heliostability. (**b**) Varying the boom flare angle

Table 2 Spacecraft data

Bus mass, kg	1
Bus size, cm	10
Boom linear density, g/m	16.3
Sail membrane areal density, g/m^2	13.2
Sail assembly mass per unit area, kg/m^2	0.050
Sail efficiency	0.85
Nominal spin rate, rph	10

3.2 Orbital Control

This mission concept is intended to allow either an increase or a decrease the semimajor axis of the orbit. This goal is achieved by opening or closing the sail booms in order to control the thrust provided by the SRP in different parts of the orbit. Assuming a zero-inclination orbit, with the sun line ideally in the orbital plane, to maximize the semimajor axis change, the acceleration should be tangential to the orbit, so the sail (in its nominal sun-pointing attitude) should be open when the spacecraft is travelling away from the Sun, and it should be closed when the spacecraft is travelling towards the Sun. In this work, we consider equatorial orbits (i.e. zero inclination) for reasons which will become clear later on.

There will only be only one true anomaly in which sun line and velocity are antiparallel, and thus the only SRP acceleration component is the tangential one; in all other locations of the orbit, the SRP acceleration will have a normal component as well. However, when the Sun is out of the orbital (equatorial) plane, then the SRP acceleration will also have an out-of-plane component. When sun-line and velocity are nearly perpendicular, an open sail provides only a small contribution to the tangential acceleration, with a large normal component that does not contribute to the semimajor axis change. For this reason, it is decided to open the sail (maximum SRP) in the range of anomaly shown in Fig. 5, where the angle $f_{control} \in [0, \ \pi/2\,]$ is a control law parameter.

By inspecting Gauss' equations again, it is found that a tangential acceleration changes the eccentricity of the orbit. The change depends on the true anomaly f, and in particular for a circular orbit ($e = 0$) is positive for half of the orbit and negative for the other half. This means that if we consider an equatorial or quasi-equatorial circular orbit, where the sun vector (apparently) rotates around the orbit once per year (see again Fig. 5), the change in eccentricity largely vanishes over the course of a full year. This justifies the choice of a circular, equatorial orbit. The net change in inclination, due to the out-of-plane acceleration component, is also almost completely cancelled out over a full year, due to opposite changes in summer and winter. In summary, this strategy results in a net increase (or decrease) of semimajor axis without substantial change to the other orbital elements, over a full year.

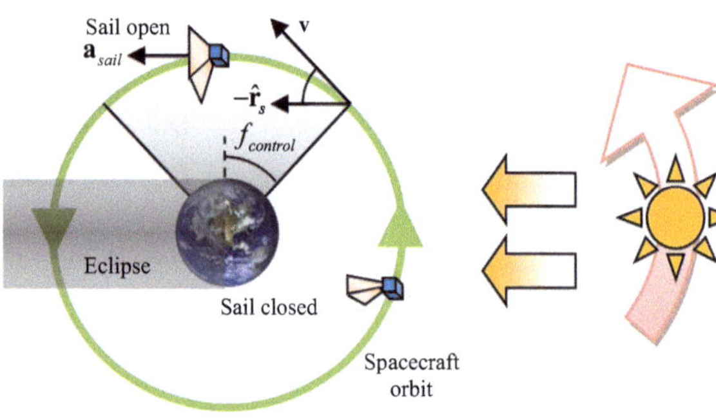

Fig. 5 Scheme of the fundamental control law

However, eclipses shall be taken into account. In fact, when the spacecraft enters the eclipse, the heliostability torque disappears and the gravity gradient becomes the dominant forcing term. Therefore only the gyro-stiffness of the spinning satellite can counteract such perturbation, to keep the sail pointing towards an inertially-fixed direction during this orbital phase. During the eclipses, the configuration of the sail does not affect the solar radiation pressure, but only the free-body motion and the gravity gradient torque, therefore it makes sense to reconfigure the spacecraft in that phase, to minimise the oscillation growth, until the heliostability is recovered at the end of the eclipse.

3.3 *Attitude Stability Analysis*

Because the spacecraft is spun, its attitude must be analysed, in order to determine the configurations (in terms of boom flare angles) that guarantee a stable spinning. This can be analysed by linearizing the Euler equations of rigid-body motion (Schaub and Junkins 2009) around the nominal configuration of the sail, and considering small displacements of the spin axis with respect to the sun direction. The result is that, in the space of possible geometries (design ratio base/length of the booms of the pyramid) and configurations (flare angles of the booms), there exist stability regions: once the geometry is fixed in the design phase, the sail should vary the flare angles such to remain in these regions, or otherwise the spinning motion will turn into tumbling (Fig. 6). Further details are in (Felicetti et al. 2016).

The solar radiation pressure produces conservative torques, therefore once an oscillation around the sun-line is started, it will continue indefinitely. In order to prevent the oscillation amplitude to grow due to perturbations such as gravity gradient, we propose to equip the spacecraft with passive nutation dampers, of the

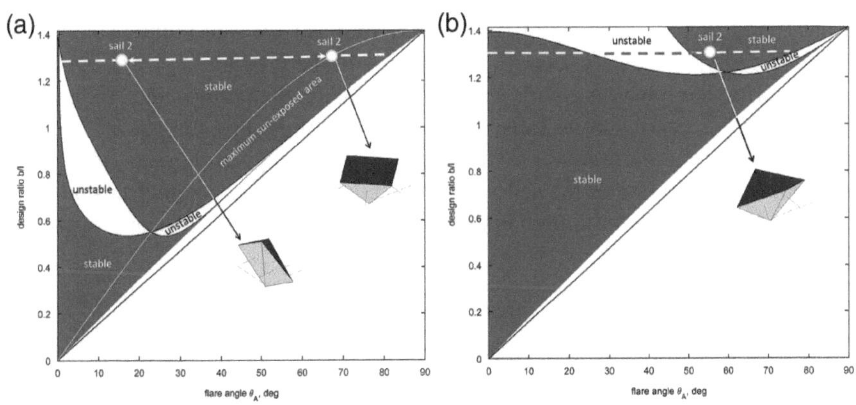

Fig. 6 Heliostable configurations for boom length $l = 2\,\text{m}$, as a function of the flare angle of the booms and the design ratio base/length of the booms. (**a**) Sun-illuminated phase. (**b**) During eclipse

type used for nutation damping. These devices usually consist of a sealed ring attached to the spacecraft bus, and they are filled with a viscous fluid. When the spacecraft experiences an angular acceleration, a viscous lag between the bulk fluid and the ring walls dissipates rotational energy as heat (Nobari and Misra 2012).

3.4 Numerical Simulations and Results

The case of the spacecraft initially orbiting in a geostationary orbit is considered, with $f_{control} = 60°$. Based on the stability analysis of the previous section, we selected a sail geometry and relative stable configurations, as in Table 3. Others are also possible, for example varying the length of the booms.

The concept was validated and assessed through numerical simulations. The fully-coupled attitude and orbital dynamic equations of motion were integrated, and included the effect of the shape change (changing the moments of inertia), nutation dampers, solar radiation pressure on the sail, eclipses, gravity gradient, and long-term effects of J_2. Figure 7 shows the working mechanism of the variable QRP sail ($l = 1$ m) during a period of 3 days. The orbit raises only during the fully-open phases (highlighted with the grey bands), when the solar radiation force is mainly directed along the velocity of the spacecraft. During the other phases the altitude is almost constant or slowly decreasing: closing the sail limits the undesired negative work produced by the solar radiation force, by minimizing its exposed area to the sun.

Year-long simulations (365 days) are used to assess the stability and the performance of the proposed concept in the long term (Felicetti et al. 2016). Figure 8 shows the pointing angle, i.e. the angle between the spin axis and the sun line. This angle should be nominally zero, as the sail stays aligned with the sun. It is known that eclipses are present only in specific seasons during the year in the geostationary orbit (around the equinoxes, when the sun is near the equatorial plane), and this is reflected in that deviations from the sun-pointing condition increase during these periods (while remaining limited), while they are extremely small during the remaining parts of the year. Finally, the performance in terms of orbit raising is presented in Table 4. It is worth to note that the use of the fluid rings for damping the transversal components of the angular velocity, also affect the long term spin rate of the sail negatively. Furthermore, a parametric study was done to determine the minimum operable orbit altitude for this concept: in fact, as

Table 3 Sail geometry and configurations

Boom length, l, m		2
Sail base length, b, m		2.6
Flare angles	Fully-closed phase, degrees	15
	Eclipse phase, degrees	55
	Fully-open phase, degrees	66.8

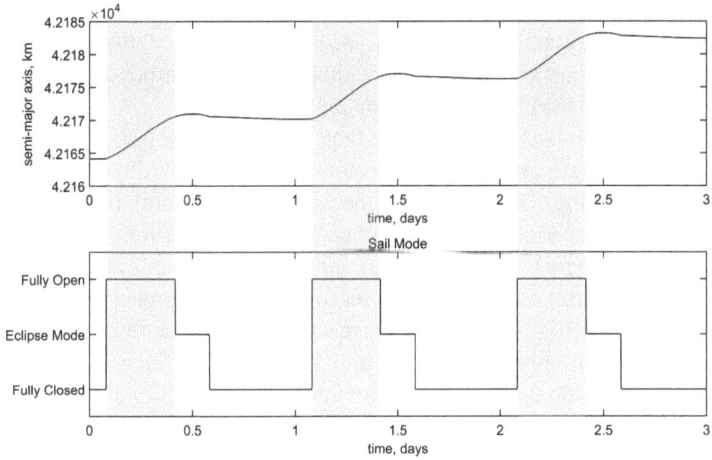

Fig. 7 Orbit raising during the fully opened phases of the sail (here $l = 1$ m). Grey bands identify fully-open phases

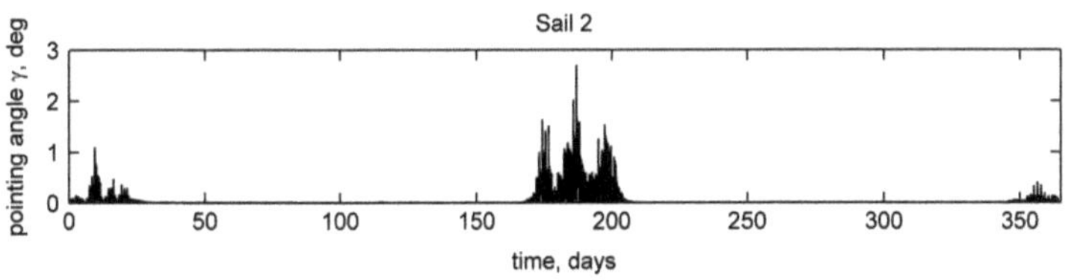

Fig. 8 Angle between spin axis and sun line for 1 year

Table 4 Sail performance over 1 year

Orbit raising rate, km/year	+8600
Maximum sun pointing deviation, degree	2.7
Spin rate of change, rad/s/year	-3.2×10^{-3}
Minimum working orbit altitude, km	28,000

the altitude decreases, both the fraction of orbit in eclipse and the gravity gradient torque magnitude increase, eventually causing instability: it resulted that the concept is stable down to an altitude of 28,000 km (for the considered sail geometry).

4 Oscillating Solar Sail

The quasi-rhombic pyramid presented in the previous section allows, through change in geometry, to avoid the rapid slew manoeuvres that would be necessary otherwise, to modify the orbital parameters. Still, electric actuation is required to change the flare angle of the booms. This leads to the next research question: is it possible to exploit a solar sail to change the orbital parameters (in some controlled way), with an entirely passive solar sail? As explained before, there is still need of a somewhat alternating acceleration throughout an orbit: is it possible to exploit the sun itself to provide the change in attitude required?

In order to answer, we first of all note that, as the QRP showed, a heliostable sail will naturally oscillate about the sun vector if perturbed. While these oscillations were undesired for the QRP (and hence the use of the dampers), they will cause the direction of the force vector to also oscillate. If these oscillations are synchronized (in period and phase) with an orbital path, the direction of the force vector will also oscillate once per orbit and thus can be set to point approximately but consistently in a particular direction, such as along the flight vector or towards zenith. This means that, once an appropriate behaviour is initiated, a force can be generated that can adjust the altitude of a spacecraft with very little energy expenditure and no propellant (Ceriotti et al. 2014b).

4.1 Orbital and Attitude Dynamics

The spacecraft is modelled by a bus from which a flat, triangular sail is deployed between two extensible booms. We consider a 1U, 1 kg CubeSat bus as before, in Table 2, deploying two, 2 m-long booms, with the triangular membrane in between them (see Fig. 9). This is essentially one side of the QRP. Figures 9 show the oscillatory motion of a sail displaced of an angle θ with respect to the sun-line, from an out-of-plane and in-plane viewpoints.

If the sail oscillation is synchronized with the orbital motion (i.e. so that they have same period), and the phase between the two is chosen such that the maximum angular displacement is reached when the spacecraft is along the sun-Earth line, then the sail provides a tangential acceleration component which has its maximum near the sun-Earth line and its null near the axis perpendicular to it (see Fig. 10). In addition, there is also a (unwanted) component of the acceleration that is normal to the orbit. In the short term, the effect is to change the eccentricity and the anomaly

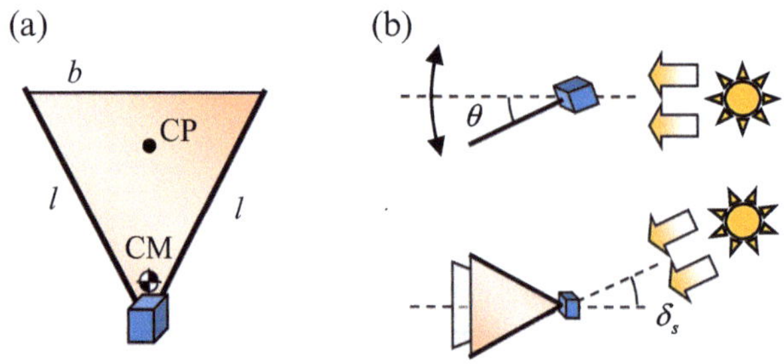

Fig. 9 (**a**) Spacecraft geometry. (**b**) Attitude oscillation: oscillation plane view (*above*) and side view, when the sun is out-of-plane (*below*)

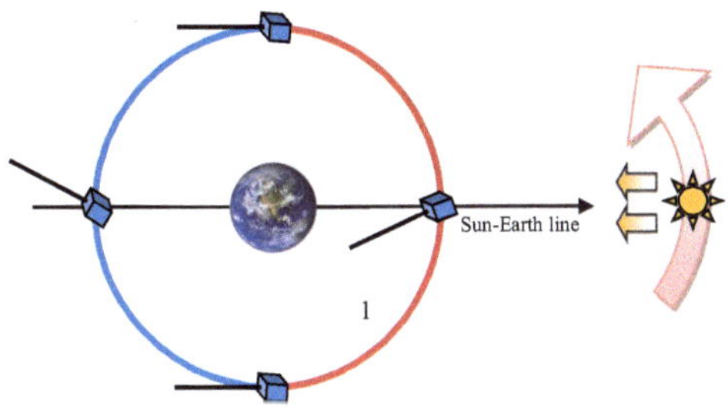

Fig. 10 Orbital plane view of sail oscillation coupled with orbital motion

of pericentre of the orbit, but the net change over 1 year is almost negligible due to the fact that the sun-Earth line rotates one full revolution in that time (as for the QRP, see Fig. 5). In fact, as the sun-Earth line rotates, the oscillation will passively stay centred around the sun-Earth line direction because of the heliostability of the sail, exactly as for the QRP, and because the system is conservative, the oscillation is undamped. However it is worth noticing that here the positive work is done in the parts of the orbit where the sail is moving perpendicularly to the sun line, while the QRP is doing positive work when moving in the direction of the sun line, and away from the sun.

However, a control is needed to maintain the phase of the attitude oscillation synchronous with the angular position on the orbit with respect to the sun line. In addition, the period of the oscillation has to be adjusted such that it matches the period of the orbit, which is increasing (or decreasing) over time following the increase in semimajor axis. To control period and phase, we assume that we can change the moment of inertia around the oscillation axis, for example by extending a boom with a tip mass. Alternatively, electronic reflectivity control could also be used.

4.2 Numerical Simulations and Results

Numerical simulations with coupled attitude and orbital dynamics were used to validate the concept, considering a similar scenario as before, i.e. circular equatorial orbit aiming to increase the semimajor axis. The orbit is set to 1000 km altitude. Further details on these simulations are in Ceriotti et al. (2014b). In this preliminary analysis, eclipses and other perturbations are neglected.

Figure 11 shows the (osculating) semimajor axis over the first orbit. The trend is the consequence of the sail oscillation discussed before. The tangential component of acceleration is the one producing the increase in semimajor axis. Note that the second well of the tangential acceleration (when the spacecraft travels toward the sun) is longer than the first one (when the spacecraft travels away from the sun), which is explained by considering the angular position of the sail with respect to the direction tangential to the orbit (see again Fig. 10).

When simulating a full year, the spacecraft achieves about 35 km of semimajor axis increase (or altitude gain), with negligible change of both eccentricity and

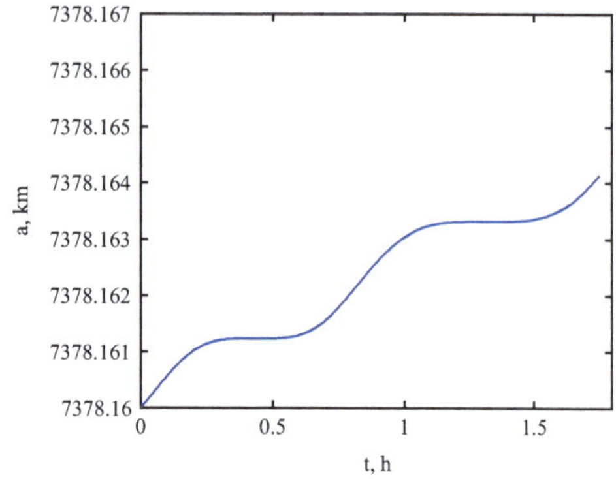

Fig. 11 Semimajor axis over the first orbit (here $l = 1$ m)

inclination. This gain increases with the initial orbital altitude: the higher the orbit, the higher the inclination change that can be obtained within the same timeframe.

5 Conclusions

This chapter has shown that novel architectures can help enhance the technology readiness level of solar sailing. This is done by enabling Earth-bound solar sailing, using the sail as adjoined propulsion system rather than the only one (hybrid propulsion), and finally using the sail as attitude and orbit control device for small satellites. Earth-bound missions are cheaper than deep space ones, and launch opportunities can usually be found—for small platforms such as CubeSats—as piggyback payload. In addition, CubeSats have very limited propulsion and attitude control capabilities: solar sailing becomes an appealing option for increasing the manoeuvrability of these satellites, and therefore expanding their range of possible applications. Low-cost, low-risk missions will be essential for a stepped entry of sail technology into future, more ambitious missions.

Acknowledgements This chapter summarises some of the work done in collaboration with many people, to whom the author is extremely thankful: Colin McInnes and Jeannette Heiligers for the hybrid propulsion (this research was funded by the European Research Council, as part of project 227571 VISIONSPACE); Patrick Harkness and Malcolm McRobb for the oscillating sail; Patrick Harkness, Leonard Felicetti and Malcolm McRobb for the quasi-rhombic pyramid.

References

Baig, S., McInnes, C.R.: Artificial three-body equilibria for hybrid low-thrust propulsion. J. Guid. Control Dyn. **31**(6), 1644–1655 (2008). doi:10.2514/1.36125

Battin, R.H.: An Introduction to the Mathematics and Methods of Astrodynamics. AIAA, New York (1999)

Biddy, C., Svitek, T.: LightSail-1 solar sail design and qualification. In: 41st Aerospace Mechanisms Symposium. Jet Propulsion Laboratory, Pasadena (2012)

Borggräfe, A.: Analysis of interplanetary solar sail trajectories with attitude dynamics. MSc, Rheinisch-Westfälische Technische Hochschule Aachen (2011)

Ceriotti, M., McInnes, C.R.: Generation of optimal trajectories for Earth hybrid pole-sitters. J. Guid. Control Dyn. **34**(3), 847–859 (2011a). doi:10.2514/1.50935

Ceriotti, M., McInnes, C.R.: Systems design of a hybrid sail pole-sitter. Adv. Space Res. **48**(11), 1754–1762 (2011b). doi:10.1016/j.asr.2011.02.010

Ceriotti, M., Heiligers, J., McInnes, C.R.: Novel pole-sitter mission concepts for continuous polar remote sensing. In: SPIE Remote Sensing, Edinburgh (2012a). doi:10.1117/12.974604

Ceriotti, M., Diedrich, B.L., McInnes, C.R.: Novel mission concepts for polar coverage: an overview of recent developments and possible future applications. Acta Astronaut. **80**, 89–104 (2012b). doi:10.1016/j.actaastro.2012.04.043

Ceriotti, M., Harkness, P.G., McRobb, M.: Variable-geometry solar sailing: the possibilities of the quasi-rhombic pyramid. In: Macdonald, M. (ed.) Advances in Solar Sailing, pp. 899–919. Springer, Berlin/Heidelberg (2014a)

Ceriotti, M., Harkness, P.G., McRobb, M.: Synchronized orbits and oscillations for free altitude control. J. Guid. Control Dyn. **37**(6), 2062–2066 (2014b). doi:10.2514/1.G000253

Ceriotti, M., Heiligers, J., McInnes, C.R.: Trajectory and spacecraft design for a pole-sitter mission. J Spacecr. Rocket. **51**(1), 311–326 (2014c). doi:10.2514/1.A32477

Driver, J.M.: Analysis of an arctic polesitter. J Spacecr. Rocket. **17**(3), 263–269 (1980). doi:10.2514/3.57736

Felicetti, L., Ceriotti, M., Harkness, P.G.: Attitude stability and altitude control of a variable-geometry earth-orbiting solar sail. J. Guid. Control Dyn. (2016). doi: 10.2514/1.G001833

Funase, R., Mori, O., Tsuda, Y., Shirasawa, Y., Saiki, T., Mimasu, Y., Kawaguchi, J.: Attitude control of IKAROS solar sail spacecraft and its flight results. In: 61st International Astronautical Congress (IAC 2010). International Astronautical Federation, Prague (2010)

Funase, R., Shirasawa, Y., Mimasu, Y., Mori, O., Tsuda, Y., Saiki, T., Kawaguchi, J.: On-orbit verification of fuel-free attitude control system for spinning solar sail utilizing solar radiation pressure. Adv. Space Res. (Special issue Solar Sailing) **48**(11), 1740–1746 (2011). doi:10.1016/j.asr.2011.02.022

Heiligers, J., Ceriotti, M., McInnes, C.R., Biggs, J.D.: Displaced geostationary orbit design using hybrid sail propulsion. J. Guid. Control Dyn. **34**(6), 1852–1866 (2011). doi:10.2514/1.53807

Heiligers, J., Ceriotti, M., McInnes, C.R., Biggs, J.D.: Design of optimal Earth pole-sitter transfers using low-thrust propulsion. Acta Astronaut. **79**, 253–268 (2012a). doi:10.1016/j.actaastro.2012.04.025

Heiligers, J., Ceriotti, M., McInnes, C.R., Biggs, J.D.: Design of optimal transfers between North and South pole-sitter orbits. In: 22nd AAS/AIAA Space Flight Mechanics Meeting. Univelt, Inc., Charleston (2012b)

Johnson, L., Whorton, M., Heaton, A., Pinson, R., Laue, G., Adams, C.: NanoSail-D: A solar sail demonstration mission. Acta Astronaut. **68**(5–6), 571–575 (2011). doi:10.1016/j.actaastro.2010.02.008

Leipold, M., Götz, M.: Hybrid Photonic/Electric Propulsion. Kayser-Threde GmbH, Munich (2002)

Macdonald, M., McInnes, C.R.: Analytical control laws for planet-centred solar sailing. J. Guid. Control Dyn. **28**(5), 1038–1048 (2005a). doi:10.2514/1.11400

Macdonald, M., McInnes, C.R.: Realistic earth escape strategies for solar sailing. J. Guid. Control Dyn. **28**(2), 315–323(2005b). doi:10.2514/1.5165

Macdonald, M., McInnes, C.R.: Solar sail mission applications and future advancement. In: 2nd International Symposium on Solar Sailing (ISSS 2010), New York (2010)

McInnes, C.R.: Solar Sailing: Technology, Dynamics and Mission Applications. Springer, Berlin (1999)

Mengali, G., Quarta, A.A.: Near-optimal solar-sail orbit-raising from low Earth orbit. J. Spacecr. Rocket. **42**(5), 954–958 (2005). doi:10.2514/1.14184

Mengali, G., Quarta, A.A.: Tradeoff performance of hybrid low-thrust propulsion system. J. Spacecr. Rocket. **44**(6), 1263–1270 (2007a). doi:10.2514/1.30298
Mengali, G., Quarta, A.A.: Trajectory design with hybrid low-thrust propulsion system. J. Guid. Control Dyn. **30**(2), 419–426 (2007b). doi:10.2514/1.22433
Nobari, N.A., Misra, A.K.: Attitude dynamics and control of satellites with fluid ring actuators. J. Guid. Control Dyn. **35**(6), 1855–1864 (2012). doi:10.2514/1.54599
Schaub, H., Junkins, J.L.: Analytical Mechanics of Space Systems, 2nd edn. AIAA, Reston (2009)
Simo, J., McInnes, C.R.: Displaced periodic orbits with low-thrust propulsion. In: 19th AAS/AIAA Space Flight Mechanics Meeting. American Astronautical Society, Savannah (2009)
Stolbunov, V., Ceriotti, M., Colombo, C., McInnes, C.R.: Optimal law for inclination change in an atmosphere through solar sailing. J. Guid. Control Dyn. **36**(5), 1310–1323 (2013). doi:10.2514/1.59931
Wie, B., Murphy, D.: Solar-sail attitude control design for a flight validation mission. J. Spacecr. Rocket. **44**(4), 809–821 (2007). doi:10.2514/1.22996
Yamaguchi, T., Mimasu, Y., Tsuda, Y., Takeuchi, H., Yoshikawa, M.: Estimation of solar radiation pressure force for solar sail navigation. In: 61st International Astronautical Congress (IAC 2010). International Astronautical Federation, Prague (2010)

Two-Body Approximations in the Design of Low-Energy Transfers Between Galilean Moons

Elena Fantino and Roberto Castelli

Abstract Over the past two decades, the robotic exploration of the Solar System has reached the moons of the giant planets. In the case of Jupiter, a strong scientific interest towards its icy moons has motivated important space missions (e.g., ESAs' JUICE and NASA's Europa Mission). A major issue in this context is the design of efficient trajectories enabling satellite tours, i.e., visiting the several moons in succession. Concepts like the Petit Grand Tour and the Multi-Moon Orbiter have been developed to this purpose, and the literature on the subject is quite rich. The models adopted are the two-body problem (with the patched conics approximation and gravity assists) and the three-body problem (giving rise to the so-called low-energy transfers, LETs). In this contribution, we deal with the connection between two moons, Europa and Ganymede, and we investigate a two-body approximation of trajectories originating from the stable/unstable invariant manifolds of the two circular restricted three body problems, i.e., Jupiter-Ganymede and Jupiter-Europa. We develop ad-hoc algorithms to determine the intersections of the resulting elliptical arcs, and the magnitude of the maneuver at the intersections. We provide a means to perform very fast and accurate evaluations of the minimum-cost trajectories between the two moons. Eventually, we validate the methodology by comparison with numerical integrations in the three-body problem.

1 Introduction

The justification of the work resides in the strong scientific interest towards the close-up observation of the moons of the giant planets. The arrival of Cassini at Saturn in 2004 marked the beginning of the era of the extended exploration of the outer planetary systems. Concerning Jupiter's system, Europa is a very attractive

E. Fantino (✉)
Space Studies Institute of Catalonia (IEEC), Polytechnic University of Catalonia (UPC), E.T.S.E.I.A.T., Colom 11, 08222 Terrassa, Spain
e-mail: elena.fantino@upc.edu

R. Castelli
VU University Amsterdam, De Boelelaan 1081, 1081 HV Amsterdam, The Netherlands
e-mail: r.castelli@vu.nl

G. Gómez, J.J. Masdemont (eds.), *Astrodynamics Network AstroNet-II*, Astrophysics and Space Science Proceedings 44,
DOI 10.1007/978-3-319-23986-6_5

target and quite a number of space missions have been proposed, in particular ESA/JUICE (Grasset et al. 2013), that will be launched in 2022 and will perform flybys of Europa, Ganymede and Callisto. A capture at a moon presents formidable challenges to traditional conic analysis since the dynamics in its vicinity is strongly influenced by Jupiter. Hence, three-body techniques are required if the transfer is to end up with a temporary or permanent capture at the moon. In the case of a multi-moon exploration mission, the cost of the transfer between moons is an issue of fundamental importance. For example, a traditional Hohmann transfer between Ganymede and Europa costs 2.8 km/s. The use of three-body dynamics may provide low-fuel solutions. Additionally, LETs add flexibility and versatility to the mission. The specialized literature on the use of three-body dynamics for trajectories among the moons of Jupiter can be divided into two main research lines: the Petit Grand Tour (Gómez et al. 2003) and the Multi-Moon Orbiter concept (Ross et al. 2003). The former consists in coupling three-body problems and seeking intersections in phase space between the corresponding invariant manifolds. This approach requires longer transfer times than Hohmann maneuvers, but allows ΔV savings at the 40 % level with respect to the classical solution. The Multi-Moon Orbiter concept consists in integrating low-energy trajectories with resonant flybys. The resulting transfers are necessarily longer than traditional conic solutions, but the costs are greatly reduced.

In this contribution, we present a method to determine the minimum-cost, direct low-energy trajectories between two moons of Jupiter. The bodies considered are Ganymede and Europa. The starting and end points of these trajectories are planar Lyapunov orbits around collinear libration points of each moon. Sections 2 and 3 illustrate the statement of the problem and the dynamical model, respectively. This is followed by a description of the approach (Sect. 4). The application is dealt with in Sect. 5 and concluding remarks are made in Sect. 6.

2 Problem Statement

The study is carried out in planar approximation, with the orbits of the moons assumed circular (Fig. 1a). This allows the adoption of the circular approximation for the Jupiter-moon-spacecraft restricted three-body problem (CR3BP). Table 1 reports the values of orbital radii, orbital periods and masses of the four Galilean moons. The transfers sought connect planar Lyapunov orbits of Jupiter-Ganymede's and Jupiter-Europa's CR3BPs. Specifically, the Europa-to-Ganymede connection consists in traveling from a planar Lyapunov orbit around the L_2 point of Jupiter-Europa (JEL_2 from now on) to a planar Lyapunov orbit around the L_1 point of Jupiter-Ganymede (JGL_1). The reverse holds for the Ganymede-to-Europa connection, which starts close to JGL_1 and ends at an orbit around JEL_2.

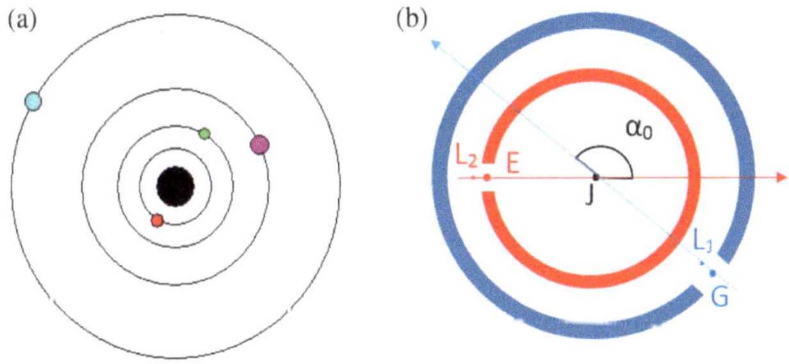

Fig. 1 (**a**) The orbits of the Galilean moons; (**b**) Schematic representation of the coupled CR3BP made up of Jupiter, Ganymede and Europa: α_0 is the angle between the x-axes of the two synodical reference frames

Table 1 Relevant data of the four Galilean moons: orbital period (days), orbital radius (km), mass (kg). The mass of Jupiter is $1898.13 \cdot 10^{24}$ kg

Moon	Period	Radius	Mass
Io	1.77	$4.218 \cdot 10^5$	$8.93 \cdot 10^{22}$
Europa	3.55	$6.711 \cdot 10^5$	$4.80 \cdot 10^{19}$
Ganymede	7.15	$1.070 \cdot 10^6$	$1.48 \cdot 10^{23}$
Callisto	16.69	$1.883 \cdot 10^6$	$1.08 \cdot 10^{23}$

3 Dynamical Model

The design is carried out in the following two models:

1. The coupled CR3BP, which results from the kinematical link between the CR3BP with Jupiter and Ganymede as primaries, and the CR3BP in which the primaries are Jupiter and Europa. The coupling is possible because the two models have a common primary, i.e., Jupiter. The kinematical connection requires the definition of a relative angle α_0 between the two synodical frames at some reference time t_0 (see Fig. 1b).
2. The Jupiter-spacecraft two-body problem.

The coupled CR3BP is used in the vicinity of the moons, i.e., when the spacecraft-to-moon distance is smaller than three times the radius of the moon's sphere of influence calculated with the Tisserand formula. At the boundary of such domain (that we called circle of influence), the two-body approximation takes over and is used to deal with the spacecraft dynamics far from the moons. This is justified by the fact that the masses of the moons are 4 to 5 orders of magnitude smaller than Jupiter's.

4 Solution Method

The solution is constructed upon a database of planar Lyapunov orbits around the two previously-mentioned libration points. The Jacobi constant levels range from 3.0024 and 3.0036 for Europa and from 3.0061 to 3.0075 for Ganymede. The y-amplitudes reach 13,600 km and 23,500 km, respectively in the two cases. Figure 2a shows the circles of influence, the Tisserand circles and a portion of the orbits of the two moons, all scale drawn. The selected ratio between the circles's of influence radius and the Tisserand radius allows to include the largest Lyapunov orbits of the database. The appropriate branch of the suitable (unstable or stable) invariant manifold is propagated to the circle of influence (Figs. 2b, c). At the intersection, the state vector of the spacecraft is transformed from the synodical frame to an inertially-oriented Jupiter-centered frame denoted (J, X, Y, Z). The transformation consists in a change of origin, a rotation around the z-axis and a change of scale (from synodical normalized units to physical units). The state vector is used to determine the orbital elements of the osculating ellipse. Specifically, since the model is two-dimensional, the semi-major axis a, the eccentricity e and the argument of pericenter ω uniquely identify the ellipse geometrically. ω is defined as the angle from the X-axis to the pericenter of the ellipse. The position on the elliptical orbit is given by the true anomaly θ. The procedure is applied to all the trajectories in a manifold for a given discretization on the original Lyapunov orbit, and to all the Lyapunov orbits of the database. In this way, a set of Keplerian orbits is obtained. Then, we form all the possible pairs of Keplerian orbits from the departure CR3BP and the arrival CR3BP, and we compute the geometrical intersections within each pair. Since the two ellipses of each pair have a common focus, the number of points of intersection cannot be higher than two. The difference $\Delta\mathbf{V}$ in velocity at each intersection represents the impulse to be applied by the propulsion system of the spacecraft to transfer from one orbit to the other. In the case of two intersections, the magnitude of the impulse is as shown by the two curves of Fig. 3b (by changing

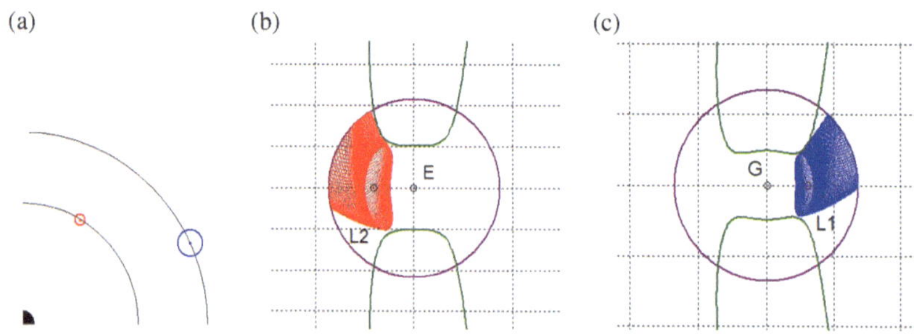

Fig. 2 (**a**) Europa (*red*), Ganymede (*blue*) and the respective circles of influence; (**b**) the circle of influence around Europa and the trajectories of an unstable invariant manifold originating from L_2; (**c**) the circle of influence around Ganymede and the trajectories of a stable invariant manifold originating from L_1

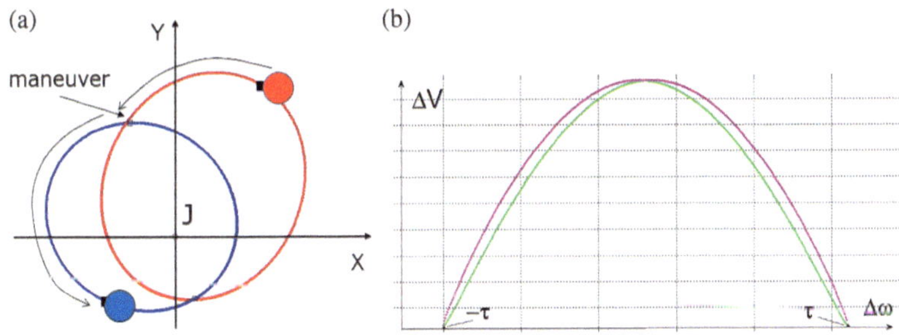

Fig. 3 (**a**) Schematic representation of the solution method; (**b**) example of the relationship between the magnitude of the impulses needed to pass from one elliptical arc to the other and the relative orientation $\Delta\omega$ between the ellipses

semi-major axis and eccentricity of the two ellipses, the shape of the curves and the difference between them may vary but the pattern followed is similar). The total transfer time is different on the two solutions. One typically chooses the shorter transfer (Fig. 3a). In principle, the procedure should be repeated for every choice of α_0 obtained by scanning the domain (from 0 to 2π) according to a selected angular resolution. However, it is straightforward to prove that two different values of α_0 provide the same ellipses rotated by α_0. Hence, there is no need to recompute the ellipses upon varying α_0, because performing the correct rotation gives the same result. Rotating an ellipse changes its argument of pericenter and the relative orientation $\Delta\omega$ with the other ellipse of the pair. Given two ellipses, we determine the interval of values of $\Delta\omega$ for which intersections exist. This is done by setting up the system of polar equations for the two curves

$$\begin{cases} r = \dfrac{a_1(1-e_1^2)}{1+e_1\cos\theta_1} \\ r = \dfrac{a_2(1-e_2^2)}{1+e_2\cos(\theta_1+\Delta\omega)} \end{cases} \tag{1}$$

and determining the values of $\Delta\omega$ for which the discriminant of the associated second-degree algebraic equation is positive or zero (the latter corresponding to the situation in which the two ellipses are tangent to one another and the two intersections reduce to one point). This range is symmetrical around $\Delta\omega = 0$, in which the two pericenters are on the same side of the same line through the focus. Let us call $-\tau$, and τ the limits of the interval, with $\tau \geq 0$. When τ is strictly positive, the two limits correspond to tangency configurations. By analysing the relationship ΔV versus $\Delta\omega$, we observed that, at least when the two orbital eccentricities are small (< 0.2), the dependence has the pattern shown in Fig. 3b, i.e., ΔV is minimum at the two limits $-\tau$ and τ. Hence, when aiming at minimum-cost connections, only the tangency configuration need to be computed, which drastically reduces the number of intersections to be tried.

5 Application: From Europa to Ganymede and Viceversa

Our database of Keplerian orbits includes 7000 orbits on each side of the transfer and for each type of stability (resulting from 70 energy levels and a discretization of 100 points on each Lyapunov orbit). This produces 49 million pairs of ellipses in each transfer direction. Figure 4 illustrates the orbital eccentricities, which are always smaller than 0.14. Figure 5 maps the pericenter and apocenter radii, which are important parameters in the a priori exclusion of non-intersecting pairs: as a matter of fact, for an intersection to exist the pericenter radius of the orbits originating from Ganymede must be smaller than the apocenter radius of the orbits originating from Europa. The number of candidate pairs left is close to 3.5 million, i.e., 7 % of the total. This subset is processed by the algorithm illustrated in Sect. 4. All combinations exhibit strictly positive τ, thus they all include the tangency, minimum-cost configurations. The minimum-cost ΔV's range from 0.97 km/s to 1.27 km/s in both directions (see Fig. 6). Eventually, by taking the absolute minima

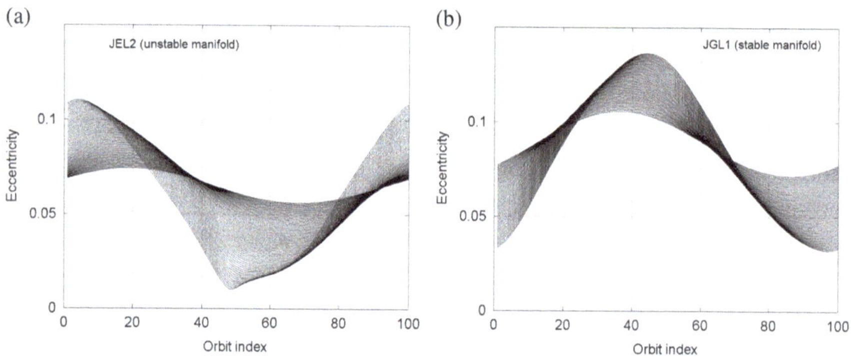

Fig. 4 Orbital eccentricities of the ellipses originating from JEL_2 (**a**) and JGL_1 (**b**)

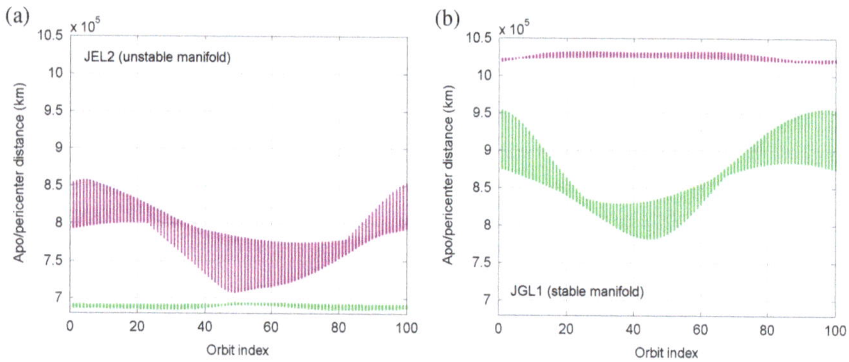

Fig. 5 Apocenter (*purple*) and pericenter (*green*) radii of the ellipses originating from JEL_2 (**a**) and JGL_1 (**b**)

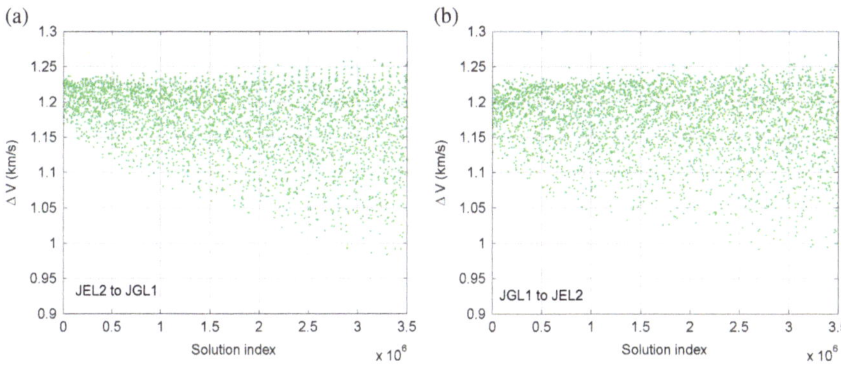

Fig. 6 Minimum-cost intersections between elliptical orbits originating from JEL_2 and JGL_1: from Europa to Ganymede (**a**) and from Ganymede to Europa (**b**)

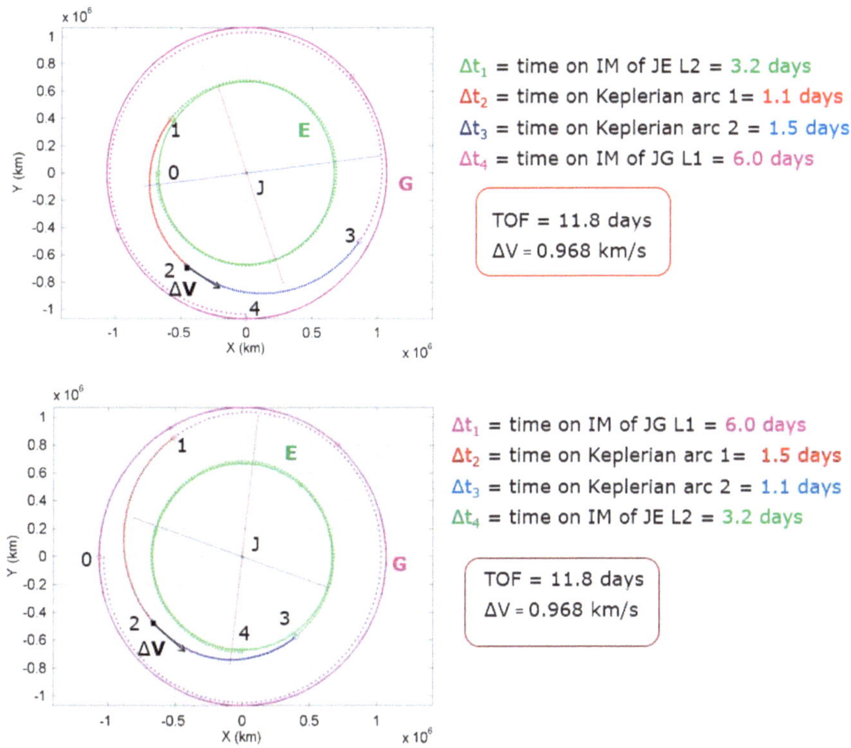

Fig. 7 Illustration of the minimum-cost Europa-to-Ganymede (*top*) and Ganymede-to-Europa (*bottom*) low-energy transfer

in the two directions, we have been able to fully characterize the corresponding trajectories (Fig. 7). These minimum-cost connections originate from invariant

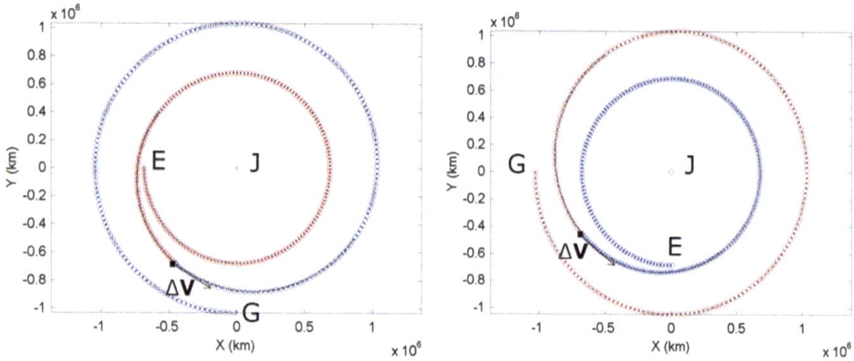

Fig. 8 The minimum-cost Europa-to-Ganymede (*left*) and Ganymede-to-Europa (*right*) low-energy transfers as obtained by the proposed approach (*red and blue dots*) and by numerical integration in the CR3BPs (*black open circles*)

manifolds of Lyapunov orbits with Jacobi constant values of 3.0024 and 3.0061, respectively in the case of JEL_2 and JGL_1. Note that the exact symmetry of the model is responsible for the two solutions having the same cost and transfer time.

A verification of the physical validity of the trajectories obtained has been carried out by propagating the state vector, taken respectively at the beginning and at the end of each transfer, in the corresponding CR3BP up to the maneuver point (Fig. 8).

6 Conclusions

In this paper a new strategy for the preliminary design of low-energy transfers between Galilean moons is presented. The advantage of adopting the two-body approximation in the region where the influence of Jupiter is largely predominant is twofold. Since the two-body problem is completely integrable, we do not need to perform costly and unnecessary numerical simulations and we can a-priori determine the set of physical parameters associated to the minimum-ΔV transfer. This allows to deal with a huge amount of possible connections in a very short time. The validity of the method is proved in the design of transfers between Europa and Ganymede. Our results are close to those published by other authors (Gómez et al. (2003) find LETs requiring an impulsive maneuver at the level of 1.2 km/s). Furthermore, the trajectory obtained is in very good agreement with the result of the numerical integration of the same initial conditions in the CR3BP.

The method is problem-independent and can be applied in different scenarios, anytime part of the force field is dominated by the gravitational influence of one body and the model is parametrized by a relative phase angle.

Acknowledgements E.Fantino acknowledges support from the Marie-Curie Research Training Network Astronet-II.

References

Gómez, G., Koon, W.S., Lo, M.W., Marsden, J.E., Masdemont, J., Ross, S.D.: Invariant manifolds, the spatial three-body problem and petit grand tour of Jovian moons. In: Libration Point Orbits and Applications. World Scientific, Singapore (2003)

Grasset, O., Dougherty, M.K., Coustenis, A., et al.: JUpiter ICy moons Explorer (JUICE): an ESA mission to orbit Ganymede and to characterise the Jupiter System. Planet. Space Sci. **78**, 1–21 (2013)

Ross, S.D., Koon, W.S., Lo, M.W., Marsden, J.E.: Design of a multi-moon orbiter. In: Proceedings of the 13th AAS/AIAA Flight Mechanics Conference, Ponce, 9–13 February 2003. Paper AAS 03–143 (2003)

Artificial Equilibria in the RTBP for a Solar Sail and Applications

Ariadna Farrés and Àngel Jorba

Abstract In this paper we focus on the motion of a solar sail in the Earth-Sun system, using as a model the Restricted Three Body Problem (RTBP) including the Solar Radiation Pressure (SRP). The system has a family of "artificial" equilibrium points parameterised by the orientation of the sail. We show how to use the information on the natural dynamics of the system to navigate around the family of equilibrium points in a controlled way.

1 Introduction

Solar Sails are a form of spacecraft propulsion that takes advantage of the Solar Radiation Pressure (SRP) to propel a satellite by providing it with large and highly reflecting ultra-thin mirrors. The impact and further reflection of the photos emitted by the Sun on this ultra-thin mirrors will accelerate the probe. This acceleration is small but unlimited, and allows new and challenging mission concepts such as GeoSail (McInnes et al. 2001), Polar Sitter (Ceriotti et al. 2014), SunJammer Mission (Heiligers et al. 2014), and cheap multi-rendezvous to different asteroids (Dachwald and Macdonald 2014) among others (McInnes 1999; Mckay et al. 2011; Macdonald and McInnes 2011).

In this paper we will focus on the motion of a Solar sail in the vicinity of the Earth, using the classical circular Restricted Three Body Problem (RTBP) as a model, taking as primaries Earth and Sun and adding the effect of the Solar Radiation Pressure (SRP) due to the sail. The acceleration given by the solar sail depends on three parameters: β the sail lightness number, which measures the effectiveness of the sail, and two angles α and δ, which define the orientation of the sail.

It is well know that the RTBP (when the SRP is not included) has five equilibrium points, $L_{1,\ldots,5}$, all of them on the ecliptic plane (Szebehely 1967). When we include the effect of the solar sail we can "artificially" displace the position of these

A. Farrés (✉) • À. Jorba
University of Barcelona, Gran Via de les Corts Catalanes 585, 08007 Barcelona, Catalunya, Spain
e-mail: ariadna.farres@maia.ub.es; angel@maia.ub.es

G. Gómez, J.J. Masdemont (eds.), *Astrodynamics Network AstroNet-II*,
Astrophysics and Space Science Proceedings 44,
DOI 10.1007/978-3-319-23986-6_6

equilibrium points by changing the sail parameters (β, α, δ) (McInnes 1999, 2000). In Sect. 2.1 we will describe the family of equilibrium points for different fixed values of β. We will see the dependence of the equilibrium points with respect to the sail orientation as well as their linear stability.

Many of the equilibrium points (specially the ones close to the Earth) are unstable and require station keeping to remain close to them for a long time. In previous papers (Farrés and Jorba 2008a,b, 2014), we discussed how to use the information on the linear dynamics of the system to control the trajectory of the sail. The main idea is to understand how the linear dynamics around an equilibrium point varies when we change the sail orientation, and use this information to find changes on the sail orientation that allow us to move around the system in a controlled way. In Sect. 3 we will describe the main ideas behind these type of strategies and see their performance on a test mission.

2 Equations of Motion for a Solar Sail in the RTBP

To model the motion of a solar sail we use the RTBP taking as primaries the Earth and Sun and including the SRP due to sail. We assume that the two primaries are point masses orbiting around their mutual centre of mass in a circular way due to their mutual gravitational attraction. The solar sail is also a point mass that does not affect the motion of the two primaries but is affected by their gravitational attraction as well as the SRP.

We will take a rotating reference frame where the origin is at the Earth-Sun centre of mass and such that the x-axis is along the line joining the two primaries, the z-axis is perpendicular to the orbital plane and the y-axis completes an orthogonal positive oriented reference frame. We will take normalised units of mass, distance and time such that the total mass of the system is 1, the Earth-Sun distance is 1, and their orbital period is 2π. In these units the universal gravitational constant $G = 1$, the mass of the Earth is $\mu = 3.00348060100486 \times 10^6$ and $1 - \mu$ corresponds to the mass of the Sun.

Taking this into account the equations of motion are:

$$\ddot{x} - 2\dot{y} = \frac{\partial \Omega}{\partial x} + a_x, \quad \ddot{y} + 2\dot{x} = \frac{\partial \Omega}{\partial y} + a_y, \quad \ddot{z} = \frac{\partial \Omega}{\partial z} + a_z, \tag{1}$$

where $\Omega(x, y, z) = \frac{1}{2}(x^2 + y^2) + \frac{1-\mu}{r_{ps}} + \frac{\mu}{r_{pe}}$, $r_{ps} = \sqrt{(x-\mu)^2 + y^2 + z^2}$, and $r_{pe} = \sqrt{(x - \mu + 1)^2 + y^2 + z^2}$ are the Sun-sail and Earth-sail distances, and $\mathbf{a} = (a_x, a_y, a_z)$ represents the acceleration due to the solar sail.

To model the acceleration of the solar sail one should take into account the absorption and reflection of the photons induced by the sail material (McInnes 1999; Dachwald et al. 2005). The force due to absorption, $\mathbf{F_a}$, is in the direction of the SRP ($\mathbf{r}_s = (x-\mu, y, z)$), while the force due to reflection, $\mathbf{F_r}$, is directed along the normal to the surface of the sail ($\mathbf{n}$):

$$\mathbf{F_a} = PA\langle \mathbf{r_s}, \mathbf{n}\rangle \mathbf{r_s}, \qquad \mathbf{F_r} = 2PA\langle \mathbf{r_s}, \mathbf{n}\rangle^2 \mathbf{n},$$

where $P = P_0(R_0/R)^2$ is the SRP magnitude at a distance R from the Sun ($P_0 = 4.563\text{N/m}^2$, the SRP magnitude at $R_0 = 1\text{AU}$) and A is the area of the solar sail.

If we denote by a the absorption coefficient and by ρ the reflectivity coefficient, we have $a+\rho = 1$. Hence, the solar sail acceleration in this simplified non-perfectly reflecting model SNPR (Dachwald et al. 2005) is given by:

$$\mathbf{a} = \frac{2PA}{m}\langle \mathbf{r_s}, \mathbf{n}\rangle \left(\rho\langle \mathbf{r_s}, \mathbf{n}\rangle \mathbf{n} + 0.5(1-\rho)\mathbf{r_s}\right). \tag{2}$$

The characteristic acceleration of the solar sail, a_0, is defined as the acceleration produced by the solar sail at 1AU when the sail is perpendicular to the Sun-sail line (i.e. $a_0 = (1+\rho)P_0A/m$). It is common to write the acceleration of the sail as a correction of the Sun's gravitational attraction:

$$\mathbf{a} = \beta\frac{Gm_s}{r_{ps}^2}\langle \mathbf{r_s}, \mathbf{n}\rangle \left(\rho\langle \mathbf{r_s}, \mathbf{n}\rangle \mathbf{n} + 0.5(1-\rho)\mathbf{r_s}\right), \tag{3}$$

where G is the universal gravitational constant, m_s is the mass of the Sun and r_{ps} is the Sun-sail distance. The constant β is defined as the **sail lightness number** and accounts for the effectiveness of the solar sail.

$$\beta = \sigma^*/\sigma, \qquad \sigma^* = \frac{2P_0R_0^2}{Gm_s} = 1.53\,\text{g/m}^2,$$

and $\sigma = m/A$ is the area-to-mass ratio of the solar sail. In Table 1 we can see for different sail lightness numbers β, the corresponding area-to-mass ration (σ), the characteristic acceleration, a_0, and the size of the solar sail, A, for 30 kg payload.

Notice that, $\rho = 1$ corresponds to a perfectly reflecting solar sail, and $\rho = 0$ to a perfect solar panel where the absorption by the panels is the only effect. According to (Dachwald et al. 2005) a solar sail with a highly reflective aluminium-coated side has an estimated reflectivity value $\rho \approx 0.88$. Throughout this paper we will consider the sail to be perfectly reflecting ($\rho = 1$).

Table 1 Relation between: the sail lightness number (β), the area-to-mass ration of the satellite (σ), its characteristic acceleration (a_0), and the required sail area for 30 kg of payload mass

β	σ (g/m^2)	a_0 (mm/s^2)	Area (m^2)
0.01	153.0	0.059647	$\approx 14 \times 14$
0.02	76.5	0.119294	$\approx 20 \times 20$
0.03	51.0	0.178941	$\approx 24 \times 24$
0.04	38.25	0.238588	$\approx 28 \times 28$
0.05	30.6	0.298235	$\approx 31 \times 31$

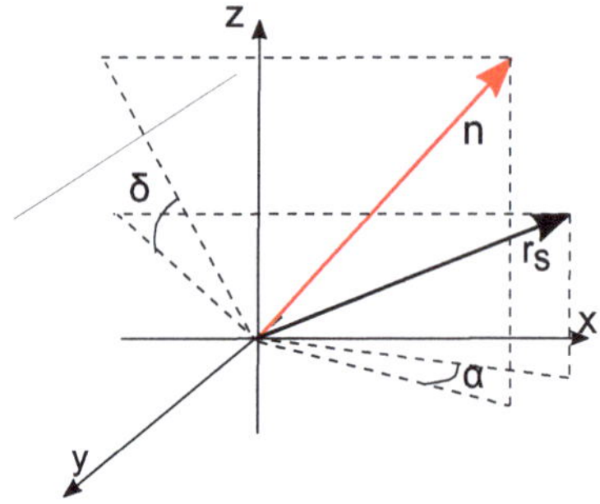

Fig. 1 Schematic representation for the definition of the angles defining the sail orientation. Where $\mathbf{r_s}$ is a unitary vector in the Sun-sail line direction, $\mathbf{n}$ is the normal vector to the surface of the sail

The **sail orientation** is defined by the normal vector to the surface of the sail, $\mathbf{n}$, which can be parameterised by two angles α and δ. The definition of these two angles will depend on the reference frame that we use, and in the literature we find different ways to define them (McInnes 2000). Here we define them as the planar and vertical displacement w.r.t the Sun-sail line r_s in a reference frame centred at the spacecraft and parallel to the rotating frame (x, y, z) defining the system. In Fig. 1 we have a schematic representation of this definition.

If we take $\mathbf{r_s}$ in polar coordinates, $\mathbf{r_s} = (\cos\phi(x,y)\cos\psi(x,y,z), \sin\phi(x,y)\cos\psi(x,y,z), \sin\psi(x,y,z))$, we have that $\mathbf{n} = (n_x, n_y, n_z)$ is

$$\begin{aligned} n_x &= \cos(\phi(x,y)+\alpha)\cos(\psi(x,y,z)+\delta), \\ n_y &= \sin(\phi(x,y)+\alpha)\cos(\psi(x,y,z)+\delta), \\ n_z &= \sin(\psi(x,y,z)+\delta), \end{aligned}$$

where $\phi(x,y) = \arctan\left(\frac{y}{x-\mu}\right)$ and $\psi(x,y,z) = \arctan\left(\frac{z}{\sqrt{(x-\mu)^2+y^2}}\right)$.

Notice that, with this definition, the sail is perpendicular to $\mathbf{r_s}$ when $\alpha = 0, \delta = 0$. Moreover, if we fix δ and vary α we are moving the sail right/left w.r.t the Sun-sail line, and if we fix α and vary δ we are moving the sail up/down w.r.t the Sun-sail line. We note that the sail acceleration cannot point towards the Sun (i.e. $\langle \mathbf{n}, \mathbf{r_s}\rangle \geq 0$), which implies $\alpha, \delta \in [-\pi/2 : \pi/2]$. Moreover, from an engineering point of view, tilting the sail more that 45° w.r.t the Sun-sail line is very hard to maintain, compromising the sail controllability and structure.

Finally, we note that when we add the effect of the solar sail to the RTBP, the system is only Hamiltonian when the sail is perpendicular to the Sun-sail line (i.e. $\mathbf{r}_s \perp \mathbf{n}$ when $\alpha = 0, \delta = 0$) or when the solar photons do not impact the surface of the sail and its effect is discarded (i.e. $\mathbf{r}_s \perp \mathbf{n}$ when $\alpha = 0, \delta = \pm\pi/2$).

2.1 *Artificial Equilibrium Points*

It is well known that when the SRP is discarded ($\beta = 0$) the Earth-Sun RTBP has five equilibrium points: three of them ($L_{1,2,3}$) are on the line joining the two primaries and are linearly unstable (saddle×centre×centre), while the other two ($L_{4,5}$) are on the ecliptic plane forming an equilateral triangle with the two primaries and are linearly stable (centre×centre×centre) (Szebehely 1967).

When the sail is perpendicular to the Sun-sail line ($\alpha = 0, \delta = 0$) we have a similar phase space portrait, notice that in this particular case we are essentially changing the magnitude of the Sun's attracting force. Hence, we also have five equilibrium points $SL_{1,...,5}$ which are slightly displaced towards the Sun w.r.t the classical Lagrangian points $L_{1,...,5}$. Three of them are also in the Earth-Sun line ($SL_{1,2,3}$) and are unstable (saddle×centre×centre), and the other two form a triangle with the two primaries ($SL_{4,5}$) and are stable (centre×centre×centre) (McInnes 1999, 2000).

When we change the sail orientation ($\alpha \neq 0$ and/or $\delta \neq 0$) we are adding an extra force in a certain direction and we can artificially displace the equilibrium point (McInnes 1999, 2000). For instance, if we take $\delta = 0$ and $\alpha \neq 0$ this extra force is in the ecliptic plane, hence we will displace the equilibrium point left/right w.r.t SL_i within the ecliptic plane. Instead, if we take $\alpha = 0$ and $\delta \neq 0$ this extra force is in the vertical direction and we will displace the equilibrium above/bellow the ecliptic plane (in the $Y = 0$ plane). Taking other sail orientations we can generate a 2D surface of equilibria homeomorphic to a sphere, where each equilibrium point corresponds to a given sail orientation. For a fixed value for δ and $\alpha \in [-\pi/2, \pi/2]$, we have a curve of equilibrium points on a plane that contains L_i and is inclined w.r.t $Z = 0$. In Fig. 2 we can see the family of artificial equilibria related to L_1 (right) and L_2 (left) for $\beta = 0.01$, where we have highlighted the equilibrium points in the family with a fixed value of δ and $\alpha \in [-\pi/2, \pi/2]$.

For small β ($< 10^{-6}$) we have five disconnected families of equilibrium points, each family parameterised by the two angles defining the sail orientation. As β grows ($\approx 10^{-6}$) the equilibria surfaces related to SL_3, SL_4 and SL_5 merge into each other, having three disconnected families of equilibria (two spheres one containing

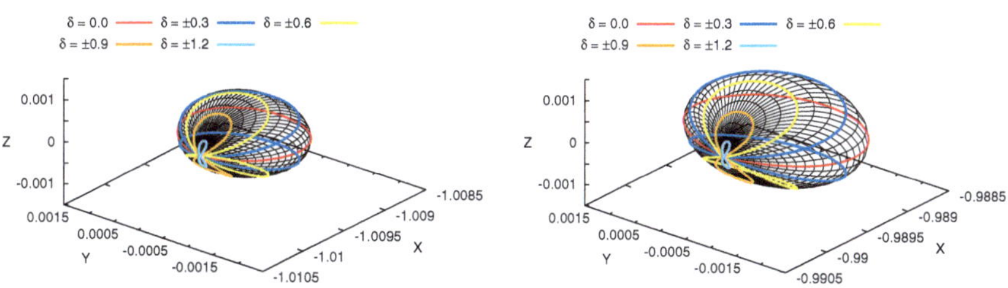

Fig. 2 Family of artificial equilibria related to L_1 (*right*) and L_2 (*left*) for $\beta = 0.01$. *Colour lines* correspond to equilibrium points for $\delta = 0, \pm 0.30, \pm 0.60, \pm 0.90, \pm 1.20$ and ± 1.5 rad

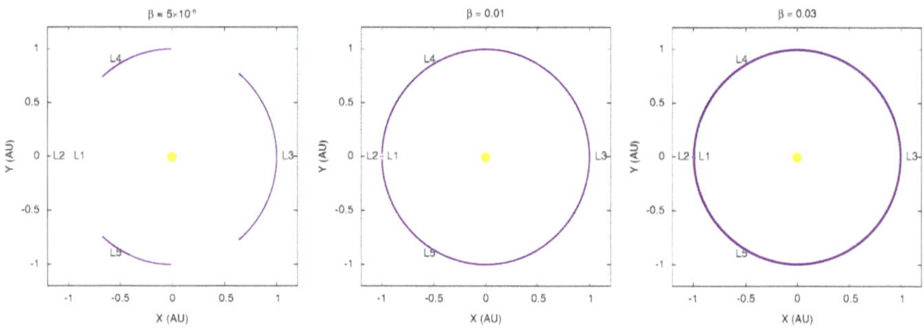

Fig. 3 Family of equilibrium points on the xy-plane ($\delta = 0$) for $\beta = 5 \times 10^{-6}, 0.01$ and 0.03

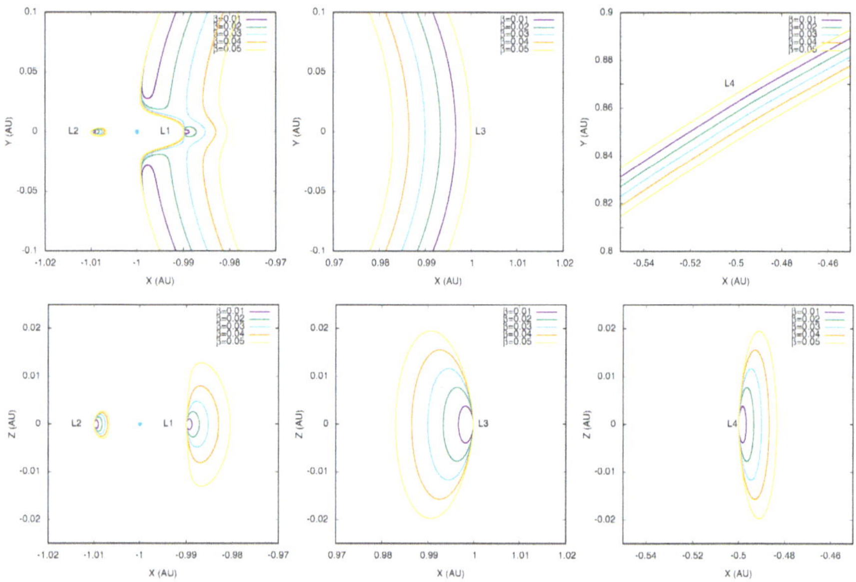

Fig. 4 Slices of the surface of equilibrium point for $\beta = 0.01, 0.02, 0.03, 0.04$ and 0.05. *Top*: xy-plane ($\delta = 0$) around $L_{1,2}$ (*left*), L_3 (*middle*), L_4 (*right*). *Bottom*: xz-plane ($\alpha = 0$) around $L_{1,2}$ (*left*), L_3 (*middle*), L_4 (*right*)

SL_1 and the other SL_2, and a banana shape surface containing $SL_{3,4,5}$). Between $\beta = 0.02$ and 0.03 the surface related to SL_1 merges with the large surface containing $SL_{3,4,5}$, having only two disconnected surfaces of equilibria (a sphere containing SL_2, and a torus containing the other four points). In Fig 3 we can see slices of these surfaces on the $Z = 0$ plane for $\beta = 5 \times 10^{-6}, 0.01$ and 0.03. The two remaining surfaces will never merge into each other, the fact that $\langle \mathbf{n}, \mathbf{r_s} \rangle > 0$ delimits the regions of possible equilibria, separating the SL_1 and SL_2 families (McInnes 1999, 2000).

In Fig. 4 we have zoomed the regions around the Lagrangian points to appreciate better the difference between different values of β. We plot slices of the surfaces

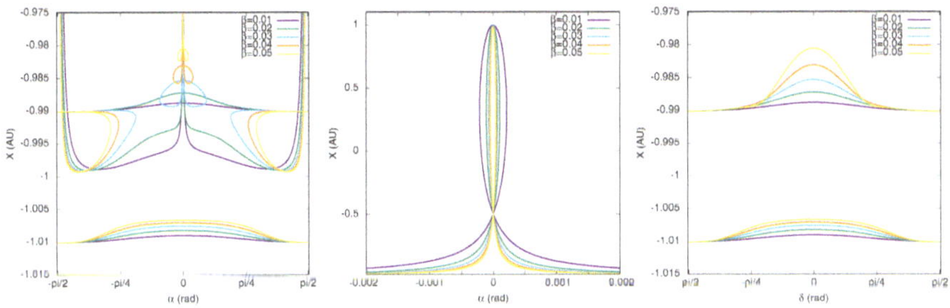

Fig. 5 For $\beta = 0.01, 0.02, 0.03, 0.04$ and 0.05 relationship between the x coordinate of the equilibrium point and the sail orientation

of equilibria for $\beta = 0.01$, 0.02, 0.03, 0.04 and 0.05. On the top we have the intersection of family of equilibria on the xy-plane (i.e. for $\alpha \in [-\pi/2 : \pi/2]$ and $\delta = 0$) and on the bottom we have the projection with the xz-plane (i.e. $\alpha = 0$ and $\delta \in [-\pi/2 : \pi/2]$). From left to right we have the L_1/L_2 region, L_3 region and L_4 region (L_5 is symmetric to L_4). On the top we have the intersection of family of equilibria on the xy-plane (i.e. for $\alpha \in [-\pi/2 : \pi/2]$ and $\delta = 0$) and on the bottom we have the projection with the xz-plane (i.e. $\alpha = 0$ and $\delta \in [-\pi/2 : \pi/2]$).

If we look at how the sail orientation varies along the different families of equilibria we can see that, when the families related to L_1, L_3, L_4 and L_5 merge into each other, there are some sail orientations that are lost. With this we mean that we no longer have for each sail orientation five equilibrium points. In Fig. 5 we plot, for $\beta = 0.01, 0.02, 0.03, 0.04$ and 0.05, the relation between the x component of the equilibrium points and the sail orientation (α or δ). On the left-hand side we have the variation of α for the equilibrium points related to L_1 and L_2 close to the Earth. Notice how for $\beta \geq 0.03$ the line joining the equilibrium points related to L_1 has split in two, "loosing" sail orientations. By this we mean that there are sail orientations that are not related to an equilibrium point. In the middle we have the variation of α for the equilibrium points related to $L_{1,3,4,5}$. As we can see as β increases the range of admissible α is smaller. Finally, on the right-hand side we have the variation of δ for the equilibrium points related to L_1 and L_2, where for the values of β that we consider we have fixed points for each $\delta \in [-\pi/2 : \pi/2]$. we recall that these equilibrium points are above and bellow the ecliptic plane.

As we have mentioned before, although $\alpha, \delta \in [-\pi/2 : \pi/2]$, orienting the solar sail more that 45° with respect to the Sun-sail line can compromise the control of a solar sail. In Fig. 6 we plot (on the plane $Z = 0$) the relation between the equilibrium position and sail orientation. In green we have the equilibrium points with $|\alpha| \leq \pi/4$ and purple those with $|\alpha| \geq \pi/4$. Notice that for a fixed β, the equilibrium points that are closer to the Sun are good candidates for placing a solar sail for mission applications.

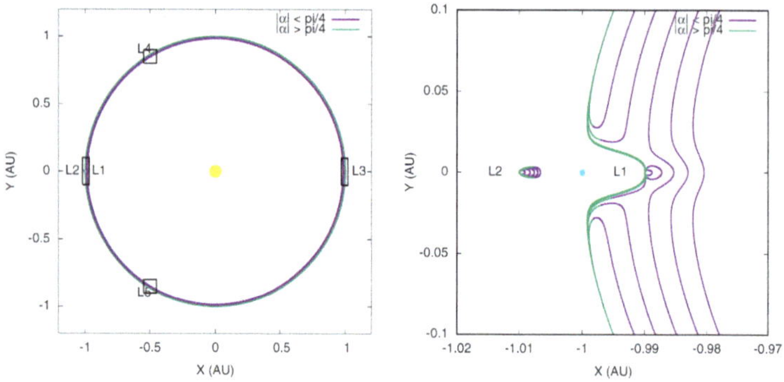

Fig. 6 Sail orientation relation between the family of equilibrium points on the xy-plane ($\delta = 0$) for $\beta = 0.01, 0.02, 0.03, 0.04$ and 0.05. *Green points* $|\alpha| \leq \pi/4$ and *Purple points* $|\alpha| \geq \pi/4$

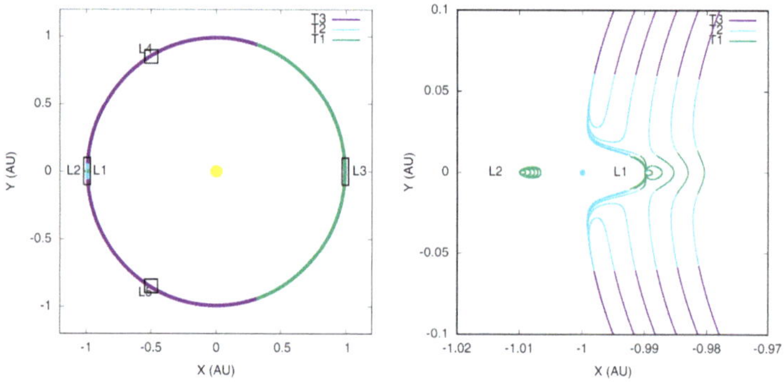

Fig. 7 Stability indicator for the family of equilibrium points on the xy-plane ($\delta = 0$) for $\beta = 0.01, 0.02, 0.03, 0.04$ and 0.05. The class $\mathcal{T}_1, \mathcal{T}_2, \mathcal{T}_3$ are represented in colour *green*, *blue* and *purple* respectively

Finally, we can classify the equilibrium points according to their stability, which is given by the eigenvalues of the linearised flow around an equilibrium point. We can distinguish three class of equilibrium points: class $\mathcal{T}_1$ that are unstable and have as eigenvalues $\lambda_1 > 0, \lambda_2 < 0, \nu_1 \pm \mathrm{i}\omega_1$ and $\nu_2 \pm \mathrm{i}\omega_2$ (here the main instability is given by the saddle as $|\nu_{1,2}| \ll |\lambda_{1,2}|$); class $\mathcal{T}_2$ are equilibria with three pair of complex eigenvalues $\nu_{1,2,3} \pm \mathrm{i}\omega_{1,2,3}$, where at least one of the real parts $|\nu_i| > 0.001$ (they have some instability given by a complex saddle); and class $\mathcal{T}_3$ also have three pair of complex eigenvalues $\nu_{1,2,3} \pm \mathrm{i}\omega_{1,2,3}$ but now $|\nu_{1,2,3}| < 0.001$ (we say that these points are almost stable as we require a long time to escape form the vicinity of the equilibria). In Fig. 7 we show the relation between the position of the equilibrium point and the class where they belong. As we can see the equilibria close to L_1 and L_2 are unstable, as well as those close to L_3.

3 Mission Applications at the Artificial Equilibria

The family of equilibrium points close to SL_1 and SL_2 offer interesting locations for placing a solar sail to perform observational missions (McInnes 1999). Recently, the SunJammer mission (Heiligers et al. 2014) was proposed, where the goal was to place a Solar sail around an equilibrium point in the SL_1 family displaced 5° from the Earth-Sun line. From this location we can observe the activity of the Sun and perform enhanced warnings of the Geomagnetic storms. Given the fact that SL_1 is closer to the Sun than L_1 (with an appropriate sail size) we can almost double the warning time from a satellite at L_1.

As we have seen in the previous section, the equilibrium points close to the Earth are unstable and of class $\mathscr{T}_1$, so the linear dynamics is close to saddle×centre×centre. This means that a solar sail that is close to equilibria will escape along the unstable direction, and a station keeping strategy is required. In previous papers (Farrés and Jorba 2008a,b, 2014) we have seen how we can use the information on the natural dynamics of the system to derive simple strategies to: (a) remain close to the unstable equilibria and (b) navigate along the family of equilibria in a controlled way. In this paper we want to give a quick overview of the main ideas behind them and show their performance on a simple test mission.

We focus only on the unstable equilibria, where the linear dynamics is close to saddle×centre×centre ($\mathscr{T}_1$ type points), and take advantage of the unstable manifolds to move around the system. We recall that each equilibrium points, p_0, is associated to a given sail orientation α_0, δ_0 (orientation required to have equilibria). If we are close to p_0 with fixed sail orientation α_0, δ_0, the trajectory will escape along the unstable manifold while rotating in the other two centre directions. If we change the sail orientation, α_1, δ_1, the position, p_1, of the equilibrium point is displaced, as well as its stable and unstable directions. Now the trajectory will escape along the new unstable manifold. If we choose an appropriate new sail orientation we can make the solar sail come back to the original equilibrium point, p_0, or to surf towards a new equilibria.

In Fig. 8 we have a schematic representation of the linear dynamics of the system on the saddle and the two centre projections of an equilibrium point p_0 for α_0, δ_0.

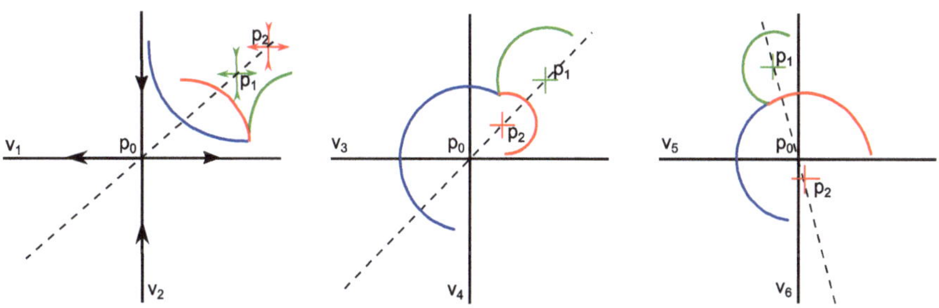

Fig. 8 Sketch for possible effects on the sail trajectory for changes on the sail orientation

Depending on the relative position between the satellite and the new equilibrium point when we change the sail orientation, the behaviour of the trajectory will be different. Here we see two different possibilities for each projection. Looking at the saddle projection (left), if the new equilibrium point, p_1, is closer to p_0 than the position of the satellite (blue line), the trajectory will continue to escape from the vicinity of p_0 (green line). On the other hand, if the new equilibrium point p_2, is further away from p_0 than the satellites position, then the trajectory will com close to p_0 (red line). Looking at the trajectory on the centre components (middle and right), we have a sequence of rotations around the different equilibrium points. Depending if p_1, p_2 are inside or outside the circle defined by the trajectory while rotating around p_0 (blue line), the trajectory will get closer or away from to p_0. This must also be controlled to forbid possible unbounded motions.

In order to decide when and how to change the sail orientation, so that the new equilibrium point, p_1 for (α_1, δ_1), and its unstable manifold drives the trajectory where we want to go, we will track the trajectory of the spacecraft using a reference frame centred at the equilibrium point p_0 and defined by the vectors $\{\mathbf{v}_1, \mathbf{v}_2, \mathbf{v}_3, \mathbf{v}_4, \mathbf{v}_5, \mathbf{v}_6\}$, hence the position of the sail at time t, is written as $\phi(t) = p_0 + \sum_{i=0}^{6} s_1(t)\mathbf{v}_i$. Where $\mathbf{v}_1, \mathbf{v}_2$ are the unstable and stable eigenvectors (defining the motion on the saddle plane), $\mathbf{v}_3, \mathbf{v}_4$ are the real and imaginary part of the first pair of complex eigenvectors (related to the motion on the first centre projection) and $\mathbf{v}_5, \mathbf{v}_6$ are the real and imaginary parts of the second pair of complex eigenvectors (related to the second centre projection). Notice that this reference frame contains information on the relative position of the trajectory w.r.t the stable and unstable manifolds, as well as the motion on the two centre projections that must remain bounded.

3.1 Station Keeping Strategies

Here we briefly describe the station keeping scheme, where the goal is to remain close to the equilibrium point p_{ini} for a long time, for further details see Farrés and Jorba (2008a).

We start with a reference frame centred around the equilibrium point we want to remain close (p_{ini}). We know that the trajectory will escape along the unstable direction ($\mathbf{v}_1$) and rotate in the centre projections. When the trajectory is far from p_{ini} (i.e. $|s_1(t^*)| > \varepsilon_{max}$) we need to change the sail orientation. We will chose the new orientation such that the new equilibrium point (q_i) satisfies: $|\bar{s}_1| > d \cdot \varepsilon_{max}$ with $d > 1$, and $||(\bar{s}_3, \bar{s}_4)||_2 < ||(s_3(t^*), s_4(t^*))||_2$ and $||(\bar{s}_5, \bar{s}_6)||_2 < ||(s_5(t^*), s_6(t^*))||_2$. In other words, the new unstable manifold must take the trajectory towards the stable manifold of p_{ini} and the centre projections must remain bounded. Once the trajectory comes close to p_{ini} (i.e. $|s_1(t^*)| < \varepsilon_{min}$) we will restore the original sail orientation. We will repeat this process as long as we want to remain close to p_{ini}.

In Fig. 9 we have a schematic representation of the control. We can play with the parameters $\varepsilon_{max}, \varepsilon_{min}$ and d to modify how far away we can get from p_{ini} and the

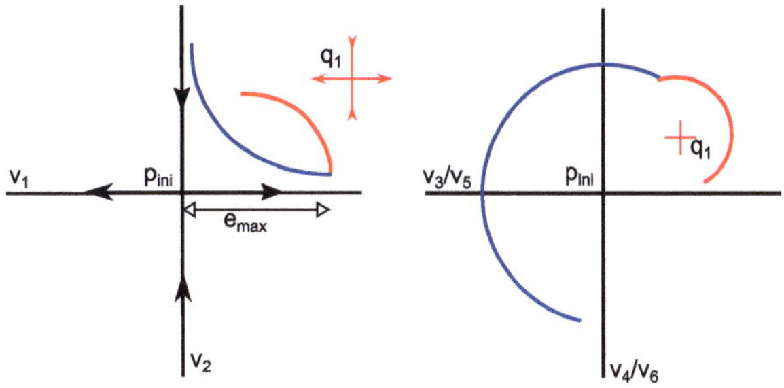

Fig. 9 Sketch on the station keeping strategy on the saddle and centre projections of the trajectory

time that it will take to come back p_{ini}. The controllability of the system can depend on the choice of these constants.

3.2 Surfing Strategies

Here we briefly describe the surfing scheme, where the goal is to find a transfer trajectory that goes from one equilibrium point p_{ini} to another p_{end}, for further details see Farrés and Jorba (2008b).

We start taking a reference frame centred at p_{ini} and find the position of the final point (p_{end}) in this reference frame. Then we draw an imaginary line joining the two points and consider q_1, an intermediate point on this imaginary line, between p_{ini} and p_{end} in each of its projections (saddle and centre). We start close to p_{ini} and escape along the unstable direction, rotation around the centre projections. When the trajectory is far away from the p_{ini} (i.e. $|s_1(t^*)| > \varepsilon_{max}$) we will change the sail orientation such that the new point q_i lies on the imaginary line joining the two point and satisfies $|\bar{s}_1| < d \cdot \varepsilon_{max}$ with $d < 1$, $||(\bar{s}_3, \bar{s}_4)||_2 > ||(s_3(t^*), s_4(t^*))||_2$ and $||(\bar{s}_5, \bar{s}_6)||_2 > ||(s_5(t^*), s_6(t^*))||_2$. In other words, the new unstable manifold takes the trajectory towards the stable manifolds of p_{end}, and the trajectory in the centre projection must also moves towards p_{end}. In Fig. 10 we have a schematic representation of these phenomena.

In many cases (as the distance between the initial and target points will be large) we will need to chose more than one intermediate point, q_i, i.e. perform several changes on the sail orientation. In order not to loose information on the dynamics of the system as we move from one point to the other, we will recompute the reference frame each time we change the sail orientation. Taking as new reference frame the one associated to the new equilibrium point q_i, and recompute the imaginary line between q_i and p_{end} to derive the new intermediate point.

Here we can also play with the parameters ε_{max} and d in order to tune the surfing speed by moving closer or further from the stable and unstable manifolds related

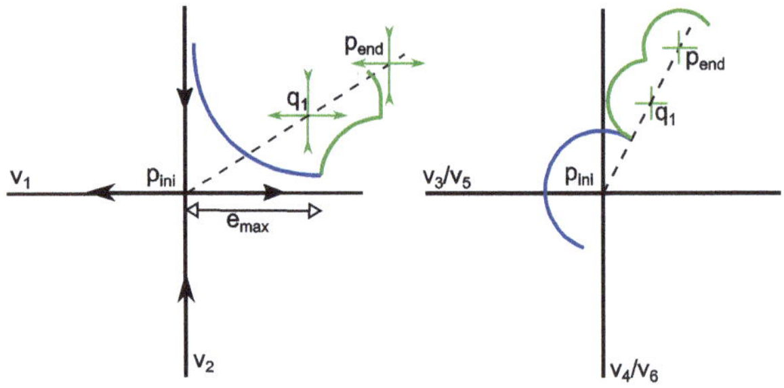

Fig. 10 Sketch of the surfing strategy on the saddle and centre projections of the trajectory

each equilibrium point. This will affect the transfer time and the controllability of the system when we get close to the target point p_{end}.

3.3 Finding an Appropriate New Sail Orientation

The key point behind these strategies is to find an appropriate sail orientation that places the new equilibrium point at the desired position. But this is not as simple as one might think. On one hand, because we do not have an explicit expression, $p(\alpha, \delta)$, for the position of the equilibrium points as a function of the sail orientation. On the other hand, because for a fixed sail lightness number, β, we have a 2D surface of equilibria in a 6D phase space. Hence, we cannot put the equilibrium point wherever we want, there are some limitations that can compromise the controllability of the system. Let us briefly see how we can overcome this issues.

It is true that we do not have an explicit expression for $p(\alpha, \delta)$, but for a given point $p(\alpha_0, \delta_0)$, if $|\alpha - \alpha_0|$ and $|\delta - \delta_0|$ are small, we can use the linear approximation (i.e. first order Taylor expansion):

$$p(\alpha, \delta) = p(\alpha_0, \delta_0) + \frac{\partial p}{\partial \alpha}(\alpha - \alpha_0) + \frac{\partial p}{\partial \delta}(\delta - \delta_0), \tag{4}$$

where $\frac{\partial p}{\partial \alpha}$ and $\frac{\partial p}{\partial \delta}$ can be easily computed numerically. We will use this equation to find the appropriate new sail orientation, where $p(\alpha_0, \delta_0) = p_0$ is the equilibrium point we have and $p(\alpha, \delta) = p_{new}$ is the desired location for the new equilibrium point.

Notice that there are more equations than unknowns (six equations vs two unknowns), which corresponds to the fact that we do not have an equilibrium point at an arbitrary place. We will use the least-squares method to solve the system and have an equilibrium point as close as possible to the desired positions. In some cases we might have to add some restrictions when we solve the system in order to guarantee that the trajectory behaves as expected. For instance, that the new unstable manifold takes us back to p_{ini} or towards p_{end}. In order to include these restrictions in an easy way, we will write this equations using the reference frame centred at p_0 used during the control schemes. This means that we only need to impose constraints to $(\bar{s}_1, \bar{s}_2, \bar{s}_3, \bar{s}_4, \bar{s}_5, \bar{s}_6)$ the coordinates of p_{new} in this reference system.

Notice that $\frac{\partial p}{\partial \alpha}$ and $\frac{\partial p}{\partial \delta}$ tell us (at first order) how the family of equilibria vary when we vary the two angles. If these two vectors have a small components on the unstable direction we will not be able to compensate the instability of the saddle, as we will not be able to have $|s_1^*| \geq \varepsilon_{max}$.

3.4 Test Mission

To illustrate these two strategies we consider a round tour visiting four points on the surface of equilibria related to SL_1. These four points are displaced 5° from the Earth-Sun line, two of them above and below the ecliptic plane (p_1, p_3) and the other two left and right from the Earth-Sun line (p_0, p_2), forming a rhomb with the Sun in the middle seen from the Earth. The mission goal is to go from one equilibrium point to the other, and once we get there remain around each of them for a long time to make observations. In this way we test both the surfing and control strategies. For the sail performance we have taken $\beta = 0.051689$, ($\approx$ 32 kg of payload mass and $38 \times 38\text{m}^2$ of sail area) the sail lightness number for the SunJammer mission. In Table 2 we have the position of the four equilibrium points we want to visit and their corresponding sail orientation.

We have divided the mission in four stages: (1) go from a vicinity of p_0 to a vicinity of p_1; (2) from p_1 to p_2; (3) from p_2 to p_3; and (4) from p_3 to p_0. Where each stage has two parts: (A) surfing from one point to the other; and (B) station keeping around the target point for 2 years. When we surf from one point to the other we use the surfing strategy described in Sect. 3.2 and when we control the trajectory to remain close to one of the equilibria we use the station keeping strategy described in Sect. 3.1.

In Fig. 11 we have the trajectory the solar sail follows throughout the mission, where each colour corresponds to the different stages. The average time to go from one equilibrium point to the other is 4 years, and as we can see we have avoided the

Table 2 (x, y, z) coordinates of the equilibrium points (p_i) we visit and their corresponding sail orientation (α_i, δ_i) in the test mission

	x	y	z	α (deg)	δ (deg)
p_1	-9.79998e-01	1.81889e-03	0.00000e+00	-0.74	0.00
p_2	-9.80036e-01	0.00000e+00	1.73948e-03	0.00	2.61
p_3	-9.79998e-01	-1.81889e-03	0.00000e+00	0.74	0.00
p_4	-9.80036e-01	0.00000e+00	-1.73948e-03	0.00	-2.61

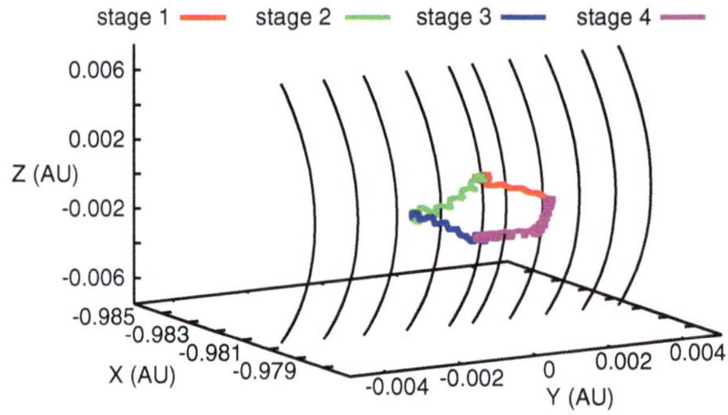

Fig. 11 XYZ projection of the solar sail trajectory during the whole mission

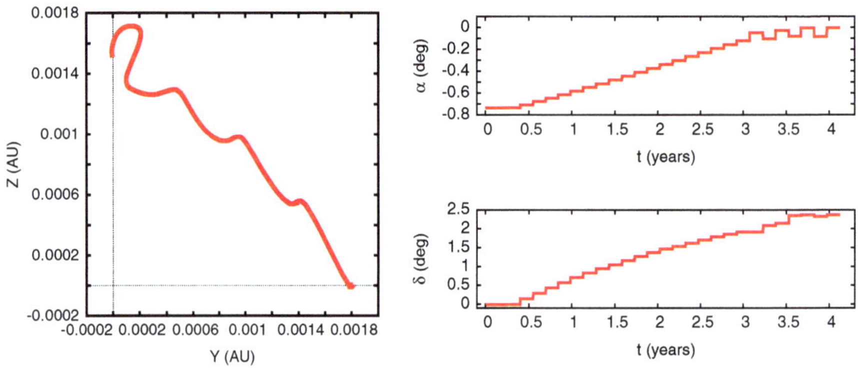

Fig. 12 Different projections of the trajectory of the first stage surfing form p_0 to p_1

Solar disc. The behaviour of each stage is very similar, and for simplicity we only present detailed results of the first stage.

On the left-hand side of Fig. 12 we have the *YZ* projection of the stage 1.A of the mission, where we clearly see how the trajectory gains altitude while moving towards the left until it reaches a vicinity of p_1. On the right-hand side of this figure we see the required control scheme, i.e. the variation of α (top) and δ (bottom) along time. In Fig. 13 we have the projection of the trajectory in the saddle and centre planes related to the initial point p_0. The black crosses correspond to the projection of the intermediate equilibria (q_i) that play a role in the surfing scheme. Noticc how the trajectory in the saddle projection (right) is a succession of saddle arcs, each of them centred around q_i. The trajectory on the other two centre directions (middle and left) also moves along the family of equilibria and rotates around them.

On the left-hand side of Fig. 14 we have the *YZ* projection of the trajectory of the solar sail during stage 1.B. As we can see, despite the instability of the region, the trajectory remains close of the equilibrium point for 2 years. On the right-hand side of this figure we see the variation of the sail orientation during the 2 years. In Fig. 15 we find the plots on the projection of the sail trajectory on the saddle

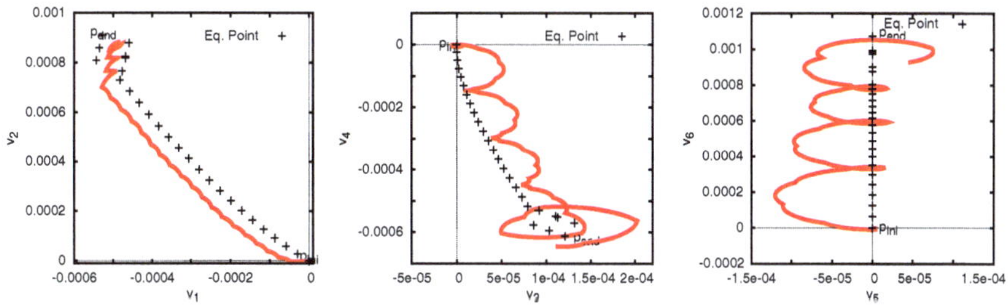

Fig. 13 Different projections of the trajectory of the first stage surfing form p_0 to p_1

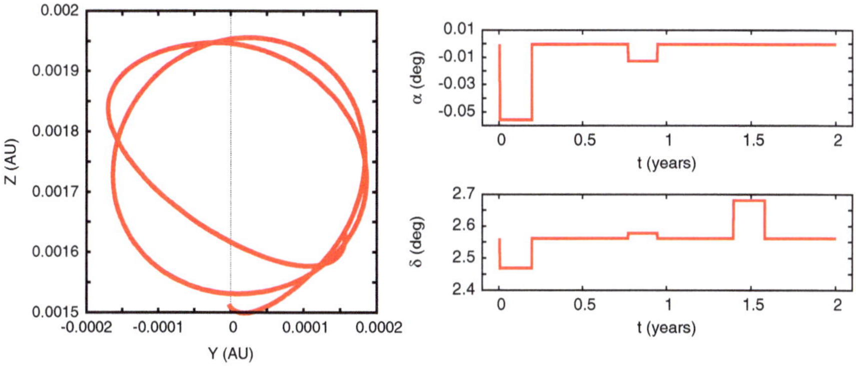

Fig. 14 Stage 1, control around p_1: α, δ variation

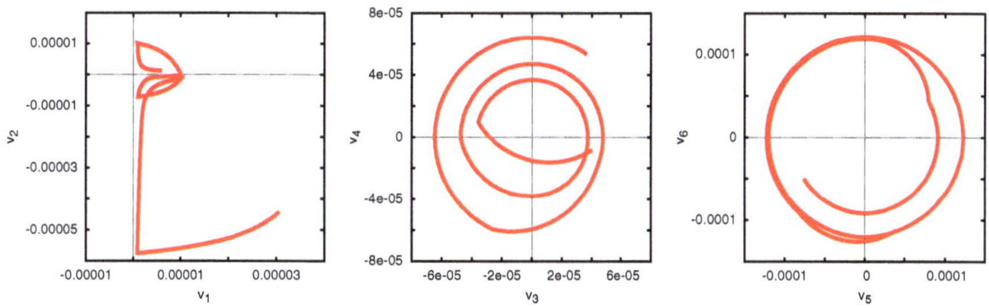

Fig. 15 Different projections of the trajectory of the first stage control around p_1

and two centre projections around the target point p_1. As we can see, in the saddle projection (right) the trajectory is a connection of saddle arcs, where each time the trajectory is escaping along the unstable direction, the sail orientation is changed and the trajectory comes back towards the stable direction of p_1 along a saddle arc. The trajectory in the two centre projections is a connection of rotations around different points, which remains bounded along time.

4 Conclusions

We know that in the Sun-Earth RTBP with a solar sail there are surfaces of equilibrium points, each point corresponding to a certain sail orientation. In this paper we have described some of the properties of these families of equilibrium points for different values of β (the sail lightness number). We have seen that some of these equilibrium points are linearly unstable and have a stable and unstable manifolds associated to them. Hence, for a fixed sail orientation, if we are close to equilibria, the trajectory will escape along the unstable manifolds. We also know that when we change the sail orientation the position of equilibria shifts, and the trajectory will escape along the new unstable manifold. If we can understand how these invariant manifolds vary with the sail orientation, we can derive schemes to make the solar sail surf around the system in a controlled way.

We have shown how to derive strategies to move along the family of equilibria in a controlled way. These strategies use information on the relative position between the solar sail and the invariant objects in the system to take decisions. Hence, they are robust when dealing with different sources of error. We have taken a test mission to illustrate the performance of these strategies.

Acknowledgements This work has been supported by the MEC grant MTM2012-32541, the AGAUR grant 2014 SGR 1145 and the AGAUR postdoctoral fellowship Beatriu de Pinós (BP-B 00142-2011).

References

Ceriotti, M., Heiligers, J., McInnes, C.: Trajectory and spacecraft design for a Pole-Sitter mission. J. Spacecr. Rocket. **51**(1), 311–326 (2014)

Dachwald, B., Seboldt, W., Macdonald, M., Mengali, G., Quarta, A.A., McInnes, C.R., et al.: Potential solar sail degradation effects on trajectory and attitude control. In: AIAA Guidance, Navigation, and Control Conference and Exhibit (AIAA 2005–6172) (2005)

Dachwald, B., Macdonald, M.: Gossamer roadmap technology reference study for a multiple NEO rendezvous mission. In: Advances in Solar Sailing, vol. 1, pp. 211–216. Springer, Berlin (2014)

Farrés, A., Jorba, À.: A Dynamical system approach for the station keeping of a solar sail. J. Astronaut. Sci. **56**(2), 199–230 (2008a)

Farrés, A., Jorba, À.: Solar sail surfing along families of equilibrium points. Acta Astronaut. **63**(1–4), 249–257 (2008b)

Farrés, A., Jorba, À.: Station keeping of a solar sail around a Halo orbit. Acta Astronaut. **94**(1), 527–539 (2014)

Heiligers, J., Diedrich, B., Derbes, B., McInnes, C.: Sunjammer: preliminary end-to-end mission design. In: Proceedings of the AIAA/AAS Astrodynamics Specialist Conference (2014)

Macdonald, M., McInnes, C.R.: Solar sail science mission applications and advancement. Adv. Space Res. **48**(11), 1702–1716 (2011)

McInnes, A.: Strategies for solar sail mission design in the circular restricted three-body problem. Master's thesis, Purdue University (2000)

McInnes, C.R.: Solar Sailing: Technology, Dynamics, and Mission Applications. Springer, Chichester (1999). ISBN 1-85233-102-X

McInnes, C.R., Macdonald, M., Angelopolous, V., Alexander, D.: GeoSail: exploring the geomagnetic tail using a small solar sail. J. Spacecr. Rock. **38**(4), 622–629 (2001)
Mckay, R., Macdonald, M., Biggs, J., McInnes, C.: Survey of highly non-Keplerian orbits with low-thrust propulsion. J. Guid. Control Dyn. **34**(3), 645–666 (2011)
Szebehely, V.: Theory of Orbits. The Restricted Problem of Three Bodies. Academic, New York (1967)

Terminal Entry Phase Trajectory Generator for Reusable Launch Vehicles

Luca De Filippis, Murray Kerr, and Rodrigo Haya

Abstract In this paper a new trajectory generator for the Terminal Area Energy Management phase of a Reusable Launch Vehicle is presented. During this phase the vehicle has to glide at low Mach to reach the point close to the runway where automatic approach and landing starts. The algorithm presented here is based on the concept of Energy Corridor management and it is composed of two main elements: a trajectory propagator and a ground track generator. Imposing a dynamic pressure profile as function of the altitude, a heading path is selected in order to steer the vehicle toward the runway, putting to zero cross and down track errors.

Symbols and Acronyms

g	=	Gravitational acceleration	$[m/s^2]$
h	=	Vehicle altitude over ground	$[m]$
m	=	Vehicle mass	$[kg]$
q	=	Dynamic pressure	$[Pa]$
sb	=	Speed-break deflection	$[rad]$
x	=	Vehicle position along X-axis	$[m]$
y	=	Vehicle position along Y-axis	$[m]$
z	=	Vehicle position along Z-axis	$[m]$

L. De Filippis (✉) • M. Kerr • R. Haya
Elecnor Deimos, Ronda de Poniente 19, 28760 Tres Cantos, Madrid, Spain
e-mail: luca.defilippis@deimos-space.com; murray.kerr@deimos-space.com; rodrigo.haya@deimos-space.com

G. Gómez, J.J. Masdemont (eds.), *Astrodynamics Network AstroNet-II*, Astrophysics and Space Science Proceedings 44,
DOI 10.1007/978-3-319-23986-6_7

α	=	Angle of attack	$[rad]$
β	=	Sideslip angle	$[rad]$
γ	=	Flight path angle	$[rad]$
χ	=	Heading angle	$[rad]$
ρ	=	Air density	$[kg/m^3]$
σ	=	Bank angle	$[rad]$
D	=	Drag	$[N]$
E	=	Total Energy	$[J]$
L	=	Lift	$[N]$
$Mach$	=	Mach number	$[-]$
S	=	Reference wing surface	$[m^2]$
V	=	Flight Speed	$[m/s]$
W	=	Vehicle weight	$[N]$
ALI	=	Approach and Landing Interface	
A&L	=	Approach and Landing	
RLV	=	Reusable Launch Vehicles	
TAEM	=	Terminal Area Energy Management	
TEP	=	Terminal Entry Point	
UAV	=	Unmanned Aerial Vehicle	

1 Introduction

The Future European Space Transportation Investigation Program of ESA was launched in the early 90's after the Hermes program was cancelled, with the idea of studying a cheaper, reliable and multipurpose Reusable Launch Vehicle (RLV). This program collected a large set of proposals and concepts that allowed the second phase of the program to develop experimental vehicles like the Hopper and IXV (Kuczera and Johnson 1999). In the United States, the Reusable Launch Vehicle Technology Program was intended to drastically reduce operational costs experienced with the Shuttle. The partnership between NASA and Lockheed Martin allowed the development of the X-33 experimental vehicle that was followed by a family of other technological demonstrators like the X-37B (Bevacqua 2004). Multiple interests motivated private companies and public organizations to design and develop RLVs. First of all, the reduced building and operational costs, together with the fact that, using the same vehicle for more than one mission, launch costs could be reduce up to one order of magnitude (Freeman et al. 1997). Costs reductions together with re-entry capabilities will allow the use of RLVs for new missions, like satellites retrieval, passengers and payload servicing and space tourism.

Autonomy of RLVs has been a key issue from the very beginning of the study and development of these systems. Tolerance to uncertainties and failures of the Guidance, Navigation and Control system (GNC), together with trajectory adaptation capabilities, has been deeply investigated (Doman et al. 2006). The GNC system needs to be designed to work on many different phases of the mission like ascent, separation, entry, approach and landing. It represents a key element for costs reduction, as its improvement directly affects flight operations (Corban et al. 2001). Current GNC systems still rely significantly on pre-mission planning, where a large and complex activity is performed before the flight in order to update the system with respect to the conditions of the day of the launch. On the other hand automatic and adaptive trajectory generation and control is mandatory to react to changing mission demands (Calise et al. 1988). According with mission objectives and vehicle characteristics, the mission planning of a RLV can vary significantly from take-off to landing. Particularly the vehicle descent can be subdivided in three main sub-phases called Entry, Terminal Area Energy Management (TAEM) and Approach & Landing (A&L) (Horneman 2010).

When vehicle speed goes down Mach 2.5 and altitude is around 30 km, TAEM phase is assumed to start. At this speed aerodynamic surfaces become the main control system and quasi-gliding flight is performed. Transition point from Entry to TAEM is called TAEM Entry Point (TEP). During TAEM the remaining total energy is controlled to put to zero the cross and down track errors (McKee 2011). The first is the distance normal to the runway axis between the vehicle position and the point where A&L starts, the Approach and Landing Interface (ALI), the second is the distance, along the runway axis, between the vehicle position and the ALI (Fig. 1). Altitude at ALI is around 3 km and Mach depends from the vehicle aerodynamic characteristics. However at this point accurate runway alignment is performed and automatic A&L is started.

The TAEM phase was first defined for the Shuttle in the early 70's and further improved with the knowledge acquired with successive tests and flights. In this same period TAEM was subdivided in shorter sub-phases and the guidance logic for each one of them was designed and implemented (Moore 1991). Few years later Russian engineers developed for the Buran TAEM a similar guidance logic, based on energy corridor generation and control. This approach, like the one developed for the Shuttle, requires a long and intensive pre-flight preparation, in order to update with day-of-launch parameters the software and the data sheets (Kirpishchikov

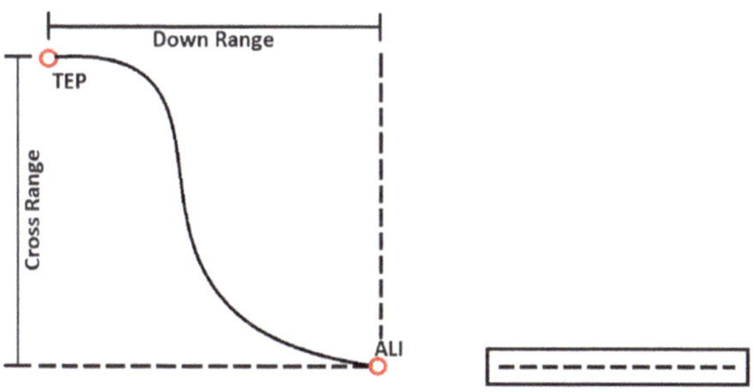

Fig. 1 Cross and down track error during TAEM

1997). With the X-34 program the problem of trying to improve the guidance logic during TAEM was reconsidered. Particularly, in 1999 Burton and Tragesser published the Auto Landing I-Load Program (ALIP), a tool for pre-mission design of the autonomous landing trajectory (Barton and Tragesser 1999). In the same years the authors started to study the possibility to apply the same techniques used for ALIP to trajectory generation of the TAEM phase. This research activity was applied in 2001 by Girerd that was able to test this method just to the subsonic portion of TAEM (Girerd 2001). In the same years Grubler described in his thesis the evolution of Girerd method, applying it to the full TAEM phase and reformulating the energy corridor approach in a more effective way. However, even if Grubler work improved the trajectory propagation algorithm of ALIP, he was unable to develop an effective algorithm to generate the ground track during TAEM (Grubler 2001). Fuzzy Logic was applied to TAEM trajectory generation by Burchett, who proposed this technique to adapt the parameters necessary to design the classic Shuttle path, according with vehicle characteristics and initial conditions. However its approach was never been verified with large test campaigns (Burchett 2004).

This paper aims to contribute to the research path opened with Groubler, Kluever and Ridder works on trajectory generation during TAEM phase of RLVs. All these authors exploit energy-corridor concepts developed for the Shuttle and Buran guidance during TAEM, to develop real-time trajectory generators where longitudinal and lateral guidance problems are split and approached separately. This method resulted of big interest here because of its simplicity and robustness with respect to more advanced and complex approaches. In this work a new guidance strategy based on aircraft gliding performance and current dynamic/kinematic conditions is presented. An optimized trajectory propagator algorithm with respect to the one described in previous works is suggested and a new ground-track generation algorithm is described. Particularly, this last is the main research contribution provided with this work. An evolution of the well known A* algorithm, called Kinematic A* (De Filippis and Guglieri 2012), has been adapted to this specific problem. Kinematic A* was developed for path planning of Unmanned Aerial Vehicles (UAVs) and in this case it has been used to plan the heading change sequence required to steer the spacecraft toward the final position and configuration.

2 Energy Corridor and Altitude Profile

The energy corridor is the basis of many guidance schemes used in TAEM and it is also fundamental to explain the method proposed here. As a matter of fact, it was already said that, being a quasi glide condition, the objective in TAEM is to control vehicle total energy. This quantity is defined as the sum of potential and kinetic energy over weight:

$$E/W = h + \frac{q}{\rho g} \tag{1}$$

Plotting vehicle total energy over weight with respect to downrange (i.e. range covered in straight flight) and defining minimum and maximum range capabilities, related to minimum and maximum dynamic pressure profiles, it is possible to obtain what is called the Energy Corridor. The trajectory generation method here described uses an altitude and a dynamic pressure profile with respect to a given downrange, in order to force the vehicle to fly a given energy profile. In other words, in order to solve the trajectory generation problem, longitudinal and lateral planes are split and treated separately. A given dynamic pressure profile, function of the altitude, is selected. This profile provides in turn a given total energy profile and a given maximum downrange. Knowing the maximum range, a ground track is selected that puts to zero cross and down track errors, following the assigned energy profile. Angle of attach and bank are the control variables used to follow the selected dynamic pressure and heading profile. The dynamic pressure is selected between maximum glide and maximum dive.

The altitude profile is a fundamental element for the implementation of this trajectory generator. The basic idea is to use altitude instead of time as independent variable to generate each vehicle state. The concept behind this change is to express each variable profile as a function of the altitude instead of time, as time is not a main constraint during this phase of the mission. As a matter of fact, the vehicle total energy is a direct function of altitude and, according with the selected atmosphere model, the same happens for dynamic pressure. Following the logic behind the energy corridor concept, it is not important when the vehicle reaches a given energy condition, but it is fundamental to link to each altitude a given energy and in turn a given flight range. The altitude profile is a discretization of the altitude between TEP and ALI. For guidance and control during TAEM, ALI altitude is a requirement coming from mission planning, on the other hand altitude at TEP is affected from errors and uncertainties coming from guidance and control along the previous phase of the mission: the entry phase. Consequently, if the trajectory generator is used in real-time, the altitude profile must be defined in flight, when TAEM trajectory is planned, just before the vehicle reaches TEP. A constant altitude step is selected in this method. This configuration parameter is used to subdivide the altitude between TEP and ALI.

3 Trajectory Generator Architecture

The ground track generator is put together with the trajectory propagator into an iterative loop, the overview of the full architecture helps to better understand how the ground track generator works and why is this iteration required.

Figure 2 shows the trajectory generator architecture and the iterative loop already mentioned. An angle of attack profile is needed to generate the dynamic pressure profile. The angle of attack profile, being a function of the Mach, is directly related to the energy profile flown inside the energy corridor. The maximum Z-acceleration profile is needed to limit maximum turning radius of the vehicle at each altitude and

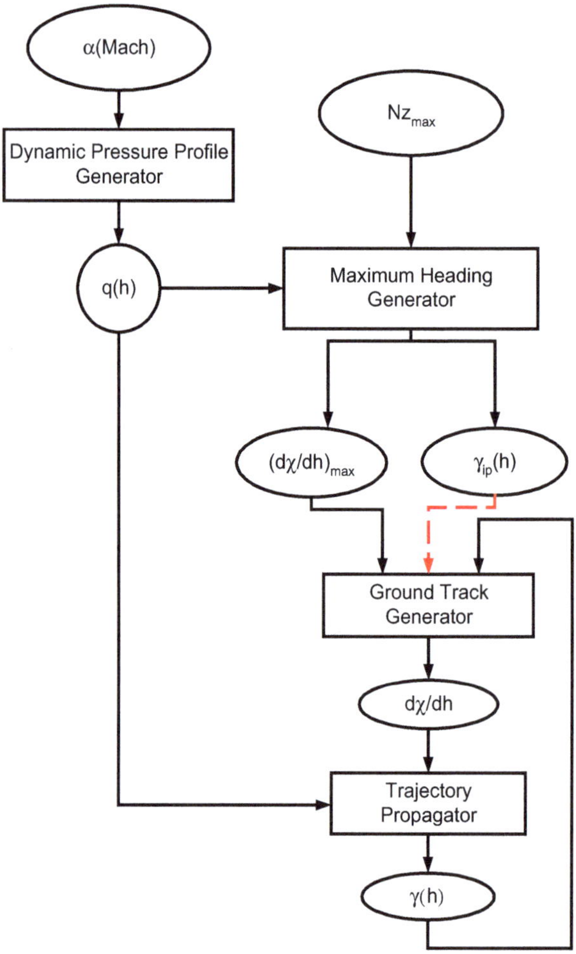

Fig. 2 Trajectory-generator architecture

speed. An iteration is required because an interconnection between the ground track generator and the trajectory propagator exists. The first block requires a flight path to build the heading angle profile. On the other hand the trajectory propagator uses the heading variation as an input to generate the flight path angle. It seems clear that iteration is required to satisfies both the blocks and converge to an acceptable solution.

The red arrow in Fig. 2 indicates that just during first iteration the flight path angle profile, coming from the maximum heading block, is used. This profile is generated assuming a constant turn at maximum bank that means a very short range, far from the one that will come out at the end of the iteration. However, this flight path is needed to start the iteration process that follows, going back to the ground track generator with the new flight path angle profile and creating a new heading profile up to converge to the solution. Convergence is assumed when the error between the flight path angles profiles obtained from two successive iterations is lower than an assigned threshold.

3.1 Trajectory Propagator

Figure 3 describes the flow diagram of the trajectory-propagation algorithm. States and commands at TEP are sent to the Step Propagator together with a constant speed-break deflection, in order to obtain a trajectory flown including the speed-break effects. This block calculates states and commands at each altitude step using as input the heading and dynamic pressure profiles. For each altitude step an iterative algorithm over the angle of attack (AoA Iteration) is used to find flight path angle, angle of attack and bank angle that respect the equation of motion and bring the

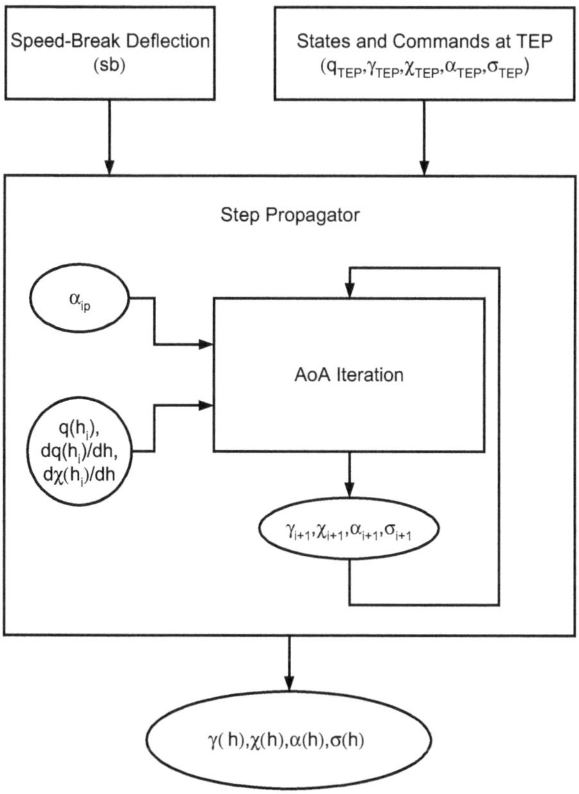

Fig. 3 Trajectory-propagator flow diagram

aircraft to the dynamic pressure and heading angle imposed from the profiles. Once convergence is reached, states and commands at the current altitude are obtained and they are used as initial condition for the next step. This process is repeated for all the altitude steps up to ALI altitude, obtaining the complete states and commands profiles and in turn a complete trajectory.

The algorithm propagates the equations of motion so that the trajectory matches the vehicle dynamics and the imposed profiles. First of all, to build the propagation algorithm a variable change is required, in order to express the equations of motion as a function of altitude instead of time:

$$\begin{cases} \dot{x} = V \cos\chi \cos\gamma \\ \dot{y} = V \sin\chi \cos\gamma \\ \dot{h} = V \sin\gamma \end{cases} \tag{2}$$

Using the third equation of system (2) to express the variable change a new system of equations of motion is obtained:

$$\begin{cases} \dfrac{dV}{dh} = -\dfrac{g}{V} - \dfrac{D}{mV\sin\gamma} \\ \dfrac{d\gamma}{dh} = -\dfrac{g}{V^2 \tan\gamma} + \dfrac{L}{mV^2}\dfrac{\cos\sigma}{\sin\gamma} \\ \dfrac{d\chi}{dh} = \dfrac{L\sin\sigma}{mV^2\cos\gamma\sin\gamma} \end{cases} \tag{3}$$

The propagation algorithm here presented is similar in concept to the one already used in previous works. However, a significant improvement has been provided, optimizing the iteration process. As a matter of fact previous implementations tend to use a set of three or two nested iterations (over a corresponding number of states) to find the desired variables at each altitude step. The propagator here described reduces to one the number of iterations, obtaining a more efficient and stable architecture for the propagator.

4 Ground-Track Generator

As early mentioned the ground track generation is based on an evolution and adaptation of a search algorithm called Kinematic A* (De Filippis and Guglieri 2012). This algorithm searches the sequence of motions that brings the vehicle from TEP to ALI with a discrete and iterative procedure. Particularly this method exploits a set of kinematic equations to generate a horizon of possible movements from one position to the next and it chooses the one that minimizes an assigned cost function.

4.1 Motion Horizon

This part of the ground track algorithm, exploits the same models described for the trajectory propagator and the same logics. Considering the discrete altitude profile, used to propagate the equations of motion, it is possible to extract from system (2) the following discrete equations:

$$\begin{cases} \dfrac{dx}{dh} = \dfrac{\cos\chi}{\tan\gamma} \\ \dfrac{dy}{dh} = \dfrac{\sin\chi}{\tan\gamma} \end{cases} \tag{4}$$

In these equations the flight path angle profile, $\gamma(h)$, is received from the maximum heading algorithm (during first execution) and from the trajectory propagator (during the following executions). The heading angle is the control variable expressed as:

$$\chi(h_i) = \chi_0 + \tfrac{d\chi}{dh} = \chi_0 + u\left(\tfrac{d\chi}{dh}\right)_{max} \cdot \Delta h \quad -1 < u < 1 \tag{5}$$

Assuming to have a set of initial conditions:

$$[x_0, y_0, \chi_0] \tag{6}$$

The control vector (5) is built summing to the initial angle, χ_0, a vector of headings bounded with minimum and maximum heading change. The discrete variable, u, is a vector of $N_u + 1$ elements given by:

$$u = [-1, \frac{1}{N_u}, 1] \tag{7}$$

substituting Eq. (7) in Eq. (5) a vector of discrete headings is obtained for a given altitude step:

$$\bar{\chi}(h_i) = [-\chi_{max}(h_i), \ldots, \chi_n(h_i), \ldots, \chi_m ax(h_i)] \tag{8}$$

In order to build the mentioned horizon of possible motions, position at the next altitude step h_i is obtained from Eqs. (4) for each one of the heading angles (χ_n) of vector (8):

$$\begin{cases} x_n(h_i) = x_0 + \dfrac{\cos\chi_n}{\tan\gamma_i} \cdot \Delta h \\ y_n(h_i) = y_0 + \dfrac{\sin\chi_n}{\tan\gamma_i} \cdot \Delta h \end{cases} \tag{9}$$

a matrix $P(R^n \times R^2)$ is obtained, where positions from (9) are placed along the rows and x and y components put as the columns:

$$\begin{cases} \bar{p}_n(h_i) = [x_n, y_n] \\ \bar{\bar{P}}(h_i) = [\bar{p}_min(h_i), \ldots, \bar{p}_n(h_i), \ldots, \bar{p}_max(h_i)] \end{cases} \tag{10}$$

This process of generating a set of new positions from a given initial one is the core of Kinematic A*.

4.2 Searching Algorithm

The searching algorithm used for this ground track generator will not be described in details, being similar to the logics of a classic search algorithm like A* (Hart et al. 1968) or one of its evolutions Theta* (Nash et al. 2007). In particular, the algorithm used here is evolved from a search algorithm called Kinematic A* (De Filippis and Guglieri 2012). Just few details will be provide here about this method, to give a general understanding to someone how is not familiar with this approaches.

The search process has to find the sequence of positions and headings that allow going from an initial position to a final one (in our case from TEP to ALI). The search process can start indistinctly from the initial or the final state that are expressed with vector (6) and continues with the generation of a horizon of new positions, according with the equations introduced in the previous subsection. Calling initial position the one from where the algorithm starts, parent position the one from where the motion horizon is generated, current position the one selected between the horizon of possibilities and final position the one that the algorithm is trying to reach, an iterative process is performed in order to find the heading sequence that connects the initial position with the final one.

The way to chose between the set of current positions that the algorithm generates from the parent one is related to the idea of minimizing an assigned cost function. This function is defined as:

$$F = H \cdot |p_{cur} - p_{fin}| + G \cdot \chi_n \tag{11}$$

It is made of two terms multiplied by two gains; the first element is called cost to go, while the second is called cost to come. The cost to go is related to the distance between the current and the final position, the cost to come is related to the effort required to reach the current position from the parent one. The gain H is chosen so that the cost to go dominates the path search and each new current position minimizes the distance from the final one. The gain G then is usually lower than the previous one and is chosen so that a smoother heading change between successive motions is provided.

Going back to the search process, when the algorithm generates a new set of current positions, they are put in a list called open list together with the associated

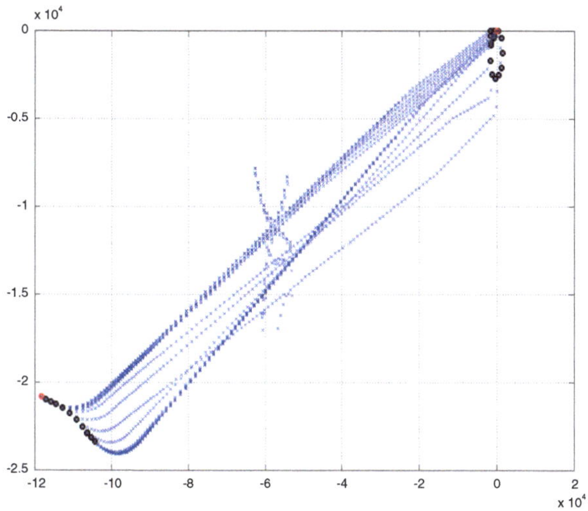

Fig. 4 Example of trajectory horizons

cost and their parent. Each time the algorithm iterates the open list is sorted and the position with minimum cost is selected. This new position becomes the new parent and it is put in another list called closed list. This list contains positions that will not be evaluated anymore and the path will be reconstructed using positions contained in this list. It should be evident that, because the current position is selected from the open list and not just from the horizon of possibilities coming from the expansion of the current parent, the algorithm jumps from one position to another according with the open list position that has the minimum cost and it stops when a position sufficiently close to the final one is reached. The path from initial to final position is then reconstructed following backward the sequence of parents that brought the algorithm to reach the last position.

In order to better understand this process, let's make an example. Assuming to start expanding the initial position, a horizon of current positions will be obtained. These positions will be put in the open list and the list will be sorted. The current position with lower cost will become the new parent and will be copied in the closed list. At this stage just the positions obtained expanding the initial state will be contained in the open list and the current position will be just the one with minimum cost. The new parent position will be expanded and its set of current positions will be copied in the open list. The list will be sorted again and a new position will be selected and copied in the closed list. The second time the algorithm sorts the open list not only the new horizon of positions will be contained in the list, also positions coming from previous expansion will be sorted. In this way, if one of the previous positions will have a cost lower than any other position in the new horizon, this position will become the new parent and will be expanded.

It should be clear from previous description that the number of positions that compose the path can vary, according with the number of evaluations required to find a current position that matches the termination criteria. As it was mentioned many times in this section, in our case the number of positions that have to compose the ground track is equal to the number of altitude steps that have been selected to generate the trajectory. This means that the trajectory generator has to start from the initial position and can select just a fixed amount of successive motions trying to reach the final one.

In order to overcome this limit and adapt to this problem the search method, a new algorithm has been designed and implemented. The main idea is that instead of going from one position to the next, every time the algorithm has to expand a new position, the search method generates a horizon of trajectories obtained with classic search approach. These trajectories are stored in the open list together with their parent position, associated cost function and final position. In Fig. 4 the horizon of two trajectories, one starting at TEP and the other starting at ALI, is shown. Defining terminal position the position associated with the last altitude step. Each time the algorithm expands a parent position, it stores a horizon of trajectories, like the one in Fig. 4, inside the open list. Then it sorts the list and it chooses the trajectory with the closer terminal position to the final one. The process then continues the same as for the classic search and the selected trajectory is put in the closed list. The process is repeated starting from the new initial position of the selected trajectory and a new expansion is performed. If the terminal position of the selected trajectory is close enough to the desired final position the algorithm stops. This condition is evaluated defining a minimum distance between the terminal position and final one.

4.3 Double Search and Boundary Conditions

Another important improvement with respect to previous implementations of the search algorithm regards the logic to perform the search. This modification has been triggered by the need to match exactly the final conditions at TEP and ALI. In fact, analyzing how the termination criterion of Kinematic A* is implemented, it should be clear that matching position and heading at TEP and ALI is not guaranteed. The algorithm expands each position generating a new set of trajectories and it iterates up to find a position close to the final one, but no exactly that one. One of the limitations of classic search is that, being the algorithm discrete, it is very difficult if not impossible to match exactly the final position. Also, if the exact final position should be matched, any constraint is imposed on the final heading. Adding this other criterion to the algorithm is hard and strongly affects convergence.

In order to have perfect matching of initial and final states a parallel search from TEP and from ALI is performed. Two parallel expansions are imposed, starting from the two bounds of the trajectory and with the identical logic. The idea is to force the algorithm to converge toward a middle position where the two trajectories meet. To do that, each time the algorithm iterates two sequential expansions are performed: the first is the one with TEP as initial position and the second is the one started from ALI. The final position for the first expansion is the terminal point of the previous

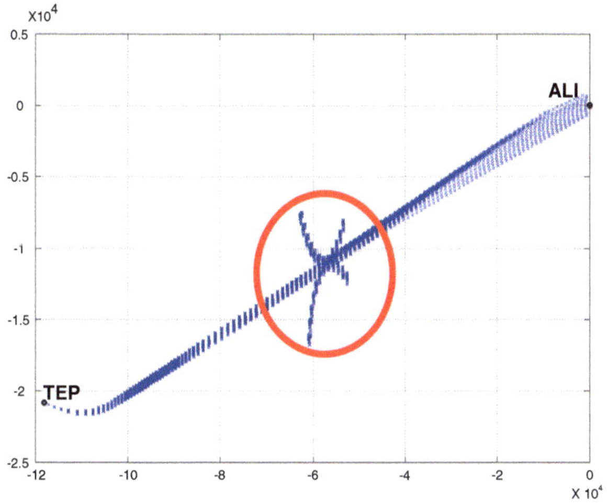

Fig. 5 Example of double search

expansion from ALI. On the other hand the final point for the second expansion is the terminal position of the last expansion from TEP. The red ellipsis in Fig. 5 shows a set of arcs in four directions. This is where the two trajectories cross and curve in order to try to reduce the distance between their terminal positions.

Figure 4 shows what happens when the algorithm expands the trajectories. The black dots close to TEP and ALI (in red) are the parent positions that the algorithm iteratively expands. Going ahead with the expansions, the algorithm tries to reduce the overshoot between the two trajectories up to reach a point where the two terminal positions are sufficiently close. The convergence criterion for this search logic is more complex then the classic one. In fact, with this approach the two trajectories are forced to find terminal points very close each other and, being these points interconnected by the fact that one is the final position for the other, heading at the two points is almost the same.

Convergence is assumed when the terminal points of two trajectories are sufficiently close and the heading changes between the parents are between their limits. Figure 6 helps to describe in details the criterion. If the starts are the terminal positions of the two trajectories and the rectangles are the relative parent positions, the distance *er* has to be lower than a given threshold to have convergence. Also, the difference between the two headings χ_1 and χ_2 has to be inside the assigned maximum heading change for the corresponding altitude step. This approach improves the previous implementations because guarantees that initial and final conditions are perfectly matched, also a faster convergence to the solution is obtained. This is due to the fact that expanding from the two boundary positions and imposing as final states the terminal positions of the two trajectories, allows forcing the algorithm to expand the positions so that the trajectories run one toward the other with close headings.

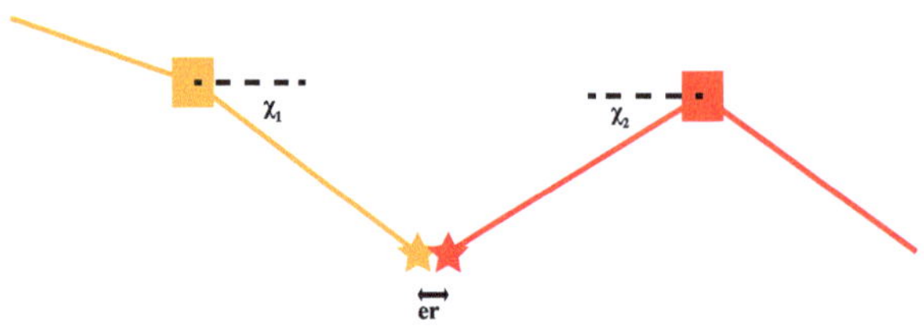

Fig. 6 Termination criterion parameters

5 Results

The trajectory generator presented here has been integrated in more complex guidance system that was tested in simulation. Two main algorithms compose the guidance block: the trajectory generator and the trajectory tracker. When the vehicle arrives at TEP, the trajectory generator receives from navigation the measure of variables and states at this point that are required to generate the path. Once the path is calculated, the trajectory is fed forward to the trajectory tracker that generates a sequence of closed-loop commands to track it. This is done receiving from navigation the measure of a set of states and parameters that are processed inside the tracker together with the feed-forward variables in order to generate angle of attack, bank angle and speed-break commands.

Some simulations made with the trajectory generator integrated with the trajectory tracker are presented here. This analysis is part of an extended test campaign that was planned to evaluate the region of positions where a TAEM phase could start according with vehicle flying performances and mission requirements. In order to study minimum and maximum cross and down track, a large number of simulations from different initial positions were run. The results of this study will not be fully reported here, but the conclusions extracted from them will be described.

Assuming to have fixed speed and flight path, it was evidenced adding dispersions on cross and down track that the relative heading between TEP and ALI, together with the relative distance between the two points, where the main variables that affect the ground track. The initial positions for the test campaign were selected choosing a decreasing distance from ALI and a set of relative headings for each distance. Figure 7 shows the results of a test campaign obtained changing the distance TEP-ALI and changing the relative heading between the two points, adding a large cross track error. The relative distance between initial and final position is varied from 140 to 100 Km with steps of 10 Km, on the other hand the relative heading is changed between three steps: $-45°$, $0°$ and $45°$.

From Fig. 7 it can be seen that not all the cases are plotted, for example the path obtained at distance 110 Km and relative heading -45 is not represented. This is due to the fact that going closer to the ALI the amount of energy that has to be dissipated with short range increases and the minimum of allowed down and cross track is reached. In these conditions not all the planned paths can be easily flown and the limits of the guidance strategy start to be approached. Figure 8 shows the results

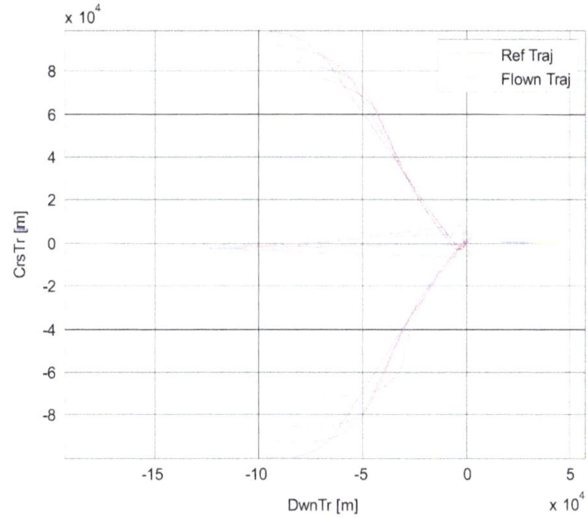

Fig. 7 Termination criterion parameters

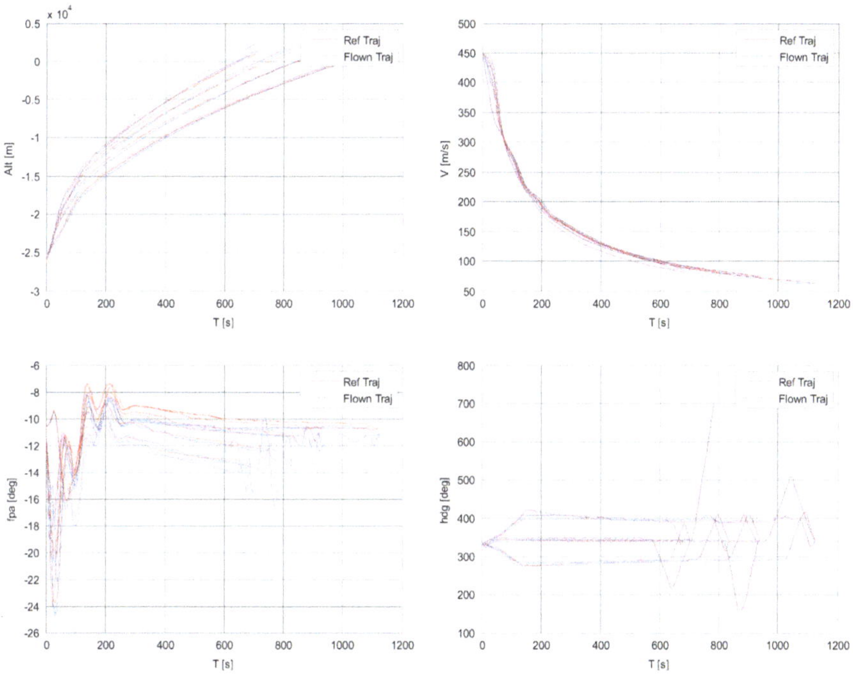

Fig. 8 Termination criterion parameters

of the simulations with the closed-loop guidance, compared with the one planned with the trajectory generator. From the figure seems that the trajectory tracker is

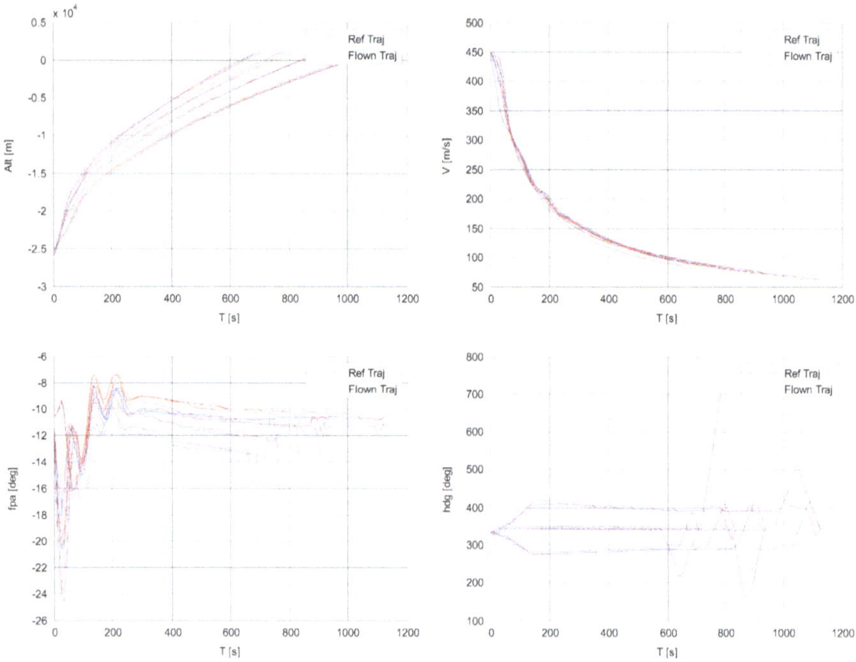

Fig. 9 Termination criterion parameters

able to follow the altitude and speed profiles planned with the trajectory generator, the same happens with the bank command represented in Fig. 9.

6 Conclusions

The results presented in the previous sections support the desired contributions and objectives that motivated the activity described in this paper. The guidance strategy showed to be a promising approach to control the RLV during its TAEM phase. Future work on this method will focus on building an accurate energy corridor, trying to simplify the strategy of choosing the dynamic pressure profile, in order to reduce the number of input parameters required to generate the desired trajectory. More work will be required also to improve the algorithm that generates the maximum heading change, in order to reduce the complexity of the trajectory tracker.

An accurate analysis will be required in order to evaluate the rate of convergence of the iterative processes that compose the algorithm. This will be an important study that will have to be performed together with robustness assessment of the algorithms. Also, the computational cost related to running the trajectory generator many times during TAEM needs to be estimated. These further analyses will be required to evidence the reliability of the presented trajectory tracker over a failure and recovery scenario.

References

Barton, G.H., Tragesser, S.G.: Autolanding trajectory design for the X-34. In: AIAA Paper 99-4161 (1999)

Bevacqua, T.R.: Advanced flight control issues for reusable launch vehicles. Master Thesis, Ohio University (2004)

Burchett, B.T.: Fuzzy logic trajectory design and guidance for terminal area energy management. J. Spacecr. Rocket. **41**(3), 444, 450 (2004)

Calise, A.J., Corban, J.E., Flandro, G.A.: Trajectory optimization and guidance law development for national aerospace plane applications. In: American Control Conference (1988)

Corban, J., Johnson, E.N., Calise, A.J.: Reusable launch vehicle adaptive guidance and control using neural networks. In: AIAA Guidance, Navigation, and Control Conference and Exhibit, pp. 1, 11 (2001)

De Filippis, L., Guglieri, G.: Advanced graph search algorithms for path planning of flight vehicles. In: Recent Advances in Aircraft Technology. Dr. Ramesh Agarwal (Ed.), InTech (2012). doi:10.5772/37033

Doman, D.B., Oppenheimer, M.W., Bolender, M.A.: Progress in guidance and control research for space access and hypersonic vehicles. In: 45th IEEE Conference on Decision and Control, pp. 6837, 6851 (2006)

Freeman Jr., D.C., Talay, T.A., Austin, R.E.: Reusable launch vehicle technology program. Acta Astronaut. **41**(11), 777–790 (1997)

Girerd, A.: Onboard trajectory generation for the unpowered landing of autonomous reusable launch vehicles. Master Thesis, Department of Aeronautics and Astronautics MIT (2001)

Grubler, A.C.: New methodologies for onboard generation of terminal area energy management trajectories for autonomous reusable launch vehicles. Master Thesis, Department of Aeronautics and Astronautics MIT (2001)

Hart, P., Nilsson, N., Raphael, B.: A formal basis for heuristic determination of minimum cost paths. IEEE Trans. Syst. Sci. Cybern. **4**(2), 100–107 (1968)

Horneman, K.R.: Automated trajectory generation and guidance for a new launch vehicle flight phases. Ph.D. Thesis, University of Missouri-Columbia (2010)

Kirpishchikov, V.P.: Descent and landing trajectories of the orbiter Buran. In: Automatic Control Algorithms: Aviation and Space Systems, pp. 46, 55 (1997)

Kuczera, H., Johnson, C.: The major results of the FESTIP system study. In: AIAA 9th International Space Planes and Hypersonic Systems and Technologies Conference, Norfolk (1999)

McKee, S.E.: Flight mechanics and control for an unpowered reusable launch vehicle. Master Thesis, University of Missouri (2011)

Moore, T.: Space shuttle entry terminal area energy management. In: NASA Technical Memorandum 104744 (1991)

Nash, A., Daniel, K., Koenig, S., Felner, A.: Theta*: any angle path planning on grids. In: Proceedings of the AAAI Conference on Artificial Intelligence, pp. 1177–1183 (2007)

Two ASRE Approaches with Application to Spacecraft Coulomb Formations

Mohammad Mehdi Gomroki, Francesco Topputo, Ozan Tekinalp, and Franco Bernelli-Zazzera

Abstract Suboptimal solutions of nonlinear optimal control problems are addressed in the present work. These suboptimal approaches are known as Approximating Sequence of Riccati Equations (ASRE) methods. In the ASRE methods, the nonlinear problem is reduced to a sequence of linear-quadratic and time-varying approximating problems. For this purpose, the nonlinear equations are written in State Dependent Coefficient (SDC) factorization form. Two different ASRE approaches are discussed and their implementation procedures will be explained. To implement and compare these two techniques, spacecraft Coulomb formations are considered. Suboptimal trajectories of formation attitude and relative position of a two-craft formation utilizing coulomb forces as well as thrusters is discussed. The effectiveness of the approaches as well as their comparison is demonstrated through numerical simulations.

1 Introduction

Many of numerical methods have been used to solve nonlinear optimal control problems in the literature. These problems have been solved by using direct and indirect methods (Betts 1998). Indirect methods stem from the calculus of variations (Bryson and Ho 1975); direct methods use a nonlinear programming optimization (Conway 2010). The computational simplicity and effectiveness of the suboptimal algorithms is an appealing alternative to the tedious task of solving the Hamilton-Jacobi-Bellman partial differential equation. These approximated methods allow circumventing the original, nonlinear Euler–Lagrange equations, and thus they do not require guessing the initial Lagrange multipliers. Nonetheless, the price to pay is the loss of optimality, since suboptimal solutions are derived. State-dependent methods (Pearson 1962; Wernli and Cook 1975) belong to this category. Çimen

M.M. Gomroki (✉) • O. Tekinalp
Dept. Aerospace Engineering, Middle East Technical University, Ankara 06800, Turkey
e-mail: mohammad.gomroki@metu.edu.tr; tekinalp@metu.edu.tr

F. Topputo • F. Bernelli-Zazzera
Dept. Aerospace Science and Technology, Politecnico di Milano, Milano 20156, Italy
e-mail: francesco.topputo@polimi.it; franco.bernelli@polimi.it

G. Gómez, J.J. Masdemont (eds.), *Astrodynamics Network AstroNet-II*,
Astrophysics and Space Science Proceedings 44,
DOI 10.1007/978-3-319-23986-6_8

and Banks (2004a,b), introduced a method known as Approximating Sequence of Riccati Equations (ASRE) which uses State Dependent Coefficient (SDC) factorization and iterative Linear Time Varying (LTV) approximations to solve Unbounded Nonlinear Optimal Control (UNOC) problem with unspecified final states (Gomroki and Tekinalp 2014b). Topputo and Bernelli-Zazzera (2012, 2013), solved UNOC problems with unspecified, fully specified, and partly specified final states by using SDC factorization and State Transition Matrix (STM) approach. The state-dependent Riccati equations (SDRE) (Cloutier et al. 1996; Mracek and Cloutier 1998; Bracci et al. 2006; Çimen 2012) is likely the most known example of approximated method due to its simplicity and effectiveness in many applications (Harman and Bar-Itzhack 1999; Bogdanov and Wan 2007; Kim et al. 2009; Gomroki and Tekinalp 2014a). This method treats the original *infinite-horizon*, nonlinear optimal control problem as an infinite-horizon linear-quadratic regulator (LQR), pointwise. A number of LQR problems are solved sequentially at each time instant in which the time domain is discretized. This is done by using state-dependent matrices, which are evaluated pointwise at each time step. With the SDRE the *closed-loop* control is treated, the control law being function of the present state. The SDRE method can also be used to solve finite-horizon optimal control (Çimen 2012); one way consists in choosing the state-dependent matrices as functions of the time-to-go (Ratnoo and Ghose 2009).

The Formation flying spacecraft for carrying out interferometry missions, making fractionated spacecraft, or patching together sensor data to obtain a higher resolution data has been envisaged in the past. Among formation control techniques, tethers to control the relative distance is also proposed. A Coulomb tether is similar to physical tether that uses coulomb forces to keep spacecraft at close proximity (Berryman and Schaub 2005). It is indicated that a coulomb tether provides an almost a propellantless formation control (Ratnoo and Ghose 2003). The charges loaded to the bodies, can create attractive and repulsive forces between these bodies. Natural charging of the spacecraft is observed at even geostationary altitudes (Mullen et al. 1986). Since the forces are relative, the total linear or angular momentum of the formation cannot be changed by coulomb forces (Schaub and Kim 2004).

The remaining part of the paper is organized as follows. In Sect. 2, the formulation of UNOC problem is given and two different approaches are recalled that solve this kind of unbounded problems. Although both methods use SDC factorization form, the first approach uses approximating sequence of Riccati equations and the second approach utilizes state transition matrix. Section 3, introduces the equations of motion for orbit radial configuration of two-craft Coulomb formation. Numerical simulations are discussed in Sect. 4, and Sect. 5 discusses about the conclusion.

2 Review of Suboptimal Solutions to UNOC Problems

An optimal control problem can be stated in a variety of forms, which differ in terms of generality. The most general definition accommodates path constraints, variable final time, control saturation, interior-point constraints, etc. The problems treated in this section are nonautonomous, nonlinear in the state, and affine (i.e., linear) in the control. The initial state is supposed given, and the final state can be either (in part) specified or unknown. The time span in which the problem is studied is fixed, and both the states and controls are unconstrained.

2.1 Statement of UNOC Problem

Consider a set of n first-order differential equations

$$\dot{\mathbf{x}} = \mathbf{f}(\mathbf{x}, t) + \mathbf{B}(\mathbf{x}, t)\,\mathbf{u} \tag{1}$$

with $\mathbf{f} : \mathbb{R}^{n+1} \to \mathbb{R}^n$ and $\mathbf{B} : \mathbb{R}^{n+1} \to \mathbb{R}^{n \times m}$. The goal is to find m control functions $\mathbf{u}(t)$ within initial t_0, and final time t_f, such that the performance index

$$J = \varphi\left(\mathbf{x}(t_f), t_f\right) + \int_{t_0}^{t_f} L\left(\mathbf{x}(t), \mathbf{u}(t), t\right) \mathrm{d}t \tag{2}$$

be minimized; $L : \mathbb{R}^{n+m+1} \to \mathbb{R}$, and $\varphi : \mathbb{R}^{n+1} \to \mathbb{R}$. We assume the initial condition to be given by the relation

$$\mathbf{x}(t_0) = \mathbf{x}_0, \tag{3}$$

whereas the final condition is allowed to take three different forms. These specify the hard constrained problem (HCP), the soft constrained problem (SCP), and the mixed constrained problem (MCP), with the final state fully specified, not specified, and partly specified, respectively.

2.2 First Approach: ASRE Method

Suppose that $\mathbf{f}(\mathbf{x}, t)$ in Eq. (1) is a continuously differentiable vector-valued function of $\mathbf{x}$ and $\mathbf{t}$ in an open set $\boldsymbol{\Gamma} \in \mathbb{R}^n$, which $\mathbf{f}(.) \in \mathscr{C}^1(\boldsymbol{\Gamma})$, and $\mathbf{B}(\mathbf{x}, t) \in \mathscr{C}^0(\boldsymbol{\Gamma})$ is a matrix-valued function. In addition, $\mathbf{f}(\mathbf{0}, t) = \mathbf{0},\ \forall t \in \mathbb{R}$. Under these conditions (Çimen 2012), the State Dependent Coefficient (SDC) factorization of Eq. (1) may be written as

$$\dot{\mathbf{x}} = \mathbf{A}(\mathbf{x}, t)\,\mathbf{x} + \mathbf{B}(\mathbf{x}, t)\,\mathbf{u} \tag{4}$$

which is a stabilizable parameterization of the nonlinear system represented in Eq. (1) in a region $\boldsymbol{\Gamma}$ if the pair $\{\mathbf{A}(\mathbf{x},t), \mathbf{B}(\mathbf{x},t)\}$ is point-wise stabilizable in the linear sense for all $\mathbf{x} \in \boldsymbol{\Gamma}$. Let also the objective function (2) be redefined in the quadratic-like form

$$J = \frac{1}{2}\mathbf{x}^T(t_f)S\left(\mathbf{x}(t_f), t_f\right)\,\mathbf{x}(t_f) + \frac{1}{2}\int_{t_0}^{t_f}\left(\mathbf{x}^T Q\,(\mathbf{x},t)\;\mathbf{x} + \mathbf{u}^T R\,(\mathbf{x},t)\;\mathbf{u}\right)\mathrm{d}t \tag{5}$$

The ASRE approach presented in Çimen and Banks (2004a,b) considers the following sequences of Time Varying Linear Quadratic Regulator (TVLQR) approximations

$$\dot{\mathbf{x}}^{[0]} = \mathbf{A}(\mathbf{x}_0)\,\mathbf{x}^{[0]}(t) + \mathbf{B}(\mathbf{x}_0)\,\mathbf{u}^{[0]}(t) \tag{6}$$

$$\dot{\mathbf{x}}^{[k]} = \mathbf{A}(\mathbf{x}^{[k-1]}(t), t)\,\mathbf{x}^{[k]} + \mathbf{B}(\mathbf{x}^{[k-1]}(t), t)\,\mathbf{u}^{[k]} \tag{7}$$

where the sequence using the iterative TVLQR approximations is denoted by a superscript $k = 0, 1, 2, \ldots$. The initial state is $\mathbf{x}^{[k]}(t_0) = \mathbf{x}_0$, and the corresponding linear-quadratic cost functional is

$$\begin{aligned} J^{[k]} = {} & \frac{1}{2}\left(\mathbf{x}^{[k]}(t_f)\right)^T \mathbf{S}(\mathbf{x}^{[k-1]}(t_f), t_f)\,\left(\mathbf{x}^{[k]}(t_f)\right) + \\ & \frac{1}{2}\int_{t_0}^{t_f}\left(\mathbf{x}^{[k]^T}\mathbf{Q}(\mathbf{x}^{[k-1]}(t), t)\mathbf{x}^{[k]} + \mathbf{u}^{[k]^T}\mathbf{R}(\mathbf{x}^{[k-1]}(t), t)\mathbf{u}^{[k]}\right)\mathrm{d}t \end{aligned} \tag{8}$$

Since each approximation is time-varying and linear-quadratic, the optimal control sequence is in the form (Çimen and Banks 2004a)

$$\mathbf{u}^{[k]}(t) = -\mathbf{R}^{-1}(\mathbf{x}^{[k-1]}(t))\mathbf{B}^T(\mathbf{x}^{[k-1]}(t))\,\mathbf{P}^{[k]}(t)\,\mathbf{x}^{[k]}(t) \tag{9}$$

where the real, symmetric and positive-definite matrix $\mathbf{P}^{[k]}(t)$ is the solution of the following differential equation

$$\begin{aligned} \dot{\mathbf{P}}^{[k]}(t) = {} & -\mathbf{Q}(\mathbf{x}^{[k-1]}(t)) - \mathbf{P}^{[k]}(t)\,\mathbf{A}(\mathbf{x}^{[k-1]}(t)) - \mathbf{A}^T(\mathbf{x}^{[k-1]}(t))\,\mathbf{P}^{[k]}(t) \\ & + \mathbf{P}^{[k]}(t)\,\mathbf{E}(\mathbf{x}^{[k-1]}(t))\,\mathbf{P}^{[k]}(t) \end{aligned} \tag{10}$$

with

$$\mathbf{P}^{[k]}(t_f) = \mathbf{S}(\mathbf{x}^{[k-1]}(t_f)) \tag{11}$$

$$\mathbf{E}(\mathbf{x}^{[k-1]}(t)) = \mathbf{B}(\mathbf{x}^{[k-1]}(t))\mathbf{R}^{-1}(\mathbf{x}^{[k-1]}(t))\mathbf{x}^B(\mathbf{x}^{[k-1]}(t)) \tag{12}$$

Consider that the differential Riccati equation in Eq. (10) has to be solved backward in time and the optimal state trajectory is obtained by integrating the following differential equation forward in time

$$\dot{\mathbf{x}}^{[k]}(t) = \left[\mathbf{A}(\mathbf{x}^{[k-1]}(t)) - \mathbf{E}(\mathbf{x}^{[k-1]}(t))\,\mathbf{P}^{[k]}(t)\right]\,\mathbf{x}^{[k]}(t) \tag{13}$$

In the present implementation, the convergence is reached when

$$\varepsilon = \|\mathbf{x}^{[k]} - \mathbf{x}^{[k-1]}\|_\infty = \max_{t\in[t_0,\,t_f]} \{|x_j^{[k]}(t) - x_j^{[k-1]}(t)|, j = 1, \ldots, n\} \leq \text{tol} \tag{14}$$

where 'tol' is a prescribed tolerance; i.e., iterations terminate when the difference between each component of the state, evaluated for all times, changes by less than 'tol' between two consecutive iterations. The sequence of solutions $\mathbf{x}^{[k]}$, $\mathbf{u}^{[k]}$ is proven to converge to the solution of the original problem (4)–(5) provided that $\mathbf{A}(\mathbf{x}, t)$ and $\mathbf{B}(\mathbf{x}, t)$ are Lipschitz continuous with respect to their arguments (Çimen and Banks 2004a).

2.3 Second Approach: Solution of TVLQR by State Transition Matrix

The sequence of TVLQR is solved by exploiting the structure of their Euler–Lagrange equations, so avoiding dealing with the matrix differential Riccati equation. This approach, in part, is described in Bryson and Ho (1975), and differs from that implemented in Çimen and Banks (2004a,b). Suppose the following dynamics are given,

$$\dot{\mathbf{x}} = \mathbf{A}(t)\,\mathbf{x} + \mathbf{B}(t)\,\mathbf{u}, \tag{15}$$

together with the quadratic objective function

$$J = \frac{1}{2}\mathbf{x}^T(t_f)\mathbf{S}\,\mathbf{x}(t_f) + \frac{1}{2}\int_{t_0}^{t_f} \left(\mathbf{x}^T\mathbf{Q}\,(t)\;\mathbf{x} + \mathbf{u}^T\mathbf{R}\,(t)\;\mathbf{u}\right)\mathrm{d}t \tag{16}$$

where $\mathbf{Q}$, $\mathbf{S}$ and $\mathbf{R}$ are positive semi-definite and positive definite time-varying matrices, respectively. The necessary conditions for this problem are

$$\dot{\mathbf{x}} = \mathbf{A}(t)\,\mathbf{x} + \mathbf{B}(t)\,\mathbf{u}, \tag{17}$$

$$\dot{\boldsymbol{\lambda}} = -\mathbf{Q}(t)\,\mathbf{x} - \mathbf{A}^T(t)\,\boldsymbol{\lambda}, \tag{18}$$

$$0 = \mathbf{R}(t)\,\mathbf{u} + \mathbf{B}^T(t)\,\boldsymbol{\lambda}. \tag{19}$$

From Eq. (19) it is possible to get

$$\mathbf{u} = -\mathbf{R}^{-1}(t)\mathbf{B}^T(t)\,\boldsymbol{\lambda}, \tag{20}$$

which can be substituted into (17)–(18) to yield

$$\begin{pmatrix} \dot{\mathbf{x}} \\ \dot{\boldsymbol{\lambda}} \end{pmatrix} = \begin{bmatrix} \mathbf{A}(t) & -\mathbf{B}(t)R^{-1}(t)\mathbf{B}^T(t) \\ -\mathbf{Q}(t) & -\mathbf{A}^T(t) \end{bmatrix} \begin{pmatrix} \mathbf{x} \\ \boldsymbol{\lambda} \end{pmatrix}. \tag{21}$$

Since (21) is a system of linear differential equations, the solution can be written as

$$\mathbf{x}(t) = \phi_{xx}(t_0, t)\,\mathbf{x}_0 + \phi_{x\lambda}(t_0, t)\,\boldsymbol{\lambda}_0, \tag{22}$$

$$\boldsymbol{\lambda}(t) = \phi_{\lambda x}(t_0, t)\,\mathbf{x}_0 + \phi_{\lambda\lambda}(t_0, t)\,\boldsymbol{\lambda}_0, \tag{23}$$

where $\mathbf{x}_0$, $\boldsymbol{\lambda}_0$ are the initial state, costate, respectively, and the functions ϕ_{xx}, $\phi_{x\lambda}$, $\phi_{\lambda x}$, and $\phi_{\lambda\lambda}$ are the components of the state transition matrix, which can be found by integrating the following dynamics

$$\begin{bmatrix} \dot{\phi}_{xx} & \dot{\phi}_{x\lambda} \\ \dot{\phi}_{\lambda x} & \dot{\phi}_{\lambda\lambda} \end{bmatrix} = \begin{bmatrix} \mathbf{A}(t) & -\mathbf{B}(t)\mathbf{R}^{-1}(t)\mathbf{B}^T(t) \\ -\mathbf{Q}(t) & -\mathbf{A}^T(t) \end{bmatrix} \begin{bmatrix} \phi_{xx} & \phi_{x\lambda} \\ \phi_{\lambda x} & \phi_{\lambda\lambda} \end{bmatrix}, \tag{24}$$

with initial conditions

$$\phi_{xx}(t_0, t_0) = \phi_{\lambda\lambda}(t_0, t_0) = \mathbf{I}_{n\times n}, \quad \phi_{x\lambda}(t_0, t_0) = \phi_{\lambda x}(t_0, t_0) = 0_{n\times n}. \tag{25}$$

If both $\mathbf{x}_0$ and $\boldsymbol{\lambda}_0$ were given, it would be possible to compute $\mathbf{x}(t)$ and $\boldsymbol{\lambda}(t)$ through (22)–(23), and therefore the optimal control function $\mathbf{u}(t)$ with (20). As only $\mathbf{x}_0$ is given, the issue is computing $\boldsymbol{\lambda}_0$ for the three problems defined previously.

2.3.1 Hard Constrained Problem

In a HCP ($\mathbf{x}_f$ fully given, $\mathbf{S}$ not defined), the value of $\boldsymbol{\lambda}_0$ can be found by writing (22) at final time

$$\mathbf{x}_f = \phi_{xx}(t_0, t_f)\,\mathbf{x}_0 + \phi_{x\lambda}(t_0, t_f)\,\boldsymbol{\lambda}_0, \tag{26}$$

and solving for $\boldsymbol{\lambda}_0$; i.e.,

$$\boldsymbol{\lambda}_0(\mathbf{x}_0, \mathbf{x}_f, t_0, t_f) = {\phi_{x\lambda}}^{-1}(t_0, t_f)\left[\mathbf{x}_f - \phi_{xx}(t_0, t_f)\,\mathbf{x}_0\right]. \tag{27}$$

2.3.2 Soft Constrained Problem

In a SCP ($\mathbf{x}_f$ not specified, $\mathbf{S}$ $n \times n$ positive definite matrix), the transversality condition $\boldsymbol{\lambda}(t_f) = \partial\varphi/\partial\mathbf{x}$ applied to (16) reads

$$\boldsymbol{\lambda}(t_f) = \mathbf{S}\,\mathbf{x}(t_f), \tag{28}$$

which can be used to find $\boldsymbol{\lambda}_0$. This is done by writing (22)–(23) at final time and using (28),

$$\mathbf{x}(t_f) = \phi_{xx}(t_0, t_f)\,\mathbf{x}_0 + \phi_{x\lambda}(t_0, t_f)\,\boldsymbol{\lambda}_0, \tag{29}$$

$$\mathbf{S}\,\mathbf{x}(t_f) = \phi_{\lambda x}(t_0, t_f)\,\mathbf{x}_0 + \phi_{\lambda\lambda}(t_0, t_f)\,\boldsymbol{\lambda}_0. \tag{30}$$

Equations (29)–(30) represent a linear algebraic system of $2n$ equations in the $2n$ unknowns $\{\mathbf{x}(t_f), \boldsymbol{\lambda}_0\}$. The system can be solved by substitution to yield

$$\boldsymbol{\lambda}_0(\mathbf{x}_0, t_0, t_f) = \left[\phi_{\lambda\lambda}(t_0, t_f) - \mathbf{S}(t_f)\phi_{x\lambda}(t_0, t_f)\right]^{-1}\left[\mathbf{S}(t_f)\phi_{xx}(t_0, t_f) - \phi_{\lambda x}(t_0, t_f)\right]\mathbf{x}_0. \tag{31}$$

This approach can be used for Mixed Constrained Problems (MCP) as well (Topputo and Bernelli-Zazzera 2013).

3 Equations of Motion of Two-Craft Coulomb Formation at Earth Circular Orbits

The nonlinear equations of motion for orbit-radial direction for a two-craft coulomb formation at Earth circular orbit may be written as

$$\ddot{L} - L\left(\dot{\boldsymbol{\theta}}^2 + \left(\dot{\boldsymbol{\psi}} + \Omega\right)^2\cos^2\boldsymbol{\theta} - \Omega^2\left(1 - 3\cos^2\boldsymbol{\theta}\cos^2\boldsymbol{\psi}\right)\right) = \frac{Q_L}{m} \tag{32}$$

$$\ddot{\boldsymbol{\psi}}\cos^2\boldsymbol{\theta} - 2\,\dot{\boldsymbol{\theta}}\sin\theta\,\cos\theta\left(\dot{\boldsymbol{\psi}} + \Omega\right) + 2\frac{\dot{L}}{L}\cos^2\boldsymbol{\theta}\left(\dot{\boldsymbol{\psi}} + \Omega\right)$$
$$+ 3\,\Omega^2\cos^2\boldsymbol{\theta}\cos\psi\,\sin\psi = \frac{Q_{\boldsymbol{\psi}}}{m\,L^2} \tag{33}$$

$$\ddot{\theta} + 2\frac{\dot{L}}{L}\dot{\theta} + \sin\theta\,\cos\theta\left(\left(\dot{\boldsymbol{\psi}} + \Omega\right)^2 + 3\,\Omega^2\cos^2\boldsymbol{\psi}\right) = \frac{Q_{\boldsymbol{\theta}}}{m\,L^2} \tag{34}$$

The constant m is defined as $m = \frac{m_1 m_2}{m_1 + m_2}$, and $Q_L, Q_{\boldsymbol{\psi}}, and\, Q_{\boldsymbol{\theta}}$ are the generalized forces associated with $L, \boldsymbol{\psi}, and\, \boldsymbol{\theta}$, respectively. L is the relative distance between two craft and the relative attitude is represented using the 3-2-1 ($\boldsymbol{\psi}$ *yaw*, $\boldsymbol{\theta}$ *pitch, and* $\boldsymbol{\phi}$ *roll*) Euler angle notation from Hill orbit frame to the formation fixed frame.

For a two spacecraft Coulomb formation, with F_{cf} being the Coulomb force acting between the two craft, $Q_L = -F_{cf}$, which

$$F_{cf} = -k_c\,\frac{q_1\,q_2}{L^2}\,\exp\left(\frac{-L}{\lambda_d}\right)\left(1 + \frac{L}{\lambda_d}\right) \tag{35}$$

where λ_d is the Debye length, and $Q_{\psi} = F_{\psi} L, and Q_{\theta} = F_{\theta} L$, where $F_{\psi} and F_{\theta}$ are the Electric Propulsion (EP) thrusting forces that introduce net formation torques in the $\psi and \theta$ directions. The state and input definitions are

$$\mathbf{x} = (\mathbf{x}_1, \mathbf{x}_2, \mathbf{x}_3, \mathbf{x}_4, \mathbf{x}_5, \mathbf{x}_6)^T = \left(L, \psi, \theta, \dot{L}, \dot{\psi}, \dot{\theta}\right)^T$$

$$\mathbf{u} = \left(\mathbf{u}_L, \mathbf{u}_{\psi}, \mathbf{u}_{\theta}\right)^T$$

where the inputs are in the form of $\mathbf{u}_L = \frac{F_{cf}}{m}, \mathbf{u}_{\psi} = \frac{F_{\psi}}{m}, and\, \mathbf{u}_{\theta} = \frac{F_{\theta}}{m}$. Nonlinear equations of motion of the formation are properly manipulated to obtain suitable state dependent coefficient factorization form for orbit-radial configuration. In the ASRE approach, the nonlinear problem is reduced to a sequence of linear-quadratic and time-varying approximating problems. The detailed procedure is expressed in Gomroki and Tekinalp (2014a,b).

4 Numerical Simulations

Comparison of two different suboptimal methods for two-craft Coulomb formation configuration is analyzed through numerical simulations. The first ASRE approach is applicable to SCP in which the final states are not specified. The second ASRE approach has the ability to be used for SCP, HCP, and MCP conditions. Simulation parameters and weighting matrices are given in Gomroki and Tekinalp (2014a), and the only difference is that the weighting matrix $R = diag(10^{21}, 10^{20}, 10^{20})$ is considered. Also, Debye length, $\lambda_d = 150$ m, is considered.

Figure 1 shows approximate trajectory solutions using two ASRE methods for SCP which the final state is not specified. It is concluded that the both curves coincide to each other, and the states are regulated. Note that the equilibrium point for the separation distance is 25 m.

The approximate control solutions are shown in Fig. 2 which demonstrates that the thrusters for formation attitude control go to zero, and the Coulomb thrust approaches a constant value since two craft are separated 25 m at equilibrium condition.

The charge product of two craft is displaced in Fig. 4 (left). It shows that the two craft have charges with opposite signs which attract each other. For the SCP case, the results of two approaches are more or less the same, and the curves coincide. Table 1 gives a comparison between two ASRE methods expressed in Sect. 2. It is demonstrated that the second approach has lower objective function.

For the HCP case in which the final states are specified, the first approach is not working and there would be just the second method. So, this is an advantage of the second method to the first one. The results are shown in Figs. 3, 4 (right), and 5, and

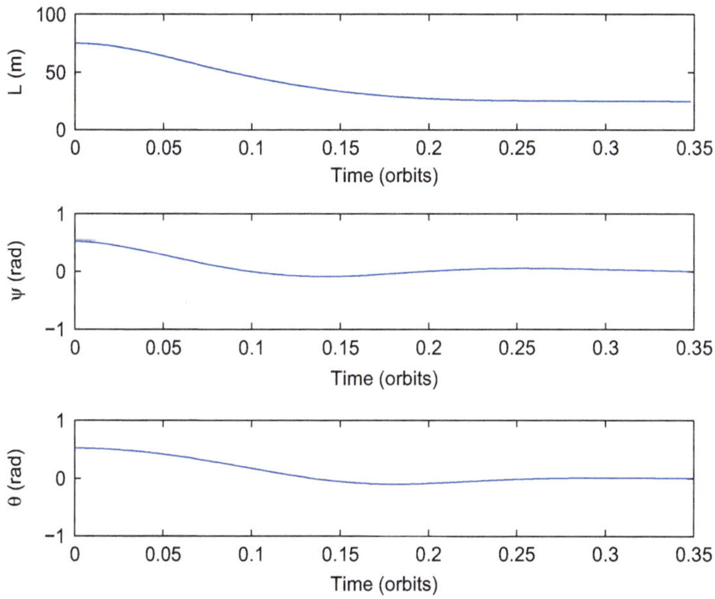

Fig. 1 Approximate trajectory solutions with ASRE approaches for SCP condition

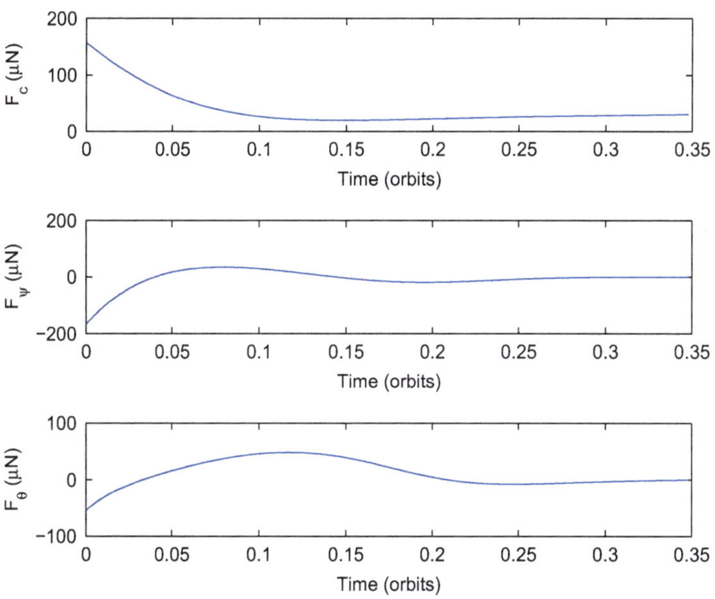

Fig. 2 Approximate control solutions with ASRE approaches for SCP condition

Table 1 Comparison of ASRE methods for SCP condition

Method	Objective function
ASRE (Çimen and Banks 2004a)	1.143735e+13
ASRE (Topputo and Bernelli-Zazzera 2013)	1.143680e+13

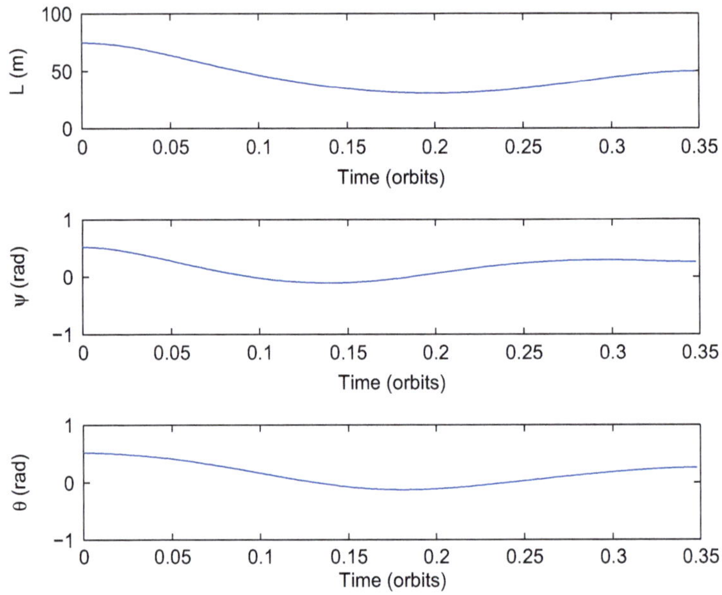

Fig. 3 Approximate trajectory solutions with ASRE approaches for HCP condition

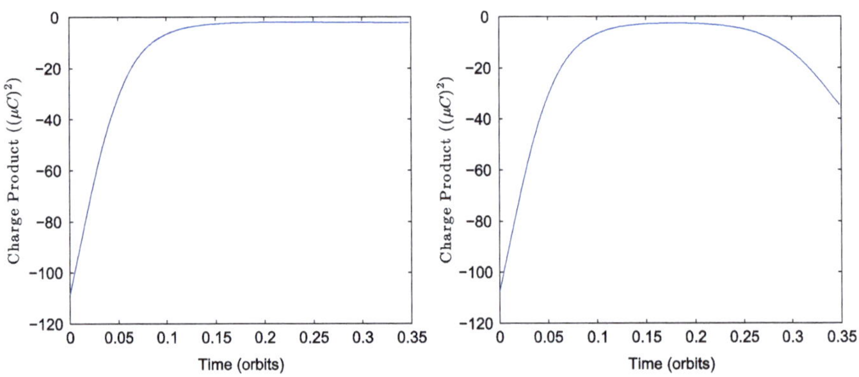

Fig. 4 Charge product for SCP condition (*left*) and HCP condition (*right*)

the formation attitude states are not going to zero anymore. The approximate control solutions are obtained and the charge product of two craft is showed. Moreover, the objective function for second ASRE approach is given in Table 2.

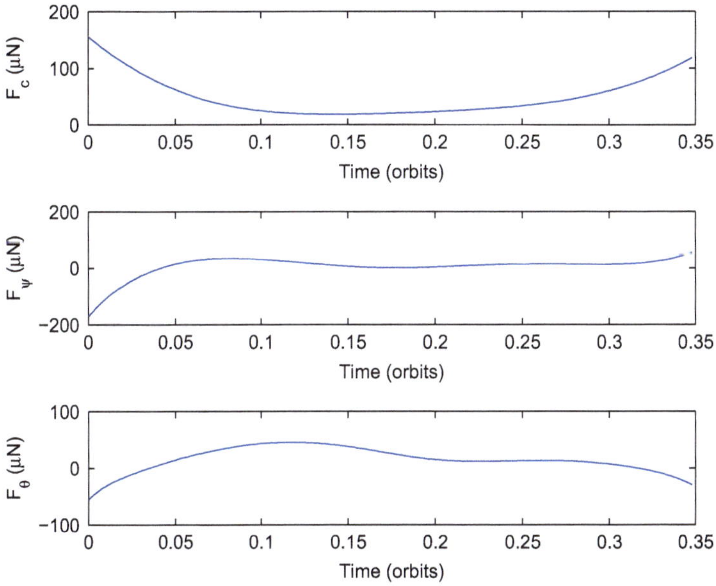

Fig. 5 Approximate control solutions with ASRE approaches for HCP condition

Table 2 ASRE method for HCP condition

Method	Objective function
ASRE (Topputo and Bernelli-Zazzera 2013)	1.518926e+13

5 Conclusions

The computational simplicity and effectiveness of the suboptimal algorithms is an appealing alternative to the tedious task of solving the Hamilton-Jacobi-Bellman partial differential equation. These approximated methods allow circumventing the original, nonlinear Euler–Lagrange equations, and thus they do not require guessing the initial Lagrange multipliers. Nonetheless, the price to pay is the loss of optimality, since suboptimal solutions are derived. This is the case of approximating sequence of Riccati equations, for which the nonlinear equations are written in state dependent coefficient factorization form. In the present paper, it has been shown that two-craft Coulomb formation example with different final state conditions is solved by the two proposed approaches. The effectiveness of the proposed techniques is demonstrated through numerical simulations.

References

Berryman, J., Schaub, H.: Static equilibrium configurations in geo Coulomb spacecraft formations. In: Paper AAS 05-104, AAS/AIAA Space Flight Mechanics Meeting, Copper Mountain (2005)
Betts, J.T.: Survey of numerical methods for trajectory optimization. J. Guid. Control Dyn. **21**, 193–207 (1998)
Bogdanov, A., Wan, E.A.: State-dependent riccati equation control for small autonomous helicopters. J. Guid. Control Dyn. **30**, 47–60 (2007)
Bracci, A., Innocenti, M., Pollini, L.: Estimation of the region of attraction for state-dependent riccati equation controllers. J. Guid. Control Dyn **29**, 1427–1430 (2006)
Bryson, A.E., Ho, Y.C.: Applied Optimal Control, pp. 65–69, 150–151. Wiley, New York (1975)
Çimen, T.: Survey of state-dependent riccati equation in nonlinear optimal feedback control synthesis. J. Guid. Control Dyn. **35**, 1025–1047 (2012)
Çimen, T., Banks, S.P.: Global optimal feedback control for general non linear system with non-quadratic performance criteria. Syst. Control Lett. **53**, 327–346 (2004a)
Çimen, T., Banks, S.P.: Global optimal feedback control for general non linear system with non-quadratic performance criteria. Automatica **40**, 1845–1863 (2004b)
Cloutier, R., D'Souza, C.A., Mracek, C.P.: Nonlinear regulation and nonlinear h-infinity control via the state-dependent riccati equation technique: part 1, theory; part 2, examples, pp. 117–141. In: Proceedings of the First International Conference on Nonlinear Problems in Aviation and Aerospace (1996)
Conway, B.: Spacecraft Trajectory Optimization, pp. 37–78. Cambridge University Press, Cambridge (2010)
Gomroki, M.M., Tekinalp, O.: Relative position control of a two-satellite formation using the SDRE control method. In: Paper AAS 14-217, 24th AAS/AIAA SpaceFlight Mechanics Meeting, pp. 235–254, Santa Fe, 26–30 January 2014a
Gomroki, M. M., Tekinalp, O.: Maneuvering of two-craft coulomb formation using ASRE method. In: Paper AIAA 2014-4164, AIAA/AAS Astrodynamics Specialist Conference, San Diego, 4–7 August 2014b
Harman, R.R., Bar-Itzhack, I.Y.: Pseudolinear and state-dependent riccati equation filters for angular rate estimation. J. Guid. Control Dyn. **22**, 723–725 (1999)
Kim, C.-J., Park, S.H., Sung, S.K., Jung, S.-N.: Nonlinear optimal control analysis using state-dependent matrix exponential and its integrals. J. Guid. Control Dyn **32**, 309–313 (2009)
Mracek, C.P., Cloutier, J.R.: Control designs for the nonlinear benchmark problem via the state-dependent riccati equation method. Int. J. Robust Nonlinear Control **8**, 401–433 (1998)
Mullen, E.G., Gussenhoven, M.S., Hardy, D.A.: SCATHA survey of high-voltage spacecraft charging in sunlight. J. Geophys. Sci. **91**, 1074–1090 (1986)
Pearson, J.D.: Approximation methods in optimal control. J. Electron. Control **13**, 453–469 (1962)
Ratnoo, A., Ghose, D.: Study of interspacecraft coulomb forces and implications for formation flying. J. Propuls. Power **19**, 497–505 (2003)
Ratnoo, A., Ghose, D.: State-dependent riccati-equation-based guidance law for impact-angle-constrained trajectories. J. Guid. Control Dyn. **32**, 320–325 (2009)
Schaub, H., Kim, M.: Orbit element difference constraints for Coulomb satellite formations. In: Paper AIAA 04-5213, AIAA/AAS Astrodynamics Specialist Conference, Providence (2004)
Topputo, F., Bernelli-Zazzera, F.: A method to solve nonlinear optimal control problems in astrodynamics. Adv. Astronaut. Sci. **145**, 1531–1544 (2012)
Topputo, F., Bernelli-Zazzera, F.: Approximate solutions to nonlinear optimal control problems in astrodynamic. ISRN Aerosp. Eng. **2013**, 1–7 (2013)
Wernli, A., Cook, G.: Suboptimal control for the nonlinear quadratic regulator problem. Automatica **11**, 75–84 (1975)

Novel Approach on the Optimisation of Mid-Course Corrections Along Interplanetary Trajectories

Elisabetta Iorfida, Phil Palmer, and Mark Roberts

Abstract The primer vector theory, firstly proposed by Lawden, defines a set of necessary conditions to characterise whether an impulsive thrust trajectory is optimal with respect to propellant usage, within a two-body problem context. If the conditions are not satisfied, one or more potential intermediate impulses are performed along the transfer arc, in order to lower the overall cost. The method is based on the propagation of the state transition matrix and on the solution of a boundary value problem, which leads to a mathematical and computational complexity.

In this paper, a different approach is introduced. It is based on a polar coordinates transformation of the primer vector which allows the decoupling between its in-plane and out-of-plane components. The out-of-plane component is solved analytically while for the in-plane ones a Hamiltonian approximation is made.

The novel procedure reduces the mathematical complexity and the computational cost of Lawden's problem and gives also a different perspective about the optimisation of a transfer trajectory.

1 Introduction

For a spacecraft which flies in deep space, most of its flight time is influenced only by the gravitational effects of a single celestial body, namely the Sun. As a consequence, the two-body problem (2BP) model is generally used in the analysis of such kind of motion. Furthermore, when the patched conics method is taken into account, the whole transfer trajectory is built up through different conic arcs connected together, through a multiple application of the 2BP.

E. Iorfida (✉) • P. Palmer
Surrey Space Centre, University of Surrey, GU2 7XH Guildford, UK
e-mail: elisabetta_iorfida@yahoo.it; p.palmer@surrey.ac.uk

M. Roberts
Department of Mathematics, University of Surrey, GU2 7XH Guildford, UK

African Institute for Mathematical Sciences, Bagamoyo, Tanzania
e-mail: m.roberts@surrey.ac.uk

G. Gómez, J.J. Masdemont (eds.), *Astrodynamics Network AstroNet-II*,
Astrophysics and Space Science Proceedings 44,
DOI 10.1007/978-3-319-23986-6_9

When gravity assists and/or mid-course corrections (called also deep space maneuvers, DSM) are added into the scenario, the complexity of the model increases.

The addition of single or multiple DSM along an interplanetary transfer has the advantage of decreasing the overall cost of a trajectory. In fact, applying a DSM modifies not only the trajectory but also the boundary conditions (BC) of the orbit (Bernelli-Zazzera et al. 2007). In an impulsive thrust case, this causes a change in both the departure and arrival $\Delta\mathbf{V}$s since the cost of a mission can be defined as sum of impulses (magnitudes of the $\Delta\mathbf{V}$s). However, including a DSM along a transfer orbit has the main drawback of increasing the search space complexity of an optimization problem (Vasile and De Pascale 2006). For this reason, it can be useful to have a method that gives a first guess on the addition of a DSM.

In the early 1960s, Lawden introduced the concept of the primer vector in (Lawden 1963). Derived with an indirect optimal control method, this variable is related to the equation of motion and it has to satisfy a set of necessary conditions (NC) for a transfer to be optimal. For a linear system, Prussing proved in Prussing (1995) that these conditions are also sufficient.

Other studies, as Lion and Handelsman (1968), Jezewski and Rozendaal (1968), Vinh (1972) and Jezewski (1975) extensively showed where and how to perturb the trajectory for the addition of a DSM through gradient based optimization methods. At present, the primer vector theory is mainly exploited in global optimization scenario as first estimation technique for determining the position of the DSM. Therefore it allows mission planners to reduce the size of the search space (Olympio et al. 2006). However, due to its computational complexity, its usage is narrowed also to simple cases as in Luo et al. (2010).

The main disadvantage of Lawden's theory is that it is based on solving a boundary value problem (BVP) that involves the state transition matrix (STM) of the motion. At present, there are no known analytic solutions to the problem and, subsequently, no insights into what determines whether a given transfer will be optimal or not. In papers like Goodyear (1965), a formal solution to the differential equations used to derive the STM is provided. Even if the method presented in Goodyear (1965) is valid for every type of Keplerian transfer, it does not give a complete understanding of the structure of the problem. In fact, it is defined in a general inertial frame and therefore no simplifications in the equations are possible.

This paper gives an overall summary of a novel approach introduced and fully treated in Iorfida et al. (2016b) and Iorfida et al. (2016a), which simplifies the structure of the primer vector's problem.

2 Polar Coordinates Transformation

This paper presents the novel procedure of Iorfida et al. (2016b) and Iorfida et al. (2016a), which is framed in a local polar coordinate system defined by $\hat{\mathbf{e}}_r$, $\hat{\mathbf{e}}_\vartheta$ and $\hat{\mathbf{e}}_h$. They are, respectively, unit vectors in the direction of the local position vector, of the derivative of $\hat{\mathbf{e}}_r$ with respect to the true anomaly (ν) and of the orbital momentum vector ($\mathbf{h}$):

$$\hat{\mathbf{e}}_r = \frac{\mathbf{r}}{\|\mathbf{r}\|}; \quad \hat{\mathbf{e}}_\vartheta = -\hat{\mathbf{e}}_r \times \hat{\mathbf{e}}_h; \quad \hat{\mathbf{e}}_h = \frac{\mathbf{h}}{\|\mathbf{h}\|}. \tag{1}$$

It can be demonstrated that this coordinate system is an eigenvector basis of the gravity gradient matrix $\mathbf{G}$ defined to first order approximation as

$$\mathbf{G} = -\frac{\mu}{r^3}\mathbf{I}_3 + \frac{3\mu}{r^5}\mathbf{r} \cdot \mathbf{r}^T. \tag{2}$$

The corresponding eigenvalues of $\mathbf{G}$, with respect to the eigen-vectors of Eq. (1), are $2\mu/r^3$, of algebraic multiplicity 1, and $-\mu/r^3$, of algebraic multiplicity 2.

Combining together the primer vector's equation [Lawden (1963)] and the eigen-decomposition of $\mathbf{G}$, it can be expressed in polar form as

$$\ddot{\mathbf{p}} = \mathbf{G}\mathbf{p} = \frac{\mu}{r^3}(2p_r\hat{\mathbf{e}}_r - p_\vartheta\hat{\mathbf{e}}_\vartheta - p_h\hat{\mathbf{e}}_h). \tag{3}$$

Finally the first derivative of the primer vector with respect to time in polar coordinates can be derived:

$$\frac{d\mathbf{p}}{dt} = \underbrace{(\dot{p}_r - \frac{h}{r^2}p_\vartheta)\hat{\mathbf{e}}_r \; + \; (\dot{p}_\vartheta + \frac{h}{r^2}p_r)\hat{\mathbf{e}}_\vartheta}_{\text{IN PLANE}} \quad + \quad \underbrace{\dot{p}_h\hat{\mathbf{e}}_h}_{\text{OUT OF PLANE}} \quad . \tag{4}$$

From Eq. (4), it is clear that the out-of-plane component is decoupled from the in-plane components.

3 Out-of-Plane Component Analysis

3.1 Analytic Solution and Independence on the Semi-major Axis

In the 2BP the equation of motion of a point mass is

$$\ddot{\mathbf{r}} = -\frac{\mu}{r^3}\mathbf{r}. \tag{5}$$

In a perifocal reference system, the motion is only on the plane (orbital plane) while the central body is the centre of the system of coordinates.

The position vector, existing only on the plane, can be defined through its components as $\mathbf{r} = [x, y]^T$. The solution of Eq. (5) expressed in a perifocal reference system (Battin 1999) is

$$\mathbf{r} = \frac{a(1-e^2)}{1+e\cos(\nu)} \begin{bmatrix} \cos(\nu) \\ \sin(\nu) \end{bmatrix}. \tag{6}$$

As from Iorfida et al. (2016b), the second order derivative with respect to time of the out-of-plane component of the primer vector, p_h, comes from Eq. (3) as

$$\ddot{p}_h = -\frac{\mu}{r^3} p_h, \tag{7}$$

which is in the same form of Eq. (5). As a consequence, the analytic solution of Eq. (7) can be written as a linear combination of x and y in the form of:

$$p_h = Ax + By, \tag{8}$$

where A and B are the integration coefficients and are

$$\begin{aligned} A &= \frac{p_{h0} y_f - p_{hf} y_0}{(x_0 y_f - x_f y_0)} \\ B &= \frac{p_{hf} x_0 - p_{h0} x_f}{(x_0 y_f - x_f y_0)} \end{aligned}. \tag{9}$$

As from Lawden (1963), the primer vector is a unit vector in the direction of the thrust when an impulse occurs. As a consequence, the values of p_{h0} and p_{hf} are non-dimensional variables, while x and y have dimension length (L) and they are linearly dependant on the value of the semi-major axis of the orbit [Eq. (6)]. Therefore, from Eq. (9) the integration coefficients have dimension of (1/L), therefore A and B are inversely proportional to the semi-major axis. Since the out-of-plane component of the primer vector is in the form of Eq. (8) and the coefficients are proportional to $(1/a)$, as a result the primer vector is independent on the value of the semi-major axis of the orbit, as shown in Iorfida et al. (2016b).

3.2 *Graphical Representation of the Out-of-Plane Case*

The analytical solution of p_h [Eq. (8)] allows an investigation into the problem of perpendicular thrust through a simple graphical approach. In the following part of the paper, the eccentricity of the transfer orbit is assumed to be $e \in [0, 1)$. Therefore, only elliptical transfer trajectories are analysed. Since open transfer orbits are not usually considered for the heliocentric phase, the approach does not lack in generality.

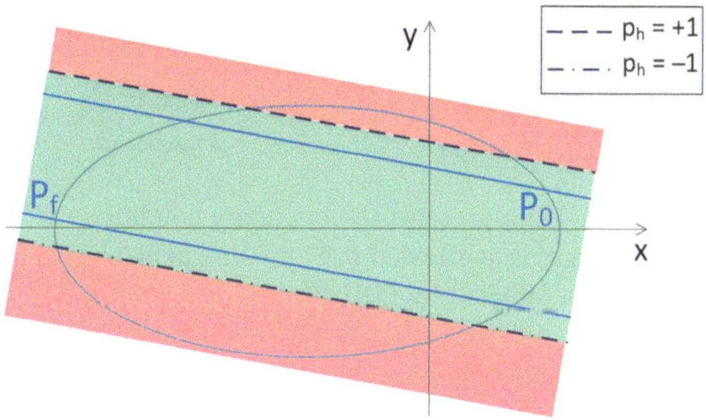

Fig. 1 Transfer ellipse with constant p_h iso-lines, general case

As demonstrated in Sect. 3.1, in a perifocal reference system p_h depends only on x and y as from Eq. (8). Consequently, the optimality of the transfer orbit can be investigated through the position's components on the orbital plane.

The solution of Eq. (8) can be related to the one of a line ($y = mx + D$) if $p_h = \text{const}$. The slope of the line (m) is $-A/B$ and the intercept (D) is $1/B$, and so Eq. (8) can be reformulated as

$$y = mx + Dp_h. \tag{10}$$

Equation 10 is the foundation of a representation of the primer vector presented in Iorfida et al. (2016b), where a set of parallel p_h-isoclines, that have same slope for same transfer orbit and boundary conditions. As a result, a graphical approach to the problem can be exploited: the 'primer vector line' of Eq. (10) can be drawn on the top of the transfer orbit with the consequence that for different values of p_h, the intercepts (Dp_h) identify distinct regions corresponding to different optimality conditions. Figure 1 gives an example of such approach.

In Fig. 1, the limiting p_h lines are plotted with a dashed segment for $p_h = +1$ and a dot-dashed segment for $p_h = -1$. The internal region bounded between those two lines (shaded in green) includes all the parallel lines with $-1 \leq p_h \leq 1$. Conversely, outside the boundary lines, p_h will be either larger than $+1$ or smaller than -1 (red shaded regions) and these areas correspond to 'non-optimal' profiles of the primer vector.

In general, the initial and final points (P_0 and P_f) always belong to the transfer orbit, but they can be associated with two different p_h-isoclines. The decoupling of the components of the primer vector permits analysis of the two cases separately. If the boundary conditions of the in-plane components are assumed to be null, there is a pure out-of-plane case, which means that only a perpendicular thrust to the orbital plane is performed.

As shown in Iorfida et al. (2016b) from the definition of the primer vector, for a pure out-of-plane thrust case, the boundary conditions for p_h are

$$p_{h0} = \pm 1; p_{hf} = \pm 1, \tag{11}$$

which, combined with Eq. (10), yields

$$\begin{array}{l} y_0 = m_0 x_0 \pm D_0 \\ y_f = m_f x_f \pm D_f \end{array}. \quad (12)$$

Equation (12) represents the boundary conditions of the p_h-isoclines which define distinct regions as showed in Fig. 1. It can be observed that the value of the slopes, $m_{0/f}$, is not affected by the value of $p_{h0/f}$. On the contrary, the intercepts $D_{0/f}$ will be either positive or negative, depending on the sign of $p_{h0/f}$.

The graphical method is extensively explained in Iorfida et al. (2016b).

3.3 Example of the Graphical Approach

In this paragraph it is presented an example where the method introduced in Sect. 3.2 and extensively presented in Iorfida et al. (2016b) is applied.

There are no assumptions on the directions of the impulses, therefore p_{h0} and p_{hf} can be both equal to ± 1. If the problem is framed in the orbital reference system, the boundary conditions are given as input of the methods. For the chosen example they are: $e = 0.7$, $a = 1.0$ DU, $\nu_0 = 10°$ and $\nu_f = 70°$. The true anomalies define the initial and final position vectors as: $\mathbf{r_0} = [0.30, 0.05, 0]^T$ DU and $\mathbf{r_f} = [0.08, 0.45, 0]^T$ DU. The slopes and intercepts of the lines to which they belong can be calculated for both the $p_{h0} = p_{hf}$ case and for different boundary conditions ($p_{h0} \neq p_{hf}$) through the equations of Iorfida et al. (2016b).

In the case when the boundary conditions are the same, that is $p_{h0} = p_{hf}$, the initial (P_0) and final (P_f) points will both belong to the same line. A second fundamental line of the problem is the one that has $p_h = -p_{0/f}$, that is a segment which goes through P_1 and P_2. From the considerations of the previous section, this line has the same slope as the one defined by P_0 and P_f, but opposite intercept.

The other possible scenario is the case when the initial and final primer vector have opposite signs. Hence, P_0 and P_f do not belong to the same line, unlike the previous case. The segment which goes through P_0 crosses the ellipse in P_1, while the one of P_f intercepts the ellipse in P_2 (as defined in Iorfida et al. 2016b).

The x-coordinates of P_1 and P_2 for the two different cases can be evaluated for both scenario and the values are summarized in Table 1.

Following the procedure of Iorfida et al. (2016b) for the considered example, Fig. 2a shows the $(D-m)$ map with a star ($\star$) which identifies the specific values of

Table 1 Parameters needed for optimality algorithm, evaluated for the considered example ($\nu_0 = 10°$, $\nu_f = 80°$, $e = 0.7$, $a = 1.0$ DU)

		$p_{h0} = p_{hf}$		$p_{h0} \neq p_{hf}$	
m	D	−1.81	0.59	1.33	−0.34
x_1	x_2	−0.03	−0.72	−0.21	−0.79

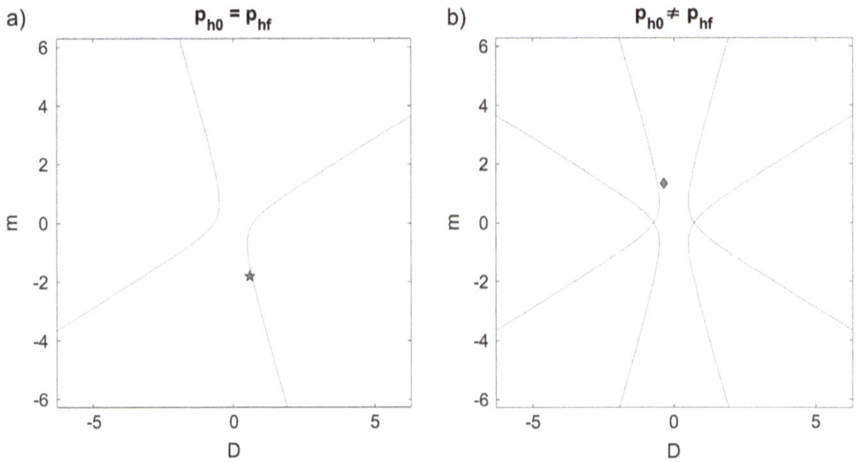

Fig. 2 $\Delta = 0$ curves in the $D - m$ plane for $e = 0.7$. **(a)** $p_{h0} = p_{hf}$ **(b)** $p_{h0} \neq p_{hf}$

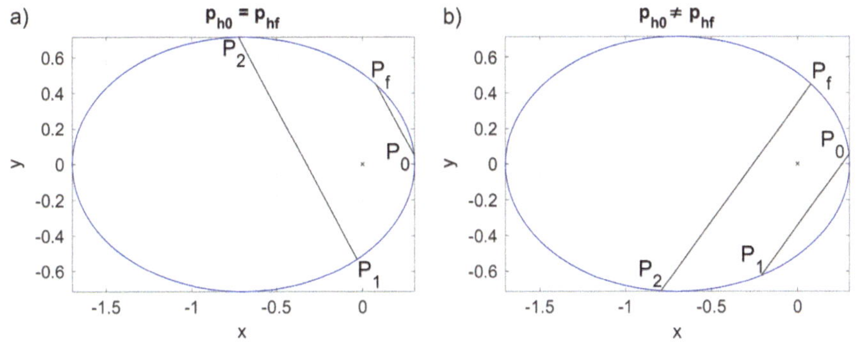

Fig. 3 Transfer ellipse and boundary p_h-isoclines for $e = 0.7$, $\nu_0 = 10°$ and $\nu_f = 80°$. **(a)** $p_{h0} = p_{hf}$, **(b)** $p_{h0} \neq p_{hf}$

D_L and m_L, which are the intercept and slope for the $p_{h0} = p_{hf}$ respectively. From the method presented in Iorfida et al. (2016b) it can be concluded that the transfer point lies in the 'non-optimal' area. In spite of the relative value of x_0 and x_f, the transfer from P_0 to P_f is either 'non-optimal' or 'sub-optimal' in both anti-clockwise and clockwise direction.

For the case when $p_{h0} \neq p_{hf}$, the $D - m$ maps (as stated in Iorfida et al. 2016b) define only the existence domain of the Δs function, Fig. 2b, where the diamond ($\diamond$) identifies the point relative to the analysed example. For this case it is necessary to verify the optimality conditions. From the values of x_1 and x_2 of Table 1, it can be concluded that the clockwise transfer is 'sub-optimal', while the anti-clockwise one is 'optimal'.

Figure 3 shows the transfer ellipse with the two combinations of lines. In particular Fig. 3a represents the case where P_0 and P_f belong to the same line ($p_{h0} = p_{hf}$), while Fig. 3b represents the $p_{h0} \neq p_{hf}$ case.

The considerations made before in the paragraph can then be graphically verified through Fig. 3 and the complete method can be found in Iorfida et al. (2016b).

4 In-Plane Components Analysis

The in-plane components of Eq. (4), p_r and p_ϑ do not have a formal analytic solution. However, an analytic approximation can be obtained, for which the time-variant variable is kept constant for small integration step size.

From Eq. (4), two new variables can be introduced:

$$\begin{cases} P_r \equiv \dot{p}_r - \dfrac{h}{r^2} p_\vartheta \\ \\ P_\vartheta \equiv \dot{p}_\vartheta + \dfrac{h}{r^2} p_r \end{cases} . \tag{13}$$

They are the components of the first derivative of the primer vector with respect to time in the local orbital reference frame. The in-plane part of Eq. (4) is then

$$\dot{\mathbf{p}} = P_r \hat{\mathbf{e}}_r + P_\vartheta \hat{\mathbf{e}}_\vartheta . \tag{14}$$

From the derivatives with respect to time of the unit vectors of Eq. (1), the second derivative of the primer vector with respect to time is

$$\ddot{\mathbf{p}} = \big(\dot{P}_r - \frac{h}{r^2} P_\vartheta\big)\hat{\mathbf{e}}_r + \big(\dot{P}_\vartheta + \frac{h}{r^2} P_r\big)\hat{\mathbf{e}}_\vartheta . \tag{15}$$

The derivatives of P_r and P_ϑ are given combining the primer vector equation in polar form [Eq. (3)] with Eqs. (13) and (15) as

$$\begin{cases} \dot{P}_r = \dfrac{2\mu}{r^3} p_r + \dfrac{h}{r^2} P_\vartheta \\ \\ \dot{P}_\vartheta = -\dfrac{\mu}{r^3} p_\vartheta - \dfrac{h}{r^2} P_r \end{cases} . \tag{16}$$

Two new time-dependant variables can be introduced in order to simplify the structure of the equations:

$$\begin{aligned} \sigma &\equiv \frac{r}{l} \equiv \frac{1}{1 + e\cos\nu} \\ \\ \omega &\equiv \frac{h}{r^2} = \frac{h}{(l\sigma)^2} \end{aligned} , \tag{17}$$

that satisfy

$$\sigma \cdot \omega^2 = \frac{r}{l}\frac{h^2}{r^4} = \frac{\mu}{r^3}\frac{r}{r}\frac{l}{l} = \frac{\mu}{r^3}. \tag{18}$$

where $l = a(1 - e^2)$ is the semi-latus rectum of the orbit, Battin (1999).
As a consequence, a new system of first order ODE in $[p_r, p_\vartheta, P_r, P_\vartheta]$ as shown in Iorfida et al. (2016a) can be obtained:

$$\begin{bmatrix} \dot{p}_r \\ \dot{p}_\vartheta \\ \dot{P}_r \\ \dot{P}_\vartheta \end{bmatrix} = \begin{bmatrix} 0 & \omega & 1 & 0 \\ -\omega & 0 & 0 & 1 \\ 2\sigma\omega^2 & 0 & 0 & \omega \\ 0 & -\sigma\omega^2 & -\omega & 0 \end{bmatrix} \begin{bmatrix} p_r \\ p_\vartheta \\ P_r \\ P_\vartheta \end{bmatrix}, \tag{19}$$

which, in compact matrix format, is

$$\dot{\mathbf{p}}_{\text{polar}}(t) = \mathbf{M}(t)\mathbf{p}_{\text{polar}}(t). \tag{20}$$

4.1 Palmer Coordinates Conversion and Integration Scheme

The Hamiltonian of the system of Eq. (19) is:

$$H_{\text{PV}} = \frac{1}{2}({P_r}^2 + {P_\vartheta}^2) + \omega(p_\vartheta P_r - p_r P_\vartheta) - \frac{1}{2}\sigma\omega^2(2{p_r}^2 - {p_\vartheta}^2). \tag{21}$$

Comparing the one of Hill's problem, H_{Hill} of Quinn et al. (2010), there is a clear similarity:

$$H_{\text{Hill}} = \frac{1}{2}({p_x}^2 + {p_y}^2) + \Omega(yp_x - xp_y) - \frac{1}{2}\Omega^2(2x^2 - y^2). \tag{22}$$

For the circular transfer case, that is Hill's problem, $\sigma = 1$ and $\omega = \Omega$, which implies a one-to-one correspondence between the two Hamiltonians. As stated in Iorfida et al. (2016a), this is an important conclusion: for a circular orbit case there is an analytic solution to the planar primer vector problem, that is precisely Hill's solution.

Palmer (2007) shows a transformation of coordinates $\tilde{\mathbf{p}} = \mathscr{P}\mathbf{z} = (p_1, p_2, p_3, p_4)^T$ on $\mathbf{z}$ that allows the separation of the motion in an oscillation and a linear drift to be explicit (as a note, the original notation of Palmer (2007) has been changed to avoid confusion). As presented in Iorfida et al. (2016a), if the transformation of coordinates of Palmer (2007) is applied to the primer vector, it becomes

$$\begin{bmatrix} p_1 \\ p_2 \\ p_3 \\ p_4 \end{bmatrix} = \begin{bmatrix} -3 & 0 & 0 & -2 \\ 0 & 1 & 0 & 0 \\ 0 & 2 & -1 & 0 \\ 2 & 0 & 0 & 1 \end{bmatrix} \begin{bmatrix} \omega p_r \\ P_r + \omega p_\vartheta \\ \omega p_\vartheta \\ P_\vartheta - \omega p_r \end{bmatrix}. \tag{23}$$

4.1.1 Integration of the Palmer Coordinates

The coordinates $\tilde{\mathbf{p}}$ (also called *Palmer coordinates*) can be expressed as functions of $(p_r, p_\vartheta, P_r, P_\vartheta)$ as shown in Eq. (23), where two couplings are noticeable: one between p_1 and p_4 and the other between p_2 and p_3.

The method presented in Iorfida et al. (2016a) starts from a combination of $\tilde{\mathbf{p}}$ with the system of Eq. (19) and then differentiated with respect to time (while σ, and therefore ω, are kept constant for small intervals of time), it yields

$$\begin{bmatrix} \dot{p}_1 \\ \dot{p}_2 \\ \dot{p}_3 \\ \dot{p}_4 \end{bmatrix} = \omega \begin{bmatrix} 0 & 4\sigma-3 & -2(\sigma-1) & 0 \\ 2\sigma-3 & 0 & 0 & 4(\sigma-1) \\ 4(\sigma-1) & 0 & 0 & 8\sigma-5 \\ 0 & -2(\sigma-1) & \sigma-1 & 0 \end{bmatrix} \begin{bmatrix} p_1 \\ p_2 \\ p_3 \\ p_4 \end{bmatrix}. \tag{24}$$

From this system of equations, it is then possible to differentiate even further to a second order system as

$$\begin{bmatrix} \ddot{p}_1 \\ \ddot{p}_2 \\ \ddot{p}_3 \\ \ddot{p}_4 \end{bmatrix} = \omega^2 \begin{bmatrix} -(2\sigma-1) & 0 & 0 & -2(\sigma-1) \\ 0 & -(2\sigma-1) & 2(\sigma-1) & 0 \\ 0 & -2(\sigma-1) & 3(\sigma-1) & 0 \\ 2(\sigma-1) & 0 & 0 & 3(\sigma-1) \end{bmatrix} \begin{bmatrix} p_1 \\ p_2 \\ p_3 \\ p_4 \end{bmatrix}. \tag{25}$$

In Eq. (25), the variables are coupled again (p_1 with p_4 and p_2 with p_3).

The structure of the problem allows to have a fourth order ODE in p_1 in the form of

$$p_1{}^{\mathrm{IV}} - \omega^2(\sigma-2)p_1{}^{\mathrm{II}} - \omega^4(\sigma-1)(2\sigma+1) = 0. \tag{26}$$

Assuming that the coefficients of Eq. (26) are constant for a small interval of time, the solution will be $p_1 \approx e^{\omega\lambda t}$.

The complete solution of Eq. (26) can be derived, and it simply is

$$p_1 = Ae^{\omega\lambda_1^+ t} + Be^{\omega\lambda_2^+ t} + Ce^{\omega\lambda_1^- t} + De^{\omega\lambda_2^- t} \tag{27}$$

where $\mathbf{A} = (A, B, C, D)^{\mathrm{T}}$ is a vector of coefficients given by the initial conditions of p_1.

The other components of $\tilde{\mathbf{p}}$ can be evaluated as for p_1, resulting in a formal matrix expression

$$\tilde{\mathbf{p}} = \boldsymbol{\Sigma}(\sigma)^{-1}\tilde{\mathbf{E}}(t,\sigma)^{-1}\mathbf{A}. \tag{28}$$

$\tilde{\mathbf{E}}(t,\sigma)$ is a diagonal matrix with exponential functions as elements of the diagonal, functions of time t and ω: while $\boldsymbol{\Sigma}(\sigma)$ is a polynomial matrix in σ see Iorfida et al. (2016a).

The propagation process presented so far, is based on time as the integration variable. In orbital dynamics it is sometimes more convenient to use true anomaly rather than time; therefore, the ODE of Eq. (24) can be converted into true anomaly as

$$\tilde{\mathbf{p}}' = \boldsymbol{\Lambda}\tilde{\mathbf{p}}. \tag{29}$$

where the symbol ($'$) represents the derivative with respect to ν.

However, the solution shown in Eq. (28) remains in the same form:

$$\mathbf{A} = \mathbf{E}(\nu, \sigma)\boldsymbol{\Sigma}(\sigma)\tilde{\mathbf{p}}, \tag{30}$$

where the only difference appears in $\mathbf{E}$. Where the exponents are explicit functions in ν.

The importance of the Palmer coordinates is easier to notice it, from the system in true anomaly. In fact, using $\tilde{\mathbf{p}}$ allows a solution which is a linear combination of the eigenvalues of the matrix $\mathbf{M}$ [Eq. (20)]. This solution depends only on σ through the polynomial matrix $\boldsymbol{\Sigma}$ and the diagonal matrix $\mathbf{E}$, with an explicit propagation variable ν.

4.1.2 Matrices Propagation Scheme

Since Eq. (30) assumes that σ is constant for small interval of ν, it is required to apply a 'double' propagation for the Palmer coordinates. This means that every time that the true anomaly is propagated to the next step, the corresponding value of σ must be updated, as shown in the scheme of Fig. 4. As explained in Iorfida et al. (2016a), for the primer vector the continuity must be guaranteed along the whole transfer arc. Therefore, a simple propagation scheme as the one of Eq. (30) guarantees that a 'continuity check' on the primer vector is executed at every update of σ.

The procedure for the propagation of the Palmer coordinates from its initial conditions $\tilde{\mathbf{p}}_0$ to a generic kth state can be expressed, as in Iorfida et al. (2016a), as

$$\tilde{\mathbf{p}}_k = \left(\prod_{i=0}^{k-1} \mathscr{I}_{\omega_i}\boldsymbol{\Sigma}^{-1}(\sigma_i)\mathbf{E}^{-1}(\nu_{i+1}, \sigma_i)\mathbf{E}(\nu_i, \sigma_i)\boldsymbol{\Sigma}(\sigma_i)\right)\tilde{\mathbf{p}}_0. \tag{31}$$

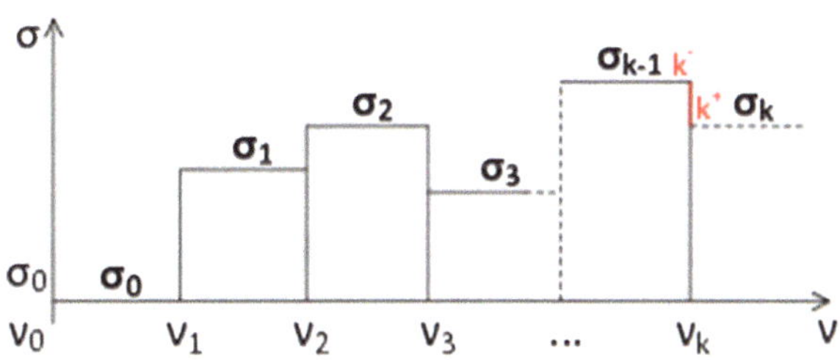

Fig. 4 Step Propagation scheme of ν and σ, with $\sigma = \text{const}$ along every interval of ν. A general kth state is highlighted with the two limiting values ($k^{\pm}$) in *red*

where $\mathscr{I}_{\omega_i}$ is the matrix which guarantees the continuity of the primer vector along the propagation. The compact matrix form of Eq. (31) is

$$\tilde{\mathbf{p}}_k = \boldsymbol{\Psi}(\nu_k, \nu_0)\tilde{\mathbf{p}}_0. \tag{32}$$

where $\boldsymbol{\Psi}(\nu_k, \nu_0)$ is defined as the propagation matrix of the Palmer coordinates.

The final step of the scheme is to convert the Palmer coordinates back into primer vector at the BC. Combining the definitions of P_r and P_ϑ [Eq. (13)] and Eq. (23), there is (where $\mathbf{p}_{tot}$ clearly refers to the primer vector and its derivative with respect to time):

$$\tilde{\mathbf{p}} = \begin{bmatrix} -3\omega & 0 & 0 & -2 \\ 0 & 0 & 1 & 0 \\ 0 & -\omega & 2 & 0 \\ 2\omega & 0 & 0 & 1 \end{bmatrix} \begin{bmatrix} p_r \\ p_\vartheta \\ \dot{p}_r \\ \dot{p}_\vartheta \end{bmatrix} \equiv \mathbf{B}(\omega)\mathbf{p}_{tot}. \tag{33}$$

Equations (32) and (33) together yield to

$$\mathbf{p}_{tot f} = \mathbf{B}^{-1}(\omega_f)\boldsymbol{\Psi}(\nu_f, \nu_0)\mathbf{B}(\omega_0)\mathbf{p}_{tot0} \equiv \tilde{\boldsymbol{\Phi}}_{(f,0)}\mathbf{p}_{tot0} \tag{34}$$

where $\tilde{\boldsymbol{\Phi}}_{(f,0)}$ has been defined as the total propagation matrix of the primer vector.

Equation (34) summarizes the whole propagation scheme for the primer vector.

For a deep understanding of the theory and of the complete scheme see Iorfida et al. (2016a), where it is moreover demonstrated that the in-plane components of the primer vector are independent on the semi-major axis of the transfer orbit.

4.2 Fixed Transfer Orbit and Variable Departure/Arrival Orbits

From the definition at the BC of the primer vector, $\mathbf{p}_0$ and $\mathbf{p}_f$ depend only on the directions of the initial and final $\Delta\mathbf{V}$s, but they are independent from their magnitudes. Therefore, as shown in Iorfida et al. (2016a), it is interesting to parametrize the $\Delta\mathbf{V}$s by their directions. When the transfer, departure and arrival orbits are coplanar, the in-plane components of the primer vector can be defined with respect to the orientation of the two local polar reference frames at initial and final position, expressed through two angles α and β:

$$\begin{aligned} \mathbf{p}_0 &= \begin{bmatrix} \cos\alpha \\ \sin\alpha \end{bmatrix} \\ \mathbf{p}_f &= \begin{bmatrix} \cos\beta \\ \sin\beta \end{bmatrix} \end{aligned} . \tag{35}$$

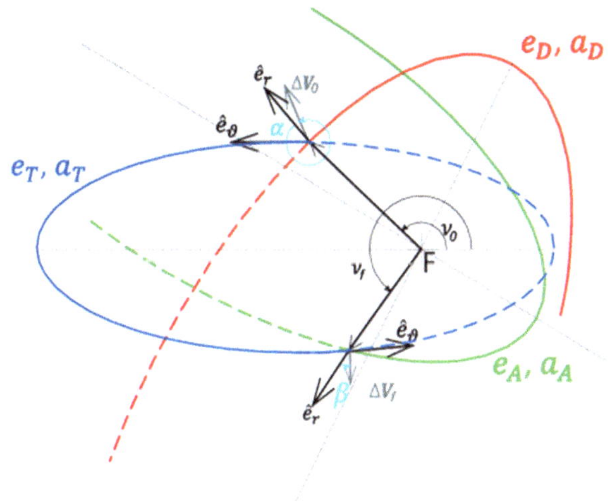

Fig. 5 Geometric representation of a co-planar departure (e_D, a_D), transfer (e_T, a_T) and arrival (e_A, a_A) orbits case (Iorfida et al. 2016a)

If the transfer orbit is fixed (therefore also the velocities $\mathbf{v}_0$ and $\mathbf{v}_f$), the departure and arrival orbits vary according to the orientation (and magnitude) of the $\Delta\mathbf{V}$s (see Fig. 5).

At the boundaries there is

$$\begin{aligned} \mathbf{v}_0 &= v_{r0}\hat{e}_{r0} + v_{\vartheta 0}\hat{e}_{\vartheta 0} \\ \mathbf{v}_f &= v_{rf}\hat{e}_{rf} + v_{\vartheta f}\hat{e}_{\vartheta f} \end{aligned}, \tag{36}$$

while the ΔVs are

$$\begin{aligned} \Delta\mathbf{V}_0 &= \Delta V_0(\cos\alpha\hat{e}_{r0} + \sin\alpha\hat{e}_{\vartheta 0}) \\ \Delta\mathbf{V}_f &= \Delta V_f(\cos\beta\hat{e}_{rf} + \sin\beta\hat{e}_{\vartheta f}) \end{aligned}. \tag{37}$$

Defining $\mathbf{v}_D$ and $\mathbf{v}_A$ respectively as the velocity at $\mathbf{r}_0$ on the departure orbit and the velocity at $\mathbf{r}_f$ on the arrival orbit, they will be

$$\begin{aligned} \mathbf{v}_D &= \mathbf{v}_0 - \Delta\mathbf{V}_0 = (v_{r0} - \Delta V_0\cos\alpha)\hat{e}_{r0} + (v_{\vartheta 0} - \Delta V_0\sin\alpha)\hat{e}_{\vartheta 0} \\ \mathbf{v}_A &= \mathbf{v}_f + \Delta\mathbf{V}_f = (v_{rf} + \Delta V_f\cos\beta)\hat{e}_{rf} + (v_{\vartheta f} + \Delta V_f\sin\beta)\hat{e}_{\vartheta f} \end{aligned}. \tag{38}$$

From Eq. (34), if the transfer orbit is fixed, the total state transition matrix, $\tilde{\boldsymbol{\Phi}}_{(f,0)}$, is fixed as well. For this scenario the optimality can vary only if α and β, directions of $\Delta\mathbf{V}$s, change.

In an optimization process, such a problem has a five dimensional search space: e, ν_0 and ν_f, which are defined on the transfer orbit, and α and β, that are the BC on the $\Delta\mathbf{V}$s.

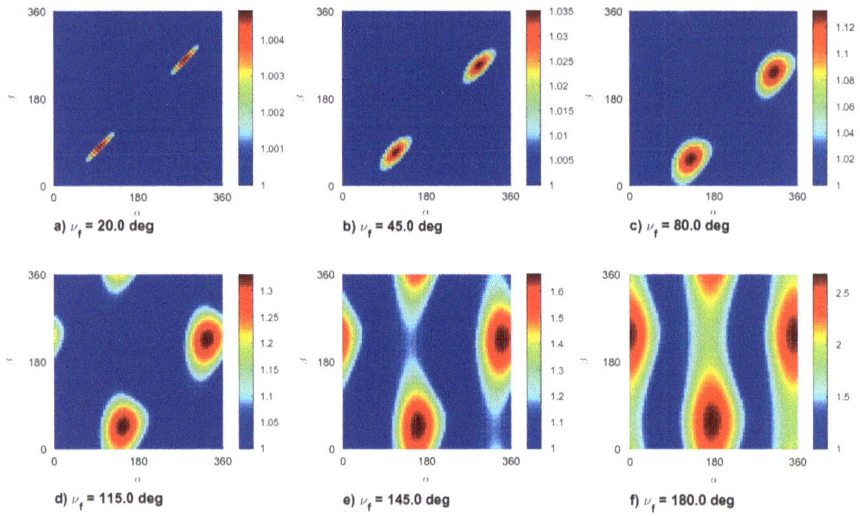

Fig. 6 Sequence of maximum values of primer vector maps for a fixed transfer orbit and increasing value of ν_f, vs α and β. The fixed parameters of the transfer orbit are e = 0.8, a = 0.95 DU, $\nu_0 = 5°$. (**a**) $\nu_f = 20.0°$. (**b**) $\nu_f = 45.0°$. (**c**) $\nu_f = 80.0°$. (**d**) $\nu_f = 115.0°$. (**e**) $\nu_f = 145.0°$. (**f**) $\nu_f = 180.0°$ (Iorfida et al. 2016a)

From Lawden's NCs, for a trajectory to be optimal, p must be always smaller than 1 along the arc and equal to 1 at the boundaries. Within this scenario, a set of transfer orbits with the same eccentricity can be parametrized by, for example, fixing the initial true anomaly and varying ν_f.

Figure 6 presents an example of Iorfida et al. (2016a), where the transfer orbit and ν_0 are considered to be fixed. The maps show the maximum value of the primer vector according to gradual increasing values of ν_f, for fixed $e = 0.8$ and $\nu_0 = 5°$. They are plotted in a complete interval (0–360°) of α and β with a computational step of 5°.

An optimal transfer has an overall maximum value of the primer vector equal to 1 (dark blues area in Fig. 6). Non-optimal trajectories have $p > 1$ and therefore in Fig. 6 they are represented in colours ('non-optimality islands'). The problem is very well structured with a defined geometry: for every specific ν_f case, the 'non-optimal' islands have always the same dimension, shape and orientation. In particular for small span of true anomalies the 'non-optimal' islands are very stretched but small in size, therefore almost the whole map of Fig. 6a represents an optimal case. In Iorfida et al. (2016a) those results are deeply analyzed, together with other cases.

The important characteristic of the results presented so far is that they encompass an entire set of orbits with same primer vector profile but different transfer orbits and BC. Specifically, what is varying are the semi-major axis of the transfer orbit (a_{transf}) and the magnitudes of the $\Delta\mathbf{V}$s at the beginning and at the end (therefore the departure and arrival orbits themselves).

In summary, this procedure gives a different approach to the primer vector theory and allows to complete an overall analysis on defined classes of departure and arrival orbits with common characteristics for specific fixed parameters of the transfer trajectory.

5 Conclusions

This paper shows how applying a reference system transformation to the primer vector equation causes the de-coupling between the in-plane and out-of-plane coordinates.

The out-of-plane component (p_h) has a simple analytic solution and it is independent from the semi-major axis of the transfer orbit.

Exploiting the simplicity of the out-of-plane analytic solution, the whole optimality conditions for a certain transfer orbit can be represented with a geometric approach, given only the eccentricity and the initial and final position vectors.

For the in-plane components it is considered the case when the transfer, departure and arrival orbits are co-planar. A further coordinates transformation is applied with the support of the Hamiltonian of the equations. The Palmer coordinates allow the simplification of the structure of the analysis and give a complete analytical approximation which, given the boundary conditions of the problem, can propagate the primer vector along the arc. The simplicity of the new procedure allows the exploitation of many other properties of the problem, not presented in this paper.

Acknowledgements This research has been funded by the European Commission through the Marie Curie Initial Training Network PITN-GA-2011-289240, AstroNet-II The Astrodynamics Network.

References

Battin, R.H.: An Introduction to the Mathematics and Methods of Astrodynamics, Revised Edition. AIAA Education Series. AIAA, Reston (1999)

Bernelli-Zazzera, F., Lavagna, M., Armellin, R., Di Lizia, P., Topputo, F., Berz, M., Vinkò, T.: Global Trajectory Optimization: can we prune the solution space when considering Deep Space Maneuvers? Technical report, ESA (2007)

Conway, B.A. (ed.): Spacecraft Trajectory Optimization. Cambridge University Press, Cambridge (2010)

Goodyear, W.H.: Completely general closed-form solution for coordinates and partial derivatives of the two-body problem. Astron. J. **70**(3), 189–192 (1965)

Iorfida, E., Palmer, P.L., Roberts, M.: A Hamiltonian approach to the planar optimization of mid-course corrections. Celest. Mech. Dyn. Astron. **124**(4), 367–383 (2016a)

Iorfida, E., Palmer, P.L., Roberts, M.: Geometric approach to the perpendicular thrust case for trajectory optimization. J. Guid. Control Dyn. **39**(5), 1059–1068 (2016b)

Jezewski, D.J.: Primer Vector Theory and Applications. Technical report, Lyndon B. Johnson Space Center, NASA (1975)

Jezewski, D.J., Rozendaal, H.L.: An efficient method for calculating optimal free-space n-impulse trajectories. AIAA J. **6**(11), 2160–2165 (1968)
Lawden, D.F.: Optimal Trajectories for Space Navigation. Butterworths Mathematical Texts. Butterworths, London (1963)
Lion, P.M., Handelsman, M.: Primer vector on fixed-time impulsive trajectories. AIAA J. **6**(1), 127–132 (1968)
Luo, Y., Zhang, J., Li, H., Tang, G.: Interactive optimization approach for optimal impulsive rendezvous using primer vector and evolutionary algorithms. Acta Astronaut. **67**, 396–405 (2010)
Olympio, J.T., Marmorat, J.P., Izzo, D.: Global Trajectory Optimization: can we prune the solution space when considering Deep Space Maneuvers? Technical report, ESA (2006)
Palmer, P.L.: Reachability and optimal phasing for reconfiguration in near circular orbit formations. J. Guid. Control Dyn. **30**(5), 1542–1546 (2007)
Prussing, J.E.: Optimal impulsive linear systms: sufficient conditions and maximum number of impulses. J. Astronaut. Sci. **48**(2), 195–206 (1995)
Quinn, T., Perrine, R.P., Richardson, D.C., Barnes, R.: A symplectic integrator for Hill's equations. Astron. J. **139**, 803–807 (2010)
Vasile, M., De Pascale, P.: Preliminary design of multiple gravity-assist trajectories. J. Spacecr. Rocket. **43**(4), 794–805 (2006)
Vinh, N.X.: Integration of the primer vector in a central force field. J. Optim. Theory Appl. **9**(1), 51–58 (1972)

Vision-Based Navigation Around Small Bodies

Pawel Kicman, Jakub Lisowski, and Ambroise Bidaux-Sokolowski

Abstract The paper is focused on the vision-based navigation around small bodies, starting with general overview of methods used in space navigation. The mission scenario is based on the latest guidelines for the ESA's Phobos Sample Return mission (until recently known as Phootprint) and the focus of the presented research is placed on the body relative navigation methods that are applicable for use around asteroids and small moons. In particular, detailed analysis of absolute navigation with reference to the body surface is performed. The results section contains analysis of the positioning accuracy achieved by the presented algorithms on a set of images generated using PANGU software.

1 Introduction

1.1 Brief Literature Review

So far there have been several successful missions to small bodies. The notable examples include NEAR-Shoemaker (Entreasian et al. 2001) that performed first soft landing on an asteroid, Hayabusa (Kominato et al. 2006) that returned a small sample of the Itokawa asteroid and most recent ROSETTA (Pardo de Santayana and Lauer 2015) mission that made first landing on the surface of a comet. All of those missions operated in a very close proximity to the bodies and used different techniques for descent and landing. The positioning systems of the satellites were designed to use vision based navigation methods. However, the level of autonomy was low in all cases. The images were processed largely on the ground, and the precise navigation information was uploaded to the spacecraft. That was possible thanks to the slow dynamics of the asteroid-spacecraft systems.

P. Kicman (✉)
GMV Spain, Calle de Isaac Newton 11, Tres Cantos, 28760 Madrid, Spain
e-mail: pkicman@gmv.com

J. Lisowski • A. Bidaux-Sokolowski
GMV Poland, Hrubieszowska 2, 01-209 Warsaw, Poland
e-mail: abidaux@gmv.com

G. Gómez, J.J. Masdemont (eds.), *Astrodynamics Network AstroNet-II*,
Astrophysics and Space Science Proceedings 44,
DOI 10.1007/978-3-319-23986-6_10

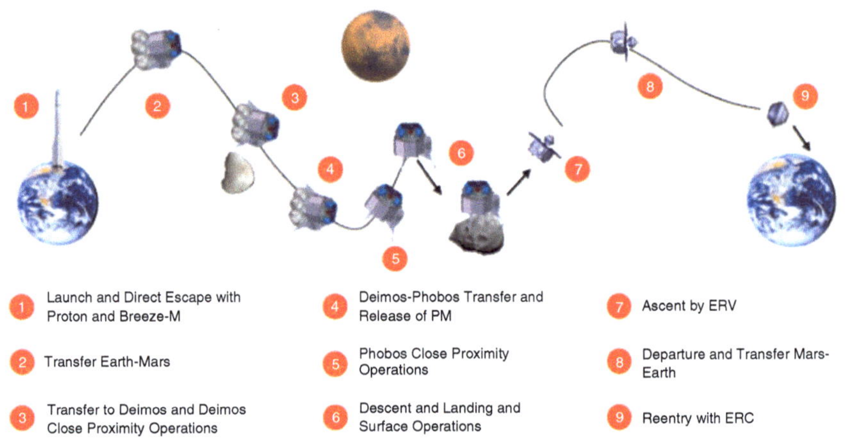

Fig. 1 Phobos Sample Return mission phases (CDF Study Report 2014)

1.2 Phobos Sample Return: Mission Scenario

Similar scenario for proximity and landing operations is envisioned for the Phobos Sample Return mission. The details are depicted in Fig. 1. The goal of the mission is to study two Martian moons—Deimos and Phobos. After several weeks of characterisation phase near Deimos, collecting measurements and images, the spacecraft transfers to Phobos and repeats the characterisation operations. When that phase is completed, landing on the surface of the moon is performed. The descent and landing operations pose high requirements on the GNC design. However, unlike in previous similar missions, high level of autonomy is required due to the long turn-around times between Earth and the spacecraft. Moreover, processing of navigation imagery on the ground is not possible due to complex dynamics of the Mars-Phobos system. Very stringent requirement on the accuracy of the landing position (50 m, 2-*sigma*) leads to even more challenges in the design of the navigation system for the spacecraft which is being analysed in more details in following paragraphs. The report on the developments of the entire GNC chain was recently reported in Gonzalez-Arjona et al. (2015).

2 System Description

The core of the navigation system consists of the relative navigation algorithm based on unknown feature tracking, supported by altimeter measurements. The architecture of the system is shown in Fig. 2. A landmark recognition algorithm (Enhanced Relative Navigation—ERN) is used to initialize the navigation filter while spacecraft is hovering at the waypoint. Then, pure relative navigation is

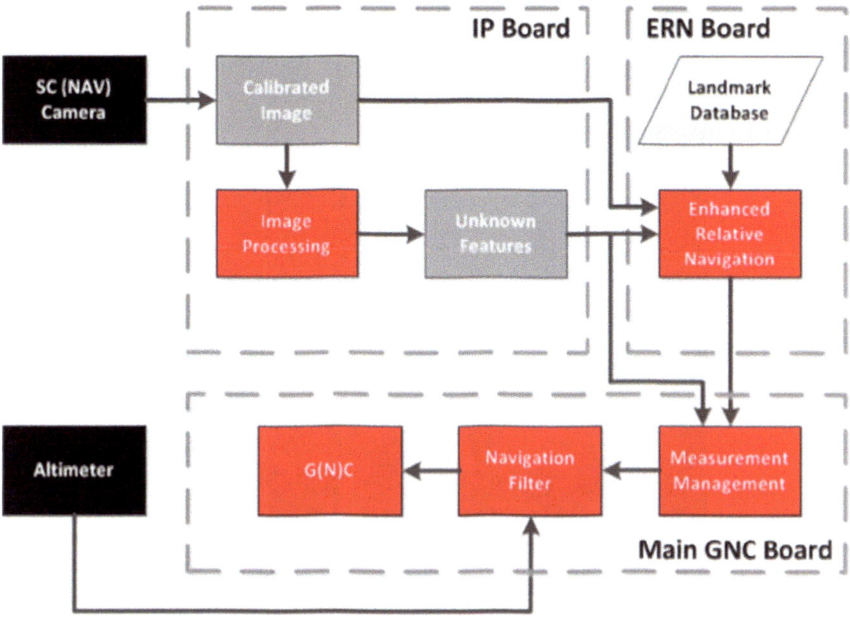

Fig. 2 Vision-based navigation system architecture

utilized during the entire D&L phase. The KLT algorithm (implemented on an FPGA board dedicated to image processing—IP Board) is used to extract and track features in the images. The positions of the features and altimeter measurements are fused within the navigation filter to provide position and velocity estimates to guidance and control modules of the spacecraft. In order to utilize ERN the database has to be prepared on the ground prior to the D&L and uploaded to the spacecraft.

2.1 Relative Navigation

2.1.1 Feature Tracking

The Kanade–Lucas–Tomasi algorithm (KLT) (Shi et al. 1994) is used to detect and track features in the sequence of images. The algorithm selects points that have high intensity gradients in both directions (corners). Such points are considered to be strong features. Once the points are selected in the first processed image the algorithm tracks them in the successive frames (Fig. 3). When a feature is lost the algorithm selects new point with high corner response to keep the number of tracked features constant. A hardware (FPGA) implementation of the algorithm was developed by GMV in order to advance TRL (technology readiness level) of the technique. The algorithm was shown to be more robust than previously used

Fig. 3 Sample KLT feature tracks during the descent simulated in a GMV's platform-art laboratory

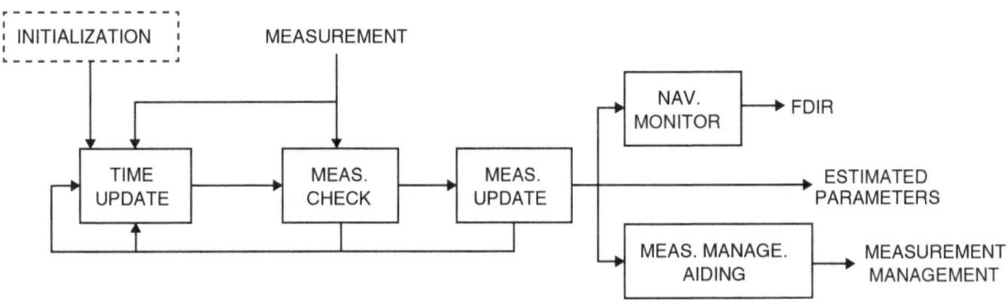

Fig. 4 Navigation filter structure

FEIC (Feature Extraction Integrated Circuit) algorithm (Dunstan et al. 2005), which directly translates to more accurate navigation estimates.

2.1.2 Navigation Filter

The navigation function provides updated values of the relative state of the spacecraft with respect to the Phobos in a Phobos-fixed frame. The functional architecture of the navigation filter is schematically shown in Fig. 4. The Initialization can be provided either by telecommand from the ground or by the Enhanced Relative Navigation.

- The time update propagates the average state and its covariance matrix from the last epoch (can be the initialization epoch and state) to the current measurement time (a priori state and covariance).
- Then, a test of hypotheses is done on the input measurements.
- If the measurements are accepted, the measurement update provides the a posteriori state and covariance. The measurement update might be iterated to smooth the non-linearities.
- The a posteriori residuals and covariance are checked against the hypotheses and the convergence criterion. The output is prepared in the proper format and some information for aiding the measurement management.

The initialization or reset of the filter with ground control involvement might consist on the epoch estimation of the spacecraft state and the parameters of the dynamics estimated during periods without manoeuvres. This approach might not be feasible for descent and landing initialization if the drift of the relative navigation is too large to achieve the landing conditions. In the enhanced relative navigation, the landmark detection function is used to start up the relative navigation.

The augmented state vector of the filter has the following components (1),

- Spacecraft relative position and velocity (6),
- Position bias of the feature location on the surface of the asteroid (3)
- Altimeter ECRV (N_{alt}),
- Position (pixel) of each tracked feature (note that the number of features changes dynamically, but their count is limited to 10 due to computational constraints) ($2N_{\text{landmark}}$).

$$\mathbf{x} = \begin{pmatrix} \mathbf{r} \\ \mathbf{v} \\ \mathbf{r}_{0\,\text{landmarks}} \\ ECRV_{\text{Alt}_i} \\ \cdots \\ ECRV_{\text{Alt}_{N\_ak}} \\ \mathbf{p}_{L_i} \\ \cdots \\ \mathbf{p}_{L_{N\_\text{landmark}}} \end{pmatrix}. \tag{1}$$

The equations of motion for time update of the above state vector are given in (2):

$$\begin{aligned} &\dot{\mathbf{r}} = \mathbf{v} \\ &\dot{\mathbf{v}} = \mathbf{a}_{g,T} + \mathbf{a}_{\text{Thrust}} \\ &\dot{\mathbf{r}}_{0\,\text{landmarks}} = 0 \\ &E\dot{C}RV_{\text{Alt}_i} = ECRV_{\text{Alt}_i}/\tau, \quad i = 1 : N_{\text{Alt}} \\ &\dot{\mathbf{p}}_{L_i} = 0, quadi = 1 : N_{\text{Landmark}} \end{aligned} \tag{2}$$

where the dots denote the derivatives with respect to the time. The dynamics of the spacecraft includes the gravity field of the Phobos-Mars system and the thruster acceleration.

The Unscented Kalman Filter (UKF) (Wan and van der Merve 2001) is a sequential filter that deals with the possible system non-linearity by replacing the analytical propagation of estimate and covariance matrix by the non-linear propagation of a set of points (sigma points). The estimate and covariance matrix of the propagated state are reconstructed as the weighted first and second moments of the sigma-point ensemble. Appropriate choice of the sigma points allows the filter to retain non-linear effects up to the third order. State estimate and its covariance are computed by means of expressions that are equivalent to those of the standard Kalman filter. A hybrid EKF-UKF is used as a navigation filter. Given the particularities of the descent and landing (D&L) problem, in which the state is quite large (e.g. 30 components if we track ten unknown features) and the dynamics of most of the states is linear (only the position and velocity of the spacecraft are non-linear), a method to accelerate the execution of the UKF without penalizing the performances has been developed at GMV. The hybrid filter considers a propagation step (time update) as in the EKF, which reduces significantly the number of propagations. Then the measurement update is performed with the UKF algorithm (i.e. using sigma points). The main advantages of such solution include:

- Significant reduction in the execution time.
- Increased numerical stability.
- Increased accuracy of measurement update due to high non-linearity of measurement equations.

2.2 *Enhanced Relative Navigation*

The Enhanced Relative Navigation computes position of the spacecraft by combining points detected in the image with previously prepared database of landmarks. In the presented implementation ORB feature descriptor (Rublee et al. 2011) is used to recognize landmark in the image. Based on obtained pairs (2D point in image matched with 3D location of a landmark) the position of the spacecraft is computed. Main steps of the process are visualised in Fig. 5.

2.2.1 Descriptor Computation

ORB feature descriptor (Rublee et al. 2011) is an improvement over the BRIEF descriptor. The detailed analysis of the applicable descriptors was reported previously in Kicman et al. (2015). The main advantage of this descriptor is very low computation time. Additionally the descriptor used is represented as set of

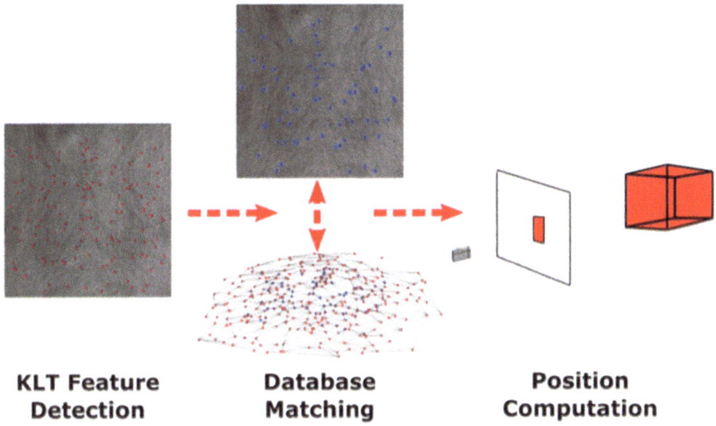

Fig. 5 Main steps of the Enhanced Relative Navigation position computation process

binary numbers which makes matching process very fast. However, the performance compared to SIFT and SURF descriptors is inferior.

The descriptor is computed comparing intensity values of pairs of pixels scattered around the detected feature. The pairs of points have been optimized by inventors of the ORB descriptor to maximize uncorrelation and in the same time to have high variance. Those properties ensure that the descriptor contains maximum amount of information and is easy to recognize.

In the implementation developed by GMV the descriptor is computed in the four main steps:

- Step 1: Blurring of the image. Convolution of the image with a Gaussian kernel.
- Step 2: Computation of the orientation of a feature. Orientation of feature is computed based on the intensity centroid.
- Step 3: Pattern rotation. The sampling pattern is rotated by the feature orientation angle.
- Step 4: Construction of the descriptor. Values in the sampling points are read and compared. Based on the results of the tests binary descriptor is constructed (Fig. 6).

The number of available features after computing the descriptor might be lower than the original list. This is due to the fact that the features detected close to the border of the image are rejected. Sampling based on the pattern is impossible in this region of image.

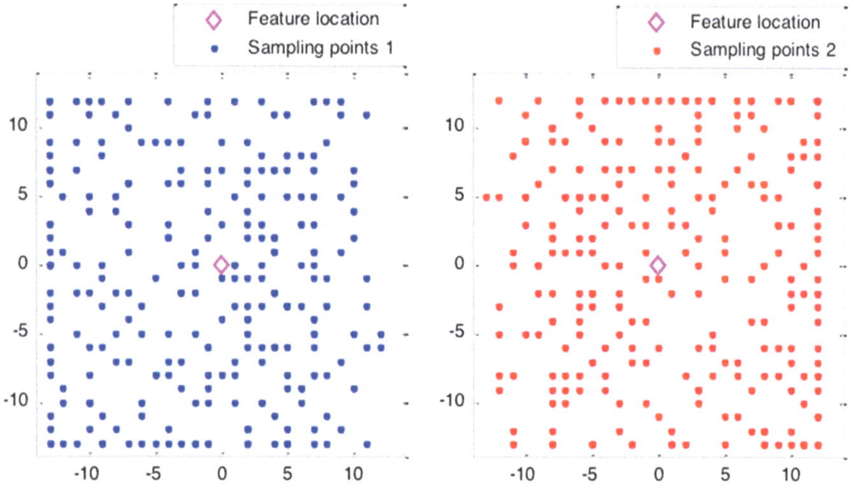

Fig. 6 Sampling pattern of ORB descriptor

2.2.2 Feature Matching

Matching of the features detected in the camera image to the database is performed using a brute-force matching algorithm that compares all descriptors. For all detected features, their descriptors ($desc_i$) are compared with the database descriptors ($desc_j$), to find the best matches.

The binary descriptors are compared using standard Hamming distance metric (3), which indicates how many elements of two binary vectors are not identical (the more different elements, the higher the distance). The Hamming distance is in the range between 0 and 1, where 0 means two identical vectors and 1 is a distance between a binary vector and its negation (i is a feature index, j is a landmark index, k is an index of element within the descriptor and n is a length of the descriptor vector).

$$d_{ij} = \left[\#(\text{desc}_i[k] \neq \text{desc}_j[k])/n\right]. \tag{3}$$

Once the detected points in the image are matched with the database we have a set of 2D points and corresponding 3D coordinates of landmarks.

2.2.3 Position Computation

PnP (Perspective-n-Point) problem is a problem of calculating a camera pose given known 3D coordinates of points (map of landmarks) and their projection on the image (KLT features). To solve this mathematical problem an EPnP algorithm (Lepetit et al. 2009) is used. It is a fast, non-iterative method that has linear computational time in terms of available 3D-2D point correspondences. The algorithm requires at least four of such correspondences to compute the solution. However, the results are degraded for the low number of available points (4–5 points). Therefore, at least seven points are required by our algorithm to compute the solution. Fortunately, the algorithm is of $\mathcal{O}(n)$ complexity and scales well with the number of points used for calculations. We have observed a fast operation of the algorithm with more than 30 matches used for calculation.

In order to eliminate any outliers that might have been falsely matched the EPnP algorithm is executed within the RANSAC framework (Fischler et al. 1981). The hypothesis of a spacecraft position is computed 500 times for seven pairs of points randomly selected among all the matches. For every hypothesis number of points that support the hypothesis is computed. As an error measure a reprojection error is used. The highest group of inliers is selected for further computation. In order to enable the execution in the processor the final solution is computed with predefined number of inliers. If number of inliers found is lower than given threshold the solution will not be computed.

3 Experimental Results

Simulated images are captured to imitate a camera that has 20° FOV and the detector with 1024 × 1024 pixels resolution. The images are taken at 5 km altitude above the surface. For such configuration image covers square are of around 1.5 km size. Previously created database covers square area of 3 km size to account for expected dispersion and pointing errors. For testing purposes a series of 100 images is used (sample image in Fig. 7).

Fig. 7 Sample image from PANGU simulated dataset

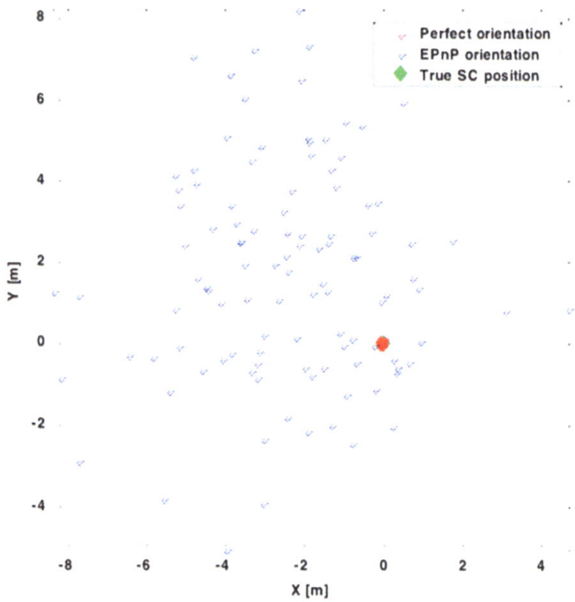

Fig. 8 Spread of positions computed using PANGU generated images. Notable difference between perfect and EPnP computed orientation

The results of the test are visualized in Fig. 8. The mean horizontal position error is equal to 4.06 m (blue points in the figure). This would indicate that the noise in the position of detected features translates to a non-negligible spread of computed positions. It shall be also noted that the images were rendered simulating also motion of the Sun. Many of the features are subject to a drift when Sun elevation changes. It was also discovered that the rate of the drift is significantly higher when Sun is low above the horizon (fast moving shadows) compared to the Sun being in a local zenith. The character of the changes is captured in the Fig. 9. This has also non-trivial effect on the positioning accuracy. Therefore it is highly recommended that images that are used for generation of the landmark database are captured in similar illumination conditions to the ones that are expected at the moment of computing position with ERN.

It was also observed that a significant contributor to the error of computed position comes from the uncertainties in rotation between camera and landing site reference frame. When the rotation is estimated by the EPnP algorithm the mean horizontal position error equals to 4.06 m. However, when perfect rotation matrix is used and only position is estimated by the EPnP algorithm the mean horizontal position error is equal to 0.07 m. This comes from the fact that in its essence the algorithm computes position of the observed object in the camera reference frame and rotation between camera and object frames. Only after rotating the position to

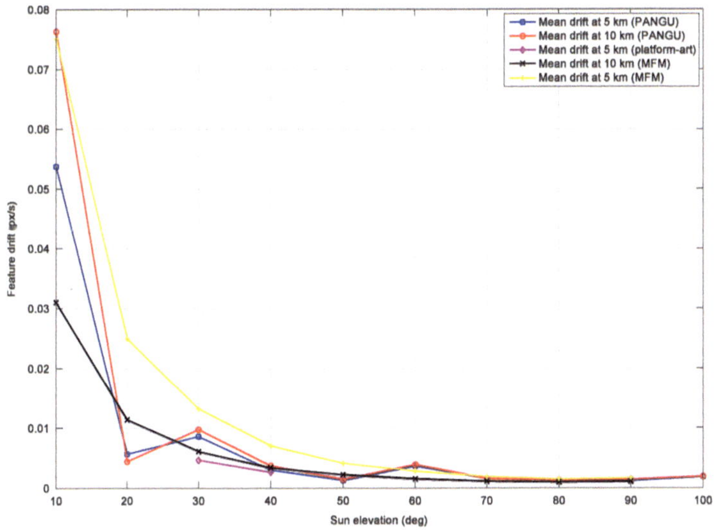

Fig. 9 Mean feature drift expressed as a function of Sun elevation

the landing site fixed frame the location of the camera and the spacecraft can be recovered. Separately, both values are computed with high accuracy. However, due to the relatively large distance between the target body and a camera even small error in the estimation of the orientation translates to large error in estimation of the position. It will be investigated further if use of the inertial orientation obtained by the star tracker, combined with the rotational state of the Phobos can provide more accurate relative rotational state, compared to the one given by the EPnP.

After initialization of the navigation filter with results computed by ERN the navigation is performed based on the relative navigation—tracking of the features and altimeter measurements. The position error for the trajectory of one of the scenarios is shown in the Fig. 10. The filter is able to successfully estimate position of the spacecraft. The deviation from initial error assumed in the simulation equals to 5.53 m. Combining that with the worst case scenario for the ERN which is equal to 8.51 m compounds to the total of 10.15 m of mean position error expected at the landing site. Further analysis with higher fidelity model and a Monte Carlo campaign are planned in the following months to validate those initial estimations.

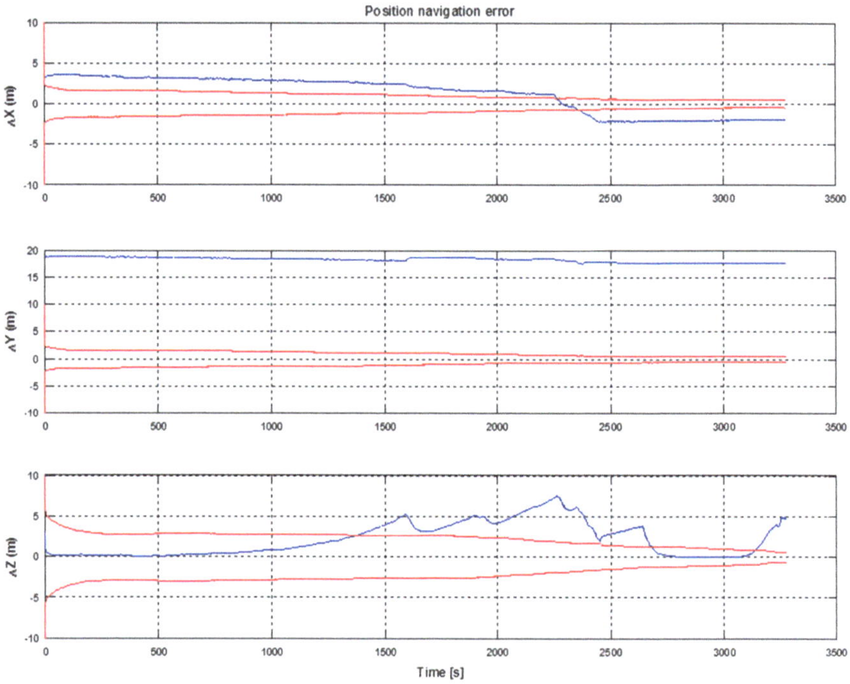

Fig. 10 Example position error in one of the simulated descent scenarios

4 Conclusions

The system capable of accurate relative navigation in proximity to the Phobos was presented in the paper. The main contribution and novelty of the presented design is integration of new absolute navigation technique (called Enhanced Relative Navigation) with well understood and tested relative navigation approach. Combination of the two techniques provides high level of autonomy for the spacecraft and enables accurate landing in a difficult 3-body dynamic system. At the moment of writing this paper the integrated version of presented system is being prepared for extensive testing campaign to analyze in-depth properties and limitations of the algorithms.

References

CDF Study Report: Phobos Sample Return (2014)

de Santayana, R.P., Lauer, M.: Optical measurements for rosetta navigation near the comet. In: Proceedings of the 25th International Symposium on Space Flight Dynamics (ISSFD), Munich (2015)

Dunstan, M., et al.: Visual Navigation Chip for Planetary Landers, Data Systems in Aerospace. DASIA (Data Systems in Aerospace), Edinburgh (2005). ISSN: 1609 042X; ISBN: 92-9092-913-8

Entreasian, P.G., et al.: The design and navigation of the NEAR-Shoemaker landing on Eros. In: AAS/AIAA Astrodynamics Specialist Conference, Quebec (2001)

Fischler, M.A., Bolles, R.C.: Random sample consensus: a paradigm for model fitting with applications to image analysis and automated cartography. Commun. ACM **24**(6), 381–395 (1981)

Gonzalez-Arjona, D., et al.: Vision-based guidance, navigation and control system for phobos sample return mission. In: 66th International Aeronautical Congress, Jerusalem (2015)

Kicman, P., et al.: Advanced vision-based navigation for approach and proximity operations in missions to small bodies. In: 6th European Conference for Aeronautics and Space Sciences (EUCASS), Karkow (2015)

Kominato, T., et al.: Optical hybrid navigation in Hayabusa – approach, station keeping & hovering. In: AAS/AIAA Space Flight Mechanics Conference, Tampa (2006)

Lepetit, V., Moreno-Noguer, F., Fua, P.: Epnp: an accurate $o(n)$ solution to the pnp problem. Int. J. Comput. Vis. **81**(2), 155–166 (2009)

Rublee, E., et al.: ORB: an efficient alternative to SIFT or SURF. In: IEEE International Conference on Computer Vision (ICCV). IEEE, Barcelona (2011)

Shi, J., Carlo, T.: Good features to track. In: IEEE Computer Society Conference on Computer Vision and Pattern Recognition, 1994. Proceedings CVPR'94. IEEE, Seattle (1994)

Wan, E.A., van der Merve, R.: The unscented Kalman. In: Hayjin, S. (ed.) Kalman Filtering and Neural Networks. Wiley, New York (2001)

Analytical and Semianalytical Propagation of Space Orbits: The Role of Polar-Nodal Variables

Martin Lara

Abstract The role of different sets of variables in analytical or semianalytical solutions to the artificial satellite theory is reviewed. In particular, it is shown that using polar-nodal variables or non-singular variables based on them is clearly advantageous in the evaluation of the transformation from mean to osculating elements.

1 Introduction

In the last two decades numerical methods seem to have completely eclipsed analytical and semianalytical solutions of the Artificial Satellite Theory (AST) for practical application. In spite of that, analytical and semianalytical integration may still play a role in the propagation of space orbits. Indeed, NORAD catalog continue to be used for many satellite operators, which must rely on the SGP4 propagator (Hoots and Roehrich 1980; Vallado et al. 2006) for programming the time of a maneuver. Unfortunately, the limited accuracy of the SGP4 orbital model and the lack of uncertainty information of the ephemeris used by SGP4 make that predictions based on the so-called Two Line Elements may lack of the required accuracy to predict some events as estimating collision probability (Kelso 2009). In this scenario the reputed Draper Semi-analytic Satellite Theory (McClain 1977) is currently being promoted as a candidate satellite theory suitable for catalogue maintenance (Setty et al. 2014), and the use of open source versions of the theory based on the independent implementation of the corresponding algorithms (Danielson et al. 1995), or open access versions of the original software is nowadays encouraged.[1] Alternatively, approaches based on the numeric-analytic integration

[1]https://www.orekit.org/, http://tastrody.unirioja.es/dsst (accessed October 22, 2015).

M. Lara (✉)
GRUCACI, University of La Rioja, 26004 Logroño, Spain
e-mail: mlara0@gmail.com

G. Gómez, J.J. Masdemont (eds.), *Astrodynamics Network AstroNet-II*,
Astrophysics and Space Science Proceedings 44,
DOI 10.1007/978-3-319-23986-6_11

of full models are claimed to provide similar accuracy to numerical integration but reducing the integration time by two orders of magnitude (Golikov 2012).

Besides, current space debris mitigation guidelines may require the design of end-of-life disposal strategies guaranteeing some safe condition for hundreds of years. In these cases, strategies based on combining semianalytical integration with global optimizers have revealed quite suitable (Armellin et al. 2015). But analytical integration is also an option for onboard orbit propagation when reduced power consumption is a critical requirement, as may be the case of micro and nanosatellite missions (Gurfil and Lara 2014).

A successful approach to finding analytical solutions of AST was the use of perturbation methods (Sanders et al. 2007). For bounded motion, solutions are assumed to be slightly distorted ellipses, and approximations to the solution are customarily obtained using canonical perturbation methods, where, in a similar fashion to the Hamilton-Jacobi method, the application of successive canonical transformations, which are given by expansions in power series of a small parameter, up to certain truncation order decouples the flow by the introduction of formal integrals, in this way simplifying the integration. Traditional approaches base on the separation of the disturbing function into short-period, long-period, and secular terms, and rely on expansions of the elliptic motion (Brouwer and Clemence 1961; Kaula 1966; Deprit and Rom 1970; Kinoshita 1977). However, by using the known differential relations of the osculating motion between the eccentric, true and mean anomalies, the closed form solution was soon accomplished—yet limited to second order secular terms and only for the lower degree zonal harmonics (Garfinkel 1959; Kozai 1959; Brouwer 1959). Achieving higher orders in the solution in closed form requires to perform highly sophisticated simplifications (Kozai 1962; Aksnes 1971) or to cope with special functions (Osácar and Palacián 1994). Eventually, the invention of the elimination of the parallax simplification (Deprit 1981; Lara et al. 2014b) eased definitively the task of achieving higher order solutions in closed form (Coffey and Deprit 1982).

It deserves to mention that the artificial satellite problem is not generally integrable although remarkable exceptions have been discovered (Vinti 1959; Kislik 1961; Aksenov et al. 1962). Hence, solutions found by the construction of formal integrals limit their application to particular regions of phase space. For instance, the usual elimination of the mean anomaly converts the semi-major axis in a formal integral, so the resulting flow only approximates those regions of phase space where the semi-mayor axis remains constant except for short-period effects, therefore clearly excluding the case of tesseral resonant orbits where the semi-major axis is affected of long-period effects (Kaula 1966). A following removal of the argument of the perigee converts the total angular momentum in a formal integral; in consequence the transformed flow is constrained to orbits whose eccentricity and inclination remain constant except for periodic effects with the period of the argument of the perigee, thus excluding the case of zonal (or inclination) resonant orbits and, in particular, the critical inclination (Jupp 1975; Coffey et al. 1986; Lara 2015b).

Efficiency in evaluating analytical expressions is closely related to the variables used in their formulation. In view of there are no singularities in the elements of an earth orbit, different efforts were made to reformulate Brouwer's gravitational solution (Brouwer 1959) in different sets non-singular variables (Lyddane 1963; Hoots 1981). Also, it deserves to mention the repeated efforts in constructing an alternative perturbation theory in polar-nodal variables (Cid and Lahulla 1969; Cid et al. 1986; Deprit and Ferrer 1989), which are free from the singularity of zero eccentricity orbits in this way avoiding the main concern on Brouwer's solution. First attempts in implementing such a theory found that, when extended further than the first order, the perturbation solution in polar-nodal variables was contaminated by mixed secular-periodic terms, which constrained the applicability of the theory for ephemeris generation to short time intervals (Cid and Lahulla 1971). However, mixed terms are nicely avoided by decomposing the perturbation theory into two different simplifications either of which is carried out in polar-nodal variables (Alfriend and Coffey 1984).

The fact that each of the canonical simplifications leading to the elimination of the argument of the latitude are radically simplified when implemented in Delaunay variables (Lara et al. 2013, 2014a,b) seems to endorse the construction of solutions to perturbed Keplerian motion in this set of canonical variables, which are the action-angle variables of the Kepler problem. Nevertheless, the canonical transformation achieving the desired Hamiltonian reduction does not need to be defined by the transformation of the Delaunay variables. On the contrary, following Izsak's suggestion of constructing the transformation from mean to osculating elements in polar-nodal variables (Izsak 1963) presents two main advantages. On the one hand, the trouble with low eccentricity orbits is completely avoided; still, the case of almost equatorial inclinations must be treated separately, but a simple set of non-singular elements which are trivially derived from the polar-nodal variables may be used for dealing properly with this case (Aksnes 1972; Lara 2015a). On the other hand, the transformation defined by the polar-nodal variables reduces spectacularly the size of the perturbation series, which adopt a compact form of straightforward evaluation, a fact that is of specific relevance when dealing with third body perturbations, a case in which the transformation from mean to osculating Delaunay variables may comprise very long Fourier series (Lara et al. 2015).

2 The Kepler Problem: Polar-Nodal and Delaunay Variables

In the Hamiltonian formalism the Kepler problem is written,

$$\mathscr{H} = \frac{1}{2}\mathbf{X}\cdot\mathbf{X} - \frac{\mu}{r}, \tag{1}$$

and the corresponding Hamilton equations are

$$\frac{\mathrm{d}\mathbf{x}}{\mathrm{d}t} = \mathbf{X}, \qquad \frac{\mathrm{d}\mathbf{X}}{\mathrm{d}t} = -\mu\frac{\mathbf{x}}{r^3}. \tag{2}$$

which is a differential system of three degrees of freedom (DOF), thus not providing too much insight into the physical properties of the Kepler problem.

The state of the satellite is conveniently given in the orbital plane by using polar variables. Thus, let the inertial frame be defined by the orthogonal unit vectors $(\mathbf{i}, \mathbf{j}, \mathbf{k})$. Let $\mathbf{x} \times \mathbf{X} = \Theta \mathbf{h}$ be the angular momentum vector (per unit of mass), where the unit vector $\mathbf{h}$ is perpendicular to the orbital plane, and let $\mathbf{l} = \mathbf{k} \times \mathbf{h} / \|\mathbf{k} \times \mathbf{h}\|$. Then, the position in the orbital plane is given by the radial distance r and the polar angle θ (the argument of the latitude), which is measured from the initial direction $\mathbf{l}$, while the velocity is given in the orbital plane by the radial and transverse components $\dot{r}$ and $r\,\dot{\theta}$, respectively, where the overdot means differentiation with respect to time. The orbital plane is then referred to the inertial frame giving the angle I between $\mathbf{k}$ and $\mathbf{h}$—its inclination with respect to the inertial $(\mathbf{i}, \mathbf{j})$ plane—and the angle ν between $\mathbf{i}$ and $\mathbf{l}$—the longitude of the ascending node.

This natural representation of the orbital motion gives rise to the introduction of the canonical set of polar-nodal variables for describing the state of the motion. This set is given by the three coordinates r, θ, and ν, and their conjugate momenta given by the radial velocity $R = \dot{r}$, the modulus of the angular momentum $\Theta = r^2\,\dot{\theta}$, and the projection of the angular momentum vector in the $\mathbf{k}$ direction $N = \Theta \cos I$, respectively. Note that this set of variables is singular for $I = 0$ or $I = \pi$, in which cases ν is not defined.

The contact transformation from polar-nodal to Cartesian variables is easily attained by applying the rotations that relate the inertial and orbital frames, the latter defined by the unit vectors $(\hat{\mathbf{x}}, \mathbf{h} \times \hat{\mathbf{x}}, \mathbf{h})$, where $\hat{\mathbf{x}} = \mathbf{x}/r$. Indeed, the position of the satellite in the orbital frame is simply $(r, 0, 0)$, whereas the velocity is $(R, \Theta/r, 0)$. Then,

$$\begin{pmatrix} x & X \\ y & Y \\ z & Z \end{pmatrix} = R_3(-\nu)\, R_1(-I)\, R_3(-\theta) \begin{pmatrix} r & R \\ 0 & \Theta/r \\ 0 & 0 \end{pmatrix} \tag{3}$$

where R_1 and R_3 are the usual rotation matrices around the x and z axes, respectively.

The transformation given by Eq. (3) is canonical, as first demonstrated by Whittaker (1904). Then, Eq. (1) is rewritten in polar-nodal variables as

$$\mathcal{H} = \frac{1}{2}\left(R^2 + \frac{\Theta^2}{r^2}\right) - \frac{\mu}{r}, \tag{4}$$

where the absence of ν, θ, and N from the Hamiltonian implies that N, Θ, and ν are constant, as easily checked from Hamilton equations. Besides, the radial motion decouples from the rest of the system:

$$\frac{\mathrm{d}r}{\mathrm{d}t} = \mathcal{H}_R = R, \qquad \frac{\mathrm{d}R}{\mathrm{d}t} = -\mathcal{H}_r = -\frac{\Theta^2}{r^3} + \frac{\mu}{r^2}. \tag{5}$$

where subindices denote partial derivation. The reduced system (5) is of 1-DOF and, therefore, integrable. After integrating it, the final coordinate is integrated by quadrature

$$\frac{d\theta}{dt} = \mathscr{H}_\Theta = \frac{\Theta}{r^2}, \qquad \theta = \theta_0 + \Theta \int \frac{1}{r(t)^2}\, dt. \tag{6}$$

Therefore, the use of polar-nodal variables and Hamiltonian mechanics directly shows the integrability of the Kepler problem, which is simplified to the integration of r from Eq. (5).

As an alternative to the integration of Eqs. (5) and (6), the Kepler problem can be solved by complete reduction of the Hamiltonian (4). Indeed, given the Hamiltonian $\mathscr{H} = \mathscr{H}(\mathbf{x}, \mathbf{X})$, if a canonical transformation

$$\mathscr{T} : (\mathbf{x}, \mathbf{X}) \longrightarrow (\mathbf{y}, \mathbf{Y}), \tag{7}$$

from old $(\mathbf{x}, \mathbf{X})$ to new variables $(\mathbf{y}, \mathbf{Y})$ is found such that the Hamiltonian in the new variables is only a function the momenta, namely

$$\mathscr{T} \circ \mathscr{H}(\mathbf{x}, \mathbf{X}) \longrightarrow \mathscr{H}(\mathbf{x}(\mathbf{y}, \mathbf{Y}), \mathbf{X}(\mathbf{y}, \mathbf{Y})) = \Phi(-, \mathbf{Y}), \tag{8}$$

then, the integration of Hamilton equations becomes trivial in the new variables, namely $\mathbf{Y} = \mathbf{Y}_0$, $\mathbf{y} = \mathbf{y}_0 + \Phi_{\mathbf{Y}} t$, and the solution in the original variables is obtained by plugging this solution into the transformation (7).

The required transformation can be computed by the Hamilton-Jacobi method (see Arnold 1989, for instance), in which the transformation is derived from a generating function in a mixed way

$$\mathbf{X} = \mathscr{S}_\mathbf{x}, \qquad \mathbf{y} = \mathscr{S}_\mathbf{Y}, \tag{9}$$

where $\mathscr{S} \equiv \mathscr{S}(\mathbf{x}, \mathbf{Y})$ depends only on the old coordinates and the new momenta. The generating function $\mathscr{S}$ must be solved from the Hamilton-Jacobi equation $\mathscr{H}(\mathbf{x}, \mathscr{S}_\mathbf{x}) = \Phi(-, \mathbf{Y})$, which is a partial differential equation in $\mathbf{x}$.

In the particular case of the Kepler problem, it seems natural to compute a canonical transformation $\mathscr{T} : (r, \theta, \nu, R, \Theta, N) \longrightarrow (\ell, g, h, L, G, H)$, such that the three known integrals of the problem from the polar-nodal variables formulation in Eq. (4) remain unaltered. This can be done choosing a generating function of the form $\mathscr{S} \equiv \mathscr{S}(r, \theta, \nu, L, G, H) = \nu H + \theta G + \mathscr{W}(r, L, G)$. In this way, from Eq. (9),

$$\Theta = \mathscr{S}_\theta = G, \qquad N = \mathscr{S}_\nu = H, \qquad h = \mathscr{S}_H = \nu, \tag{10}$$

as desired, and the rest of the transformation

$$R = \mathscr{S}_r = \mathscr{W}_r, \qquad \ell = \mathscr{S}_L = \mathscr{W}_L, \qquad g = \mathscr{S}_G = \theta + \mathscr{W}_G, \tag{11}$$

will only require to compute the function $\mathscr{W}$.

Then, the Hamilton-Jacobi equation of Eq. (4) is

$$\frac{1}{2}\left[(\mathscr{W}_r)^2 + \frac{G^2}{r^2}\right] - \frac{\mu}{r} = \Phi(L, G, -), \tag{12}$$

where Φ has been chosen independent of $N = H$ in view of this variable does not appear in the original Hamiltonian in Eq. (4). Therefore, from Eq. (12),

$$\mathscr{W} = \int \sqrt{2\Phi + 2\mu/r - G^2/r^2}\,\mathrm{d}r, \tag{13}$$

and, in consequence, Eq. (11) is written

$$R = \sqrt{2\Phi + 2\mu/r - G^2/r^2}, \tag{14}$$

$$\ell = \Phi_L \int \frac{1}{R(r)}\,\mathrm{d}r, \tag{15}$$

$$g = \theta + G \int \frac{1}{R(s)}\,\mathrm{d}s + \Phi_G \int \frac{1}{R(r)}\,\mathrm{d}r, \tag{16}$$

where $s = 1/r$. The solutions of the quadratures that appear in Eqs. (15) and (16) are achieved by making standard changes of variables (see Deprit and Rom 1970, for instance). Indeed, introducing the notation

$$p = G^2/\mu, \qquad a = -\mu/(2\Phi), \qquad e^{**}2 = 1 - p/a, \tag{17}$$

the change $r = a(1 - e\cos u)$ from the radius to the eccentric anomaly u, solves the quadrature in Eq. (15) to give

$$\ell = \mu\Phi_L(-2\Phi)^{-3/2}(u - e\sin u). \tag{18}$$

On the other hand, the change $r^{-1} = (1 + e\cos f)/p$ from the radius to the true anomaly f, converts Eq. (16) into

$$g = \theta - f + \mu\Phi_G(-2\Phi)^{-3/2}(u - e\sin u). \tag{19}$$

Therefore, Eqs. (10), (14), (18) and (19) conform a *family* of canonical transformations parameterized by Φ that carry out the complete reduction of the Kepler Hamiltonian (Ferrer and Lara 2010).

Note that, by solving Φ from Eq. (17), the reduced Hamiltonian can be written in the standard form

$$\Phi \equiv \Phi(L, G) = -\frac{\mu^2}{2G^2}(1 - e^2). \tag{20}$$

Then, the partial derivatives of Φ with respect to L and G in Eqs. (18) and (19) can be replaced by corresponding partial derivatives of e as derived from Eq. (20), in this way the family of canonical transformations is parameterized by e. Besides, since $e \equiv e(L, G)$ is non-dimensional, its formal dependence on G and L can be reduced to a dependence on the single parameter $\eta = G/L$. In particular, Eq. (18) is rewritten

$$\ell = -\eta^2 \frac{\mathrm{d}e}{\mathrm{d}\eta} \frac{e}{(1-e^2)^{3/2}} (u - e \sin u). \tag{21}$$

Any formal selection of $e = e(\eta)$ does solve the Kepler problem by complete reduction. However, a case of particular interest is that who makes ℓ to be an angle. The required form of e is obtained by imposing to Eq. (21) the angle condition $\oint \mathrm{d}\ell = 2\pi$, which states that a variation of ℓ along a closed trajectory in the original phase space must be 2π. Namely, $\ell(u = 2\pi) - \ell(u = 0) = 2\pi$, from which

$$-\frac{e\eta^2}{(1-e^2)^{3/2}} \frac{\mathrm{d}e}{\mathrm{d}\eta} = 1, \tag{22}$$

which is in separate variables and can be solved by quadrature to give

$$e^2 = 1 - \eta^2 = 1 - G^2/L^2, \tag{23}$$

a relation that further imposes the dynamical constrain $G \leq |L|$.

When e is chosen from Eq. (23), then Eqs. (18) and (19) particularize to

$$\ell = u - e \sin u, \tag{24}$$

$$g = \theta - f, \tag{25}$$

and the reduced Hamiltonian (20) in the new, action and angle variables adopts the form

$$\Phi = -\mu^2/(2L^2), \tag{26}$$

which leads to the trivial solution of the Kepler problem in the new variables:g, h, L, G, and H are constant, and $\ell = \ell_0 + (\mu^2/L^3)\,t$. As expected, this solution unavoidably requires dealing with Kepler's equation, Eq. (24). This is the famous solution given by Delaunay (1860). After him, the action-angle variables of the Kepler problem are called "Delaunay variables", viz. ℓ is the mean anomaly, g is the argument of the pericenter, h is the argument of the node, $L = \sqrt{\mu a}$ is the Delaunay action, $G = L\eta$ is the total angular momentum, and $H = G \cos I$ is the projection of the angular momentum vector along the inertial z axis. Since the Delaunay variables are the canonical counterpart of the usual Keplerian elements, they share the singularities for the cases of circular and equatorial orbits.

3 Perturbation Solutions: Brouwer's Approach

The Kepler problem is a central forces problem, and hence the orbital plane remains constant and the motion is planar. In real cases the gravitational potential is non-central, and is customarily given by the usual expansion in spherical harmonics. For the earth, the zonal harmonic coefficient of the second order $C_{2,0} = -J_2 = \mathcal{O}(10^{-3})$, whereas the rest of harmonic coefficients are of $\mathcal{O}(J_2)^2$, a reason that makes quite relevant the study of the so-called "main problem" of AST (Brouwer 1959), where all harmonic coefficients are neglected except J_2. However, for earth artificial satellites second order effects are soon apparent, thus making unsatisfactory the simplifications provided by the main problem. To better approximate the real behavior, second order effects of the earth gravitational potential must be taken into account. Still, approximate analytical solutions can be computed using perturbation theory.

Among the different perturbation methods (Sanders et al. 2007), one particularly successful in AST is the method of averaging, which was first used by Lagrange in celestial mechanics and much later by van der Pol in relation to perturbed harmonic oscillators (Mitropolsky 1967). It is quite intuitive that short-period effects add "noise" over the long-term evolution of a dynamical system, so this noise can be "filtered" when concerned about the long-term dynamics. This is illustrated in Fig. 1, where the disturbing effects of the earth's oblateness and the moon's mass point attraction on the eccentricity of a fictitious GNSS-type satellite with $a = 28,560$ km and $I = 56°$ are displayed. As shown in the figure, a clear trend in the evolution of the eccentricity uncovers from the "noise". Indeed, when the high frequencies of the motion are filtered, a simple oscillatory behavior with small amplitude and half the period of the moon orbit is clearly observed (the white dashed line). Besides, a

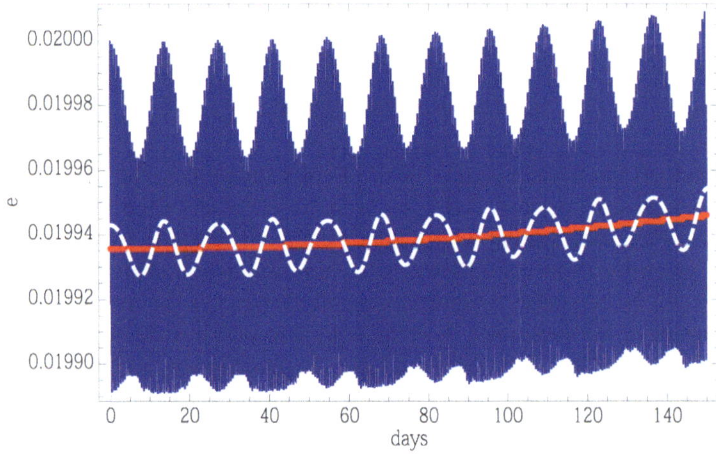

Fig. 1 Sample eccentricity time history (*blue line*) of a GNSS-type satellite under the earth's oblateness perturbation and the moon's disturbance effects

secular increase of the eccentricity along the half year propagation is also apparent (the full, red line) in Fig. 1.

From a mathematical perspective, this "noise filtering," or averaging, can be rigorously made by means of a transformation of variables. In the Hamiltonian frame, the method of Poincaré (1892) computes this transformation from a generating function in mixed coordinates, in the style of the Hamliton-Jacobi method, which is given by a (not necessarily convergent) series expansion in powers of a small parameter. Because of the implicit character of the generating function computed in Poincaré's method, the computations are generally difficult to extend further than the first order, although these difficulties are claimed to be overcame by using sophisticated inversion techniques (Deprit and Deprit 1999). Alternatively, the use of the Lie transforms method is customary these days, which provides the transformation explicitly in the new variables and is easily automated to compute higher orders of the solution by machine (Hori 1966; Deprit 1969).

A paradigmatic perturbation solution to AST was given by Brouwer, who relied in von Zeipel's modifications to the method of Poincaré (von Zeipel 1916, 1917, 1918), and worked in Delaunay variables (Brouwer 1959). Brouwer started from a gravitational model that consists of the first few zonal harmonics, thus enjoying axial symmetry and, for this reason, accepting the polar component of the angular momentum as an integral.

By carrying out two consecutive canonical transformations, Brouwer computed a solution that comprises the secular terms up to the second order of J_2 and the periodic terms limited to the first order. The first transformation, from osculating Delaunay variables to mean variables $(\ell', g', h', L', G', H')$, eliminates the short-period terms from the Hamiltonian. Paying attention only to the first two zonal harmonics of Brouwer's solution, it is found

$$\begin{aligned}\langle\mathscr{H}\rangle_\ell = &-\frac{\mu}{2a} - \frac{\mu}{2a}\frac{1}{4}J_2\frac{\alpha^2}{p^2}(4-6s^2)\eta - \frac{\mu}{2a}\frac{3}{4}J_3\frac{\alpha^3}{p^3}(4-5s^2)s\,\eta\,e\sin\omega \\ &-\frac{\mu}{2a}\frac{3}{64}J_2^2\frac{\alpha^4}{p^4}\eta\left[m_{0,0}+m_{0,1}\eta+m_{0,2}\eta^2-(28-30s^2)s^2e^2\cos 2\omega\right],\end{aligned} \tag{27}$$

where $m_{0,0} = 5(8-16s^2+7s^4)$, $m_{0,1} = (4-6s^2)^2$, $m_{0,2} = -8c^2+5s^4$, and a, p, s, η, e, and ω are assumed to be functions of the mean variables.

Brouwer only computed first order corrections of the transformation from osculating to mean Delaunay variables, which can be expressed as multivariate Fourier series of the form

$$\Delta = \sum_j\sum_k P(e)Q(I)\,\mathrm{trig}(jf+kg), \qquad \mathrm{trig}\equiv\sin,\cos, \tag{28}$$

where $P(e)$ and $Q(I)$ are eccentricity and inclination polynomials, respectively, contrary to truncated power series, because the transformation is computed in closed form. When expressed in this form, the short-period corrections in Eq. (28) need to evaluate 20 different trigonometric terms. For instance, the typical output of a computer algebra system for the first order correction to ℓ is

$$\Delta\ell = -\frac{1}{4}\frac{\alpha^2}{a^2}J_2\frac{1}{\eta^3}\left\{(2-3s^2)\left[\left(\frac{3}{e}-\frac{3}{4}e\right)\sin f+\frac{3}{2}\sin 2f+\frac{e}{4}\sin 3f\right]\right. \tag{29}$$
$$+s^2\left[-\frac{3}{8}e\sin(-f+2g)-\frac{9}{4}\sin 2g-\frac{3}{2}\left(\frac{5}{4}e+\frac{1}{e}\right)\sin(f+2g)\right.$$
$$\left.\left.-\left(\frac{e}{8}-\frac{7}{2e}\right)\sin(3f+2g)+\frac{9}{4}\sin(4f+2g)+\frac{3}{8}e\sin(5f+2g)\right]\right\}.$$

However, Brouwer cleverly arranged the transformation using $r = p/(1+e\cos f)$ as an auxiliary function, which is just a variable of the polar-nodal set, in this way reducing the number of circular functions to be evaluated to just 8. In particular, Brouwer's expression of the first order correction to ℓ simplifies to

$$\Delta\ell = -\frac{1}{2}\frac{\alpha^2}{a^2}J_2\frac{1}{e\eta}\left\{\left(1-\frac{3}{2}s^2\right)\left(\frac{a^2}{r^2}\eta^2+\frac{a}{r}+1\right)\sin f\right. \tag{30}$$
$$\left.+\frac{3}{4}s^2\left[\left(-\frac{a^2}{r^2}\eta^2-\frac{a}{r}+1\right)\sin(f+2g)+\left(\frac{a^2}{r^2}\eta^2+\frac{a}{r}+\frac{1}{3}\right)\sin(3f+2g)\right]\right\},$$

which is obviously equivalent to Eq. (29).

A drawback of the transformation of the Delaunay variables is the appearance of the eccentricity in denominators in the correction $\Delta\ell$ and Δg, with the consequent deterioration of the precision when the corrections are applied to the lower eccentricity orbits.

As a final remark, it deserves mentioning that Brouwer's elegant simplifications are further shortened when expressed in the full set of polar-nodal variables. Indeed, a standard simplification with the help of a computer algebra system gives

$$\Delta\ell = -\frac{1}{4}\frac{\alpha^2}{a^2}J_2\frac{1}{e\eta^3}\left[\frac{\sigma}{e}\left(3+3\kappa-\sigma^2\right)\left(2-3s^2\right)\right. \tag{31}$$
$$\left.+\frac{\sigma}{e}\left(5+9\kappa+4\kappa^2+\sigma^2\right)s^2\cos 2\theta+2\frac{\kappa}{e}\eta^2 s^2\sin 2\theta\right],$$

where $e=\sqrt{\kappa^2+\sigma^2}$, with $\sigma = pR/\Theta$, $\kappa = -1+p/r$, and $p = \Theta^2/\mu$. As expected, this new expression neither avoids the trouble with low eccentricity orbits, but it reduces the number of trigonometric expressions to be evaluated in the computation of short-period corrections to just 4.

Because the mean anomaly is ignorable in $\langle\mathscr{H}\rangle_\ell$, the mean L is constant (as well as the mean H, which was already constant in the original Hamiltonian). Hence, the second order truncation given in Eq. (27) is integrable. Brouwer integrated it by means of a second canonical transformation, from mean variables to action-angle variables $(\ell'', g'', h'', L'', G'', H'')$, that removes the long-period terms associated with the argument of the perigee dynamics. The resulting, secular Hamiltonian

$$\langle \mathscr{H} \rangle_{\ell,g} = -\frac{\mu}{2a} - \frac{\mu}{2a}\frac{1}{4}J_2\frac{\alpha^2}{p^2}(4-6s^2)\eta - \frac{\mu}{2a}\frac{3}{64}J_2^2\frac{\alpha^4}{p^4}\eta\left(m_{0,0}+m_{0,1}\eta+m_{0,2}\eta^2\right), \tag{32}$$

where, now, all the symbols, viz. a, p, s, η, e, and ω are assumed to be functions of the constant actions L'', G'', H'', is trivially integrable.

Again, the eccentricity appears in denominators of the first order, long-period corrections of the mean anomaly and the argument of the perigee, now called $\delta\ell$ and δg, respectively. Besides, the sine of the inclination now appears in denominators of δg, and δh. Furthermore, the denominator $1-5c^2$ appears in all the corrections. For instance, the long-period correction to the argument of the perigee is given by

$$\begin{aligned}\delta g &= \frac{\alpha^2}{p^2}\frac{J_2}{1-5c^2}\left[\left(\frac{17}{20}-\frac{17}{8}s^2+\frac{45}{32}s^4-\frac{1}{10(1-5c^2)}\right)e^2-\frac{14-15s^2}{16}s^2\right]\sin 2\omega \\ &\quad -\frac{\alpha}{p}\frac{J_3}{J_2}\frac{1}{2}\left(\frac{s}{e}-\frac{e}{s}c^2\right)\cos\omega,\end{aligned} \tag{33}$$

where all the three mentioned singularities are clearly apparent.

Note, however, the different character of theses singularities: while e and s are *virtual* singularities associated to the variables used in the transformation (Henrard 1974), the critical divisor $1-5c^2$ is an essential singularity associated to a physical resonance (Jupp 1975; Coffey et al. 1986; Lara 2015b). Thus, the trouble of the deterioration of the periodic corrections for low eccentricity orbits and also for low inclination orbits is easily solved by reformulating Brouwer's corrections in Poincaré canonical elements, which are non-singular variables (Lyddane 1963). Quite on the contrary, the fact that the second transformation converts the modulus of the angular momentum in a formal integral prevents its application to orbits whose total angular momentum is affected of secular or very long-period effects, which is precisely the case of orbits at the critical inclination. Therefore, orbits close to the critical inclination simply fall out of Brouwer's approach, and their integration requires specific treatment (Garfinkel 1966; Jupp 1972; Henrard and Wauthier 1988).

By recalling the invariance of Poisson brackets, in which canonical perturbation theory relies, Izsak realized that Brouwer's short-period transformation can be implemented in unspecified variables (Izsak 1963). In particular, he pointed out the benefits of reformulating the generating function of the short-period elimination in polar-nodal, contrary to Delaunay variables. The Hamiltonian in mean elements remains in Delaunay (mean) variables, in this way being released of short-period effects, but now the short-period corrections of the polar-nodal variables are computed.

Izsak's way of proceeding provides two main advantages: On the one hand the trouble with low eccentricity orbits is completely avoided. On the other hand, the short-period corrections adopt a compact form that only needs the evaluation of two different trigonometric functions: $\sin 2\theta$ and $\cos 2\theta$, cf. Eqs. (17)–(22) in Lara (2015a).

Later, Aksnes used Izsak's approach to compute the long-period transformation in polar-nodal variables (Aksnes 1972), thus removing any trouble with the low eccentricity orbits from Brouwer's gravitational solution. Concern with low inclination orbits still remains, but Aksnes suggested to treat this particular case computing directly the corrections to the geographic latitude and longitude. Alternatively, since satellite ephemeris do not need to be computed in canonical variables (Deprit and Rom 1970), periodic corrections of Brouwer's solution can be computed in the set of non-canonical, non-singular variables $\psi = \theta + \nu$, $\xi = s \sin\theta$, $\chi = s\cos\theta$, r, R, Θ, which keep similar evaluation performances to the polar-nodal corrections and hence can be competitive to the other formulations in non-singular variables (Lara 2015a).

4 Lunisolar Perturbations

The efficient evaluation of the short-period corrections is of much more concern when dealing with third-body perturbations. In the mass-point approximation the disturbing function is given by the usual expansion in Legendre polynomials, viz.

$$\mathscr{V}_3 = -\beta \frac{n_3^2 a_3^3}{r_3} \sum_{j\geq 2} (r/r_3)^j P_j(\cos\psi), \tag{34}$$

where ψ is the angle encompassed by the directions of the satellite, $\mathbf{x}$, and the third body, $\mathbf{r}_3$, and $\beta = m_3/(m_3 + m)$ is the third-body reduced mass. However, the convergence of this expansion when applied to lunar perturbations may be very slow for high altitude orbits. Indeed, a theory valid for highly eccentric orbits may require taking into account up to the sixth degree Legendre polynomial or even higher degrees, the corresponding short-period corrections being made of very long series.

For instance, even when dealing only with the lowest degree of Eq. (34), $j = 2$, the corresponding term of the generating function in Delaunay variables in closed form, up to the first order of the theory, is

$$W_{1,2} = L \frac{n_3^2}{n^2} \beta \frac{a_3^3}{r_3^3} \sum_{j=0}^{1} \sum_{l=-2}^{2} \sum_{\substack{m=-3 \\ m\neq 0}}^{3} C_{2,j,m}(e)\, Q_{2,j,l}(I) \tag{35}$$

$$\times \left[T_{2,l}(\hat{\mathbf{r}}_3) \cos(2j\omega + l\Omega + mu) - S_{2,l}(\hat{\mathbf{r}}_3) \sin(2j\omega + l\Omega + mu) \right],$$

Table 1 Polynomials $T_{2,m}$, $S_{2,m}$, $Q_{2,j,m}$, and $C_{2,j,m}$ in Eq. (35), cf. Lara et al. (2015)

m	0	± 1	± 2	± 3
$S_{2,m}$	$-1 + 3w_3^2$	$-v_3\,w_3$	$u_3^2 - v_3^2$	
$T_{2,m}$	0	$\pm u_3\,w_3$	$\pm 2u_3\,v_3$	
$Q_{2,0,m}$	$-(1/48)(1-3c^2)$	$(1/8)c\,s$	$-(1/32)s^2$	
$Q_{2,1,m}$	$-(1/16)s^2$	$\pm(1/8)(1\pm c)\,s$	$(1/32)(1\pm c)^2$	
$C_{2,0,m}$		$\pm(3/4)e(5+3\eta^2)$	$\mp(9/4)e^2$	$\pm(1/4)e^3$
$C_{2,1,m}$		$\mp(15/4)e(1\pm\eta)^2$	$\pm(3/4)(1\pm\eta)^2(3\mp 2\eta)$	$\mp(1/4)e(1\pm\eta)^2$

where $\hat{\mathbf{r}}_3 = (u_3, v_3, w_3) = (1/r_3)(x_3, y_3, z_3)$, is the third-body direction, u is the eccentric anomaly, and $C_{2,j,m}$, $Q_{2,j,l}$, and $T_{2,l}$ and $S_{2,l}$ are the eccentricity, inclination, and moon direction polynomials shown in Table 1. Recall that the symbols n, e, η, s, c, ω, Ω, and u, are not variables but corresponding functions of Delaunay variables.

As in Brouwer's solution, the short-period corrections to the Delaunay variables derived from W_1 may suffer again troubles for low e and I. Besides, due to the third-body disturbances the perturbation model does not enjoy axial symmetry any longer, and W_1 is a long Fourier series with 81 different trigonometric terms, resulting in a similar number of terms for each of the correction series. However, notably savings are obtained if W_1 is reformulated in polar-nodal variables.

Indeed, for a given degree n of the Legendre polynomial of the third-body expansion, P_n, the general form of the first order terms of the generating function is

$$V_{1,n} = \Theta \frac{a_3^3}{a^3}\beta \frac{n_3^2}{n^2}\left(\frac{r}{r_3}\frac{1}{\eta^2}\right)^{n+1} \sum_{j=0}^{m}\sum_{l=-n}^{n} Q_{n,j,l} \times\Big[(k_{n,j}\,T_{n,l} - q_{n,j}\,S_{n,l})\cos(2j\theta + l\nu) - (k_{n,j}\,S_{n,l} + q_{n,j}\,T_{n,l})\sin(2j\theta + l\nu)\Big], \tag{36}$$

where m is the integer part of $n/2$; now a, n, s, c, and η are functions of the polar-nodal variables. For $n = 2$, $V_{1,2}$ is made only of 14 trigonometric terms, and in spite of the eccentricity-type polynomials $k_{2,j}$ and $q_{2,j}$ in Table 2 are now much more involved than the eccentricity polynomials $C_{2,j,m}$ in Table 1, since they only involve arithmetic operations their evaluation is very fast. In consequence, the effort in evaluating the short-period corrections in polar-nodal variables, as derived from a generating function of terms $V_{1,n}$, reduces dramatically. Besides, there is no loose of precision in the short-period corrections of low eccentricity orbits, although trouble for low I may remain. Anyway, the alternatives mentioned in the previous section can be used to avoid the singularity of the polar-nodal variables for equatorial orbits.

Table 2 Eccentricity-type polynomials in Eq. (36); $k_{2,0} = 0$, cf. Lara et al. (2015)

$q_{2,0} = \eta^2\,\sigma\left[3(1+\kappa)(4+\kappa) - 2\sigma^2\right]$
$q_{2,1} = \sigma\left[(1+\kappa)^2\left(8+29\kappa+3\kappa^2\right) - (1+\kappa)(6+19\kappa)\sigma^2 - 2\sigma^4\right]$
$k_{2,1} = (1+\kappa)\left[(1+\kappa)^2\left(3-16\kappa+3\kappa^2\right) + 3(1+\kappa)(3+7\kappa)\sigma^2 - 2\sigma^4\right]$

5 Conclusions

Despite claims to the contrary, semianalytic propagation of orbital motion may be still useful in current engineering problems, where the important role played by the (instantaneous) plane of motion makes that canonical sets which explicitly include the total angular momentum among the variables are specially useful in artificial satellite theory. This is exactly the case of polar-nodal and Delaunay variables. In particular, analytic and semianalytic theories are properly constructed in Delaunay variables, but to avoid inaccurate evaluation of the short period corrections, which may happen in the case of low eccentricity orbits, satellite ephemeris are better computed in polar-nodal variables. Formulation of the periodic corrections in polar-nodal variables has the added benefit of dramatically reducing the size of the perturbation series, with the consequent improvement of the performances of the theory.

The propagation of very low inclination orbits is not generally a major concern in artificial satellite theory. When it is, the singularities which still remain in the polar-nodal variables formulation can be avoided using different sets of nonsingular variables. In particular, a simple combination of the polar-nodal variables results in a nonsingular set of non-canonical variables with similar evaluation performances that the polar-nodal set.

Acknowledgements Partial support from projects ESP2013-41634-P and ESP2014-57071-R of the Ministry of Economic Affairs and Competitiveness of Spain is recognized.

References

Aksenov, Y.P., Grebenikov, Y.A., Demin, V.G.: General solution of the problem of the motion of an artificial satellite in the normal field of the earth's attraction. Planet. Space Sci. **9**(8), 491–498 (1962)

Aksnes, K.: A note on 'The main problem of satellite theory for small eccentricities, by A. Deprit and A. Rom, 1970'. Celest. Mech. **4**, 119–121 (1971)

Aksnes, K.: On the use of the Hill variables in artificial satellite theory. Astron. Astrophys. **17**, 70–75 (1972)

Alfriend, K.T., Coffey, S.L.: Elimination of the perigee in the satellite problem. Celest. Mech. **32**(2), 163–172 (1984)

Armellin, R., San-Juan, J.F., Lara, M.: End-of-life disposal of high elliptical orbit missions: the case of INTEGRAL. Adv. Space Res. **56**(3), 479–493 (2015)

Arnold, V.I.: Mathematical Methods of Classical Mechanics, 2nd edn. Springer, New York (1989)
Brouwer, D.: Solution of the problem of artificial satellite theory without drag. Astron. J. **64**, 378–397 (1959)
Brouwer, D., Clemence, G.M.: Methods of Celestial Mechanics. Academic, New York (1961)
Cid, R., Lahulla, J.F.: Perturbaciones de corto periodo en el movimiento de un satélite artificial en función de las variables de Hill. Publ. Rev. Acad. Cienc. Zaragoza **24**, 159–165 (1969)
Cid, R., Lahulla, J.F.: Perturbaciones de segundo orden y corto periodo, para el movimiento de un satélite artificial, en las variables de Hill. Publ. Rev. Acad. Cienc. Zaragoza **26**(2), 333–343 (1971)
Cid, R., Ferrer, S., Sein-Echaluce, M.L.: On the radial intermediaries and the time transformation in satellite theory. Celest. Mech. **38**(2), 191–205 (1986)
Coffey, S., Deprit, A.: Third-order solution to the main problem in satellite theory. J. Guid. Control Dyn. **5**(4), 366–371 (1982)
Coffey, S.L., Deprit, A., Miller, B.R.: The critical inclination in artificial satellite theory. Celest. Mech. **39**(4), 365–406 (1986)
Danielson, D.A., Sagovac, C.P., Neta, B., Early, L.W.: Semianalytic Satellite Theory. Technical Report NPS-MA-95-002, Naval Postgraduate School, Monterey, CA. Dep of Math (1995)
Delaunay, C.E.: La Théorie du Mouvement de la Lune, Premier volume. Mem. Acad. Sci. Inst. Imp. Fr., vol. 28. Mallet-Bachellier, Paris (1860)
Deprit, A.: Canonical transformations depending on a small parameter. Celest. Mech. **1**(1), 12–30 (1969)
Deprit, A.: The elimination of the parallax in satellite theory. Celest. Mech. **24**(2), 111–153 (1981)
Deprit, A., Ferrer, S.: Simplifications in the theory of artificial satellites. J. Astronaut. Sci. **37**, 451–463 (1989)
Deprit, A., Rom, A.: The main problem of artificial satellite theory for small and moderate eccentricities. Celest. Mech. **2**(2), 166–206 (1970)
Deprit, E., Deprit, A.: Poincaré's méthode nouvelle by skew composition. Celest. Mech. Dyn. Astron. **74**, 175–197 (1999)
Ferrer, S., Lara, M.: Families of canonical transformations by Hamilton-Jacobi-Poincaré equation. Application to rotational and orbital motion. J. Geom. Mech. **2**(3), 223–241 (2010)
Garfinkel, B.: The orbit of a satellite of an oblate planet. Astron. J. **64**(9), 353–367 (1959)
Garfinkel, B.: Formal solution in the problem of small divisors. Astron. J. **71**, 657–669 (1966)
Golikov, A.R.: THEONA—a numerical-analytical theory of motion of artificial satellites of celestial bodies. Cosm. Res. **50**(6), 449–458 (2012)
Gurfil, P., Lara, M.: Satellite onboard orbit propagation using Deprit's radial intermediary. Celest. Mech. Dyn. Astron. **120**, 217–232 (2014)
Henrard, J.: Virtual singularities in the artificial satellite theory. Celest. Mech. **10**(4), 437–449 (1974)
Henrard, J., Wauthier, P.: A geometric approach to the ideal resonance problem. Celest. Mech. **44**, 227–238 (1988)
Hoots, F.R.: Reformulation of the Brouwer geopotential theory for improved computational efficiency. Celest. Mech. **24**, 367–375 (1981)
Hoots, F.R., Roehrich, R.L.: Models for Propagation of the NORAD Element Sets. Project SPACETRACK, Rep 3, US Air Force Aerosp Def Command, Colorado Springs, CO (1980)
Hori, G.: Theory of general perturbation with unspecified canonical variables. Publ. Astron. Soc. Jpn. **18**(4), 287–296 (1966)
Izsak, I.G.: A note on perturbation theory. Astron. J. **68**, 559–561 (1963)
Jupp, A.H.: A second-order solution of the ideal resonance problem by Lie series. Celest. Mech. **5**, 8–26 (1972)
Jupp, A.H.: The problem of the critical inclination revisited. Celest. Mech. **11**(3), 361–378 (1975)
Kaula, W.M.: Theory of Satellite Geodesy. Applications of Satellites to Geodesy. Blaisdell, Waltham (1966)
Kelso, T.J.: Analysis of the Iridium 33–Cosmos 2251 collision. In: Paper AAS 09-368, Am. Astronaut. Soc. (2009)

Kinoshita, H.: Third-Order Solution of an Artificial-Satellite Theory. SAO Special Rep 379 (1977)
Kislik, M.D.: The path of an artificial earth satellite in the normal gravitational field of the earth. Planet. Space Sci. **8**, 86–96 (1961)
Kozai, Y.: The motion of a close earth satellite. Astron. J. **64**(11), 367–377 (1959)
Kozai, Y.: Second-order solution of artificial satellite theory without air drag. Astron. J. **67**(7), 446–461 (1962)
Lara, M.: Efficient formulation of the periodic corrections in Brouwer's gravity solution. Math. Probl. Eng. **2015**, Article ID 980652 (2015a)
Lara, M.: On inclination resonances in artificial satellite theory. Acta Astronaut. **110**, 239–246 (2015b)
Lara, M., San-Juan, J.F., López-Ochoa, L.M.: Averaging tesseral effects: closed form relegation versus expansions of elliptic motion. Math. Probl. Eng. **2013**, Article ID 570127 (2013)
Lara, M., San-Juan, J.F., López-Ochoa, L.M.: Delaunay variables approach to the elimination of the perigee in artificial satellite theory. Celest. Mech. Dyn. Astron. **120**, 39–56 (2014a)
Lara, M., San-Juan, J.F., López-Ochoa, L.M.: Proper averaging via parallax elimination. Adv. Astronaut. Sci. **150**, 315–331 (2014b)
Lara, M., de Moraes, R.V., Sanchez, D.M., Prado, A.F.B.A.: Efficient computation of short-period analytical corrections due to third-body effects. Adv. Astronaut. Sci. **155**, 437–455 (2015)
Lyddane, R.H.: Small eccentricities or inclinations in the Brouwer theory of the artificial satellite. Astron. J. **68**, 555–558 (1963)
McClain, W.D.: A Recursively Formulated First-Order Semianalytic Artificial Satellite Theory Based on the Generalized Method of Averaging, Volume 1: The Generalized Method of Averaging Applied to the Artificial Satellite Problem, 2nd edn. NASA CR-156782. NASA, Greenbelt (1977)
Mitropolsky, I.: Averaging method in non-linear mechanics. Int. J. Non Linear Mech. **2**, 69–96 (1967)
Osácar, C., Palacián, J.F.: Decomposition of functions for elliptical orbits. Celest. Mech. Dyn. Astron. **60**, 207–223 (1994)
Poincaré, H.: Les methodes nouvelles de la mecanique celeste. Gauthier-Villars, Paris (1892)
Sanders, J.A., Verhulst, F., Murdock, J.: Averaging Methods in Nonlinear Dynamical Systems, 2nd edn. Springer, New York (2007)
Setty, S.J., Cefola, P.J., Montenbruck, O., Fiedler, H., Lara, M.: Investigating the suitability of analytical and semi-analytical satellite theories for space object catalogue maintenance in geosynchronous regime. Adv. Astronaut. Sci. **150**, 1027–1042 (2014)
Vallado, D.A., Crawford, P., Hujsak, R., Kelso, T.S.: Revisiting spacetrack report #3. In: Paper AIAA 2006-6753, Am. Inst. Aeronaut. Astronaut. (2006)
Vinti, J.P.: New method of solution for unretarded satellite orbits. J. Res. Natl. Bur. Stand. **62B**(2), 105–162 (1959)
von Zeipel, H.: Research on the motion of minor planets (recherches sur le mouvement des petites planètes). NASA Transl: NASA TT F-9445 (1965) (1916, 1917, 1918).
Whittaker, E.T.: A Treatise on the Analytical Dynamics of Particles and Rigid Bodies, 1st edn. Cambridge University Press, Cambridge (1904)

Small Spacecraft Formation Flying Using Solar Radiation Pressure

Junquan Li

Abstract Solar sails have long been considered as a viable mechanism for interplanetary travel by using light from the Sun to accelerate a spacecraft without propellant. They can be used for missions studying astrophysics, heliophysics and planetary science. A formation flying system using solar radiation pressure is presented in this paper. The proposed formation flying mission uses sails due to the limited volume and power of small spacecraft. Station keeping, pitch attitude slew and yaw attitude slew are accomplished using reflectivity control devices. A nonlinear non-affine controller that provides smooth control performance is applied for formation flying of small spinning spacecraft while considering gravity gradient torques for the attitude dynamics. The performance of the formation flying is numerically demonstrated.

1 Introduction

Solar sailing has been considered as a sustainable method of propulsion for enabling a highly non-Keplerian orbit about an asteroid or other displaced orbits, a sample return mission for deep space exploration, and spacecraft formation flying operations. The NASA Magnetospheric Multiscale (MMS) mission uses four spinning spacecraft in formation to study the heliophysics of magnetic reconnection (NASA 2015). Data from the MMS spacecraft could solve unanswered questions of how the magnetic field realigns and releases energy (NASA 2015). Such missions usually require propellant and consequently have a limited mission life. Sailcraft formation flying has been proposed for achieving long term orbit times in the Earth's magnetotail (Guzman and Edery 2004) using small non-spinning sailcraft. The system design of the small MMS sailcraft are described in Lappas et al. (2007), Macdonald et al. (2003), Lapps et al. (2009), Macdonald et al. (2007), and Cay et al. (2015). In future, constellations of low cost smaller spacecraft carrying electric field instruments, magnetorquers and particle detectors will be needed for

J. Li (✉)
Clyde Space Ltd., Glasgow, UK
e-mail: junquan.li@clyde-space.com

G. Gómez, J.J. Masdemont (eds.), *Astrodynamics Network AstroNet-II*,
Astrophysics and Space Science Proceedings 44,
DOI 10.1007/978-3-319-23986-6_12

the exploration of a broader region of space (Norton et al. 2014). Given the limited capacities of small spacecraft, formation flying or constellations are usually needed. In our research, a system uses two small spinning spacecraft (23×24×36 cm, 12U CubeSat form factor) performing formation flying using Solar Radiation Pressure for propulsion and control. The concept of Solar Radiation Pressure (SRP) was first raised by Sohn (Sohn 1959). Many researchers have done active attitude control research for satellites using SRP (Joshi and Kumar 1979; Kumar et al. 2006). The spinning sail described in this paper uses a reflectivity control method similar to that tested in the IKAROS mission (Tsuda et al. 2013), in which variable reflectivity materials were used to control the attitude of a spinning sail (Funase et al. 2011). A solar sail with a variable surface reflectivity distribution was also investigated in Borggrafe et al. (2013). In Mu et al. (2015), formation flying station keeping was completely controlled by reflectivity control devices (RCD) and Solar Radiation Pressure (SRP) using an LQR controller. In this paper, we present the mission concept of a 15.6 kg class small spacecraft using solar radiation pressure in a scenario similar to the MMS mission. The use of small spacecraft with reflectivity control devices (RCD) and using Solar Radiation Pressure (SRP) instead of fuel could lead to a mission life time of much more than 5 years. The proposed sail with reflective control devices (RCD) can allow a small spacecraft to study the space environment using on-board magnetometers and plasma analyzer instruments.

2 Spacecraft System Design

The proposed small spacecraft with solar sails is designed to have a spin stabilization before starting a controllable sail deployment. The sail actuator for control of the spin axis is a reaction wheel, as a fixed sail cannot directly control rotation about the spin axis. The other two axes are controlled by RCDs on the sail for the formation flying. The sail folding and controlled deployment using motors is assumed, and aluminized Kapton film is chosen for the sail instead of Mylar due to the space environment conditions. Figure 1 shows the solar sail spacecraft formation flying for Magnetospheric Multiscale Mission. The Earth centered inertial (ECI) frame, denoted by $E-XYZ$, with its origin located at the center of the Earth is as shown in Fig. 1. ω and f are the argument of perigee and true anomaly, respectively. The local vertical local horizontal (LVLH) orbital reference frame is $L - x_0\, y_0\, z_0$. $B - x_b\, y_b\, z_b$ denotes the corresponding principal body-fixed coordinate axes of the satellite. The sunlight pointing reference frame is $L - x_s\, y_s\, z_s$. Spacecraft attitude is represented using the Euler angles, $(\alpha,\ \phi,\ \gamma)$ in Fig. 2.

The sail area and the total mass of the small spacecraft relation can be written as

$$A = \frac{a_c m_{p+b}}{\eta P_s - a_c \sigma_f} \tag{1}$$

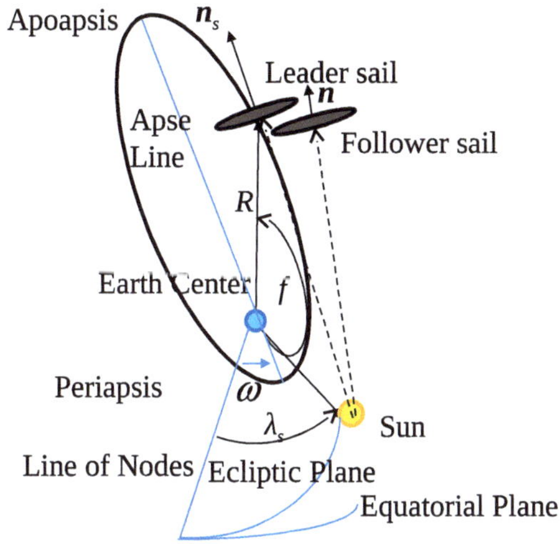

Fig. 1 Earth magnetic tail monitoring mission formation flying

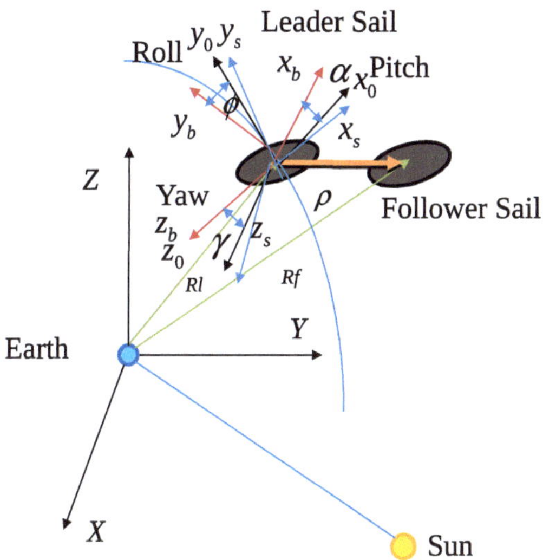

Fig. 2 Earth magnetic tail monitoring mission formation flying reference frame

where A is the area of the sail required to achieve an acceleration for the platform mass m_{p+b} (boom and structure not the sail film). $\eta = 0.9 \times 2$ is the sail efficiency. σ_f is the area density of the film. P_s is the solar radiation pressure at 1AU (4.563×10^{-6} N/m^2). a_c is the acceleration. The proposed sail system has the characteristics in Table 1.

Table 1 Sail characteristics

Sail parameters	Values
Sail area	262.43 m^2
Sail film mass	3.10 kg
Boom length	9.14 m
Boom total mass	2.56 kg
Bus and payload	9.944 kg
Total spacecraft mass	15.6 kg
Sail and boom moment of inertia	[118.12 118.12 236.24] kg/m^2
Acceleration	0.138 mm/s^2

3 Solar Radiation Pressure Model with Reflectivity Control Devices

The solar radiation pressure exerted on a non-perfect solar sail is obtained by considering reflection, absorption and re-radiation by the sail and is also parametrized by a number of coefficients representing the optical properties of the sail (Schaub and Junkin 2003). On a spinning absorption mode sailcraft, the RCD is assumed to consist of many cells with switch on and off independently (Mu et al. 2015). The amount of control that can be exerted on the sail is determined by the amount of sail area covered with reflectivity control devices (RCDs) relative to the total sail area. This is quantified by the maximum reflectivity modulation ratio u_{max} and the total area of the sail covered with RCDs. In this work we assume that the sail has ideal specular reflectivity, and the RCDs are normally assumed to be switched on, in which state they have the same reflectivity as the rest of the sail. Switching off a given area of RCDs in uniform rings or balanced patterns with respect to the center of the sail has the effect of reducing the reflective area and the corresponding sail thrust without applying a torque, and this reduced reflective area is quantified by the control parameter u, where $0 \leq u \leq u_{max}$.

For the sail with the area of the switched on surface RCD, the SRP force can be expressed as

$$\mathbf{F}_{SRPon} = 2p\Delta s\,|(\mathbf{s}\cdot\mathbf{n})\,|\,\{b_{aon}\mathbf{s} + [b_{bon}\,(\mathbf{s}\cdot\mathbf{n}) + b_{con}]\,\mathbf{n}\} \tag{2}$$

where $\mathbf{s}$ represents the Sun vector, Δs is the normal part of the sail area, $\mathbf{n}$, is the solar sail normal vector, b_a, is the absorption factor, b_b, is the specular reflection factor, and b_c, is the diffusion reflection factor. b_{ion} is the factor when the RCD is on. b_{ioff} is the factor when the RCD is off. When the parts of the RCD are switched off, the SRP force can be expressed as

$$\mathbf{F}_{SRPoff} = 2p\Delta s\,|(\mathbf{s}\cdot\mathbf{n})\,|\,\left\{b_{aoff}\mathbf{s} + \left[b_{boff}\,(\mathbf{s}\cdot\mathbf{n}) + b_{coff}\right]\mathbf{n}\right\} \tag{3}$$

The force difference between the off and on states is

$$\Delta\mathbf{F}_{SRP} = 2p\Delta s\,|(\mathbf{s}\cdot\mathbf{n})|\,\{\Delta b_a\mathbf{s} + [\Delta b_b\,(\mathbf{s}\cdot\mathbf{n}) + \Delta b_c]\,\mathbf{n}\} \tag{4}$$

Since the sail area with all RCDs switched on is the same as a normal sail, the total force on the sail can be written as

$$\begin{aligned}\mathbf{F}_{SRP} &= 2upA_s\,|(\mathbf{s}\cdot\mathbf{n})|\,\left\{b_{aoff}\mathbf{s} + \left[b_{boff}\,(\mathbf{s}\cdot\mathbf{n}) + b_{coff}\right]\mathbf{n}\right\}\\ &\quad + 2(1-u)pA_s\,|(\mathbf{s}\cdot\mathbf{n})|\,\{\Delta b_a\mathbf{s} + [\Delta b_b\,(\mathbf{s}\cdot\mathbf{n}) + \Delta b_c]\,\mathbf{n}\}\end{aligned} \tag{5}$$

To control the attitude of the spacecraft, the sail is divided into four quadrants centrally aligned with the x and y axes in the spacecraft body frame. For each quadrant, an area ratio $u_{x+}, u_{x-}, u_{y+}, u_{y-}$ is defined as the area of all switched-off RCDs relative to the total area of all RCDs in the quadrant. This ratio must be weighted for each RCD with the distance function r_d of the RCD to the sail axis about which torque is created. $\tilde{r}_d$ is the average distance of all RCDS in each half plane to the y axis.

$u_x = u_{x+} - u_{x-}$ and $u_y = u_{y+} - u_{y-}$ are control variables determined by the difference between these ratios opposite each other on the sail, and each quadrant can contribute up to a switched-off area of $1/4u_{max}$, so that $-1/4u_{max} \le u_x \le 1/4u_{max}$ and $-1/4u_{max} \le u_y \le 1/4u_{max}$. When the quadrant representing u_{x-} is made low reflectivity the control variable u_x becomes $-1/4u_{max}$ as shown in Fig. 3, and when the RCDs in the quadrant representing u_{x+} are made low reflectivity then u_x becomes $1/4u_{max}$ as shown in Fig. 3. The sail is symmetric, so the same is true of u_y, and combined control is possible by simply combining u_x and u_y as shown in Fig. 3. Fine pointing control can be achieved by controlling the proportion of sail area that is made non-reflective.

Since two opposing RCDs are located with respect to one axis of the sail at a distance r_d in opposite directions, the difference in solar radiation pressure generated between them creates a torque on the sail

$$\mathbf{T}_d = (\Delta\mathbf{F}_{SRP}\cdot\mathbf{n})r_d = 2p\Delta s cos\alpha_s\left[cos\alpha_s(\Delta b_a + \Delta b_b + \Delta b_c)\right]\mathbf{r}_d \tag{6}$$

where α_s is the sun angle, and r_d is the distance of the switched on RCD with respect to the axis.

Considering the proposed system configuration along with Eqs. (6) and the absorption sail, it can be inferred that control torques are available only about the pitch ($T_{s\alpha}$) and yaw ($T_{s\phi}$) axes by RCDs.

These torques are defined as

$$T_{s\alpha} = 2PA_s cos^2(\alpha_s) \times (\int_0^{u_{x+}} r_{dx+}(u)du - \int_0^{u_{x-}} r_{dx-}(u)du) \tag{7}$$

$$T_{s\gamma} = -2PA_s cos^2(\alpha_s) \times (\int_0^{u_{y+}} r_{dy+}(u)du - \int_0^{u_{y-}} r_{dy-}(u)du) \tag{8}$$

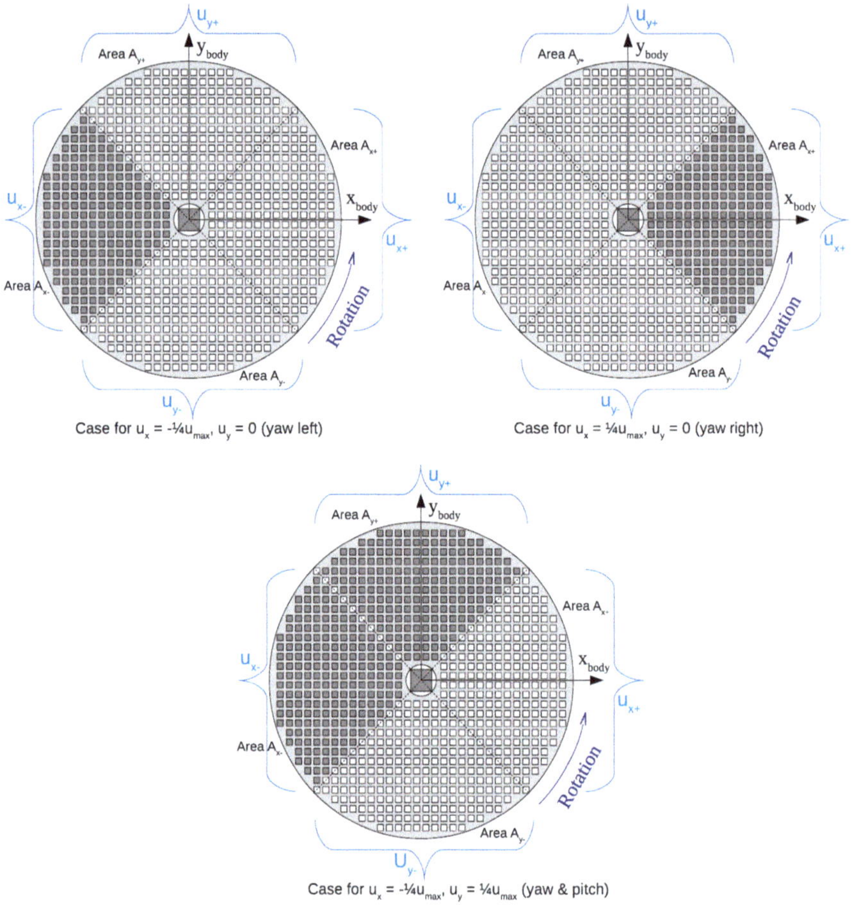

Fig. 3 RCDS sail attitude case 1 (*top left*), RCDS sail attitude case 2 (*top right*), and RCDS sail attitude case 3 (*bottom*)

where r_{dx+}, and r_{dx-} are the distance functions of each cell of RCDS to y axis. r_{dy+} and r_{dy-} are the distance functions of each cell of RCDS to x axis.

Define

$$u_x = -(\frac{1}{4}u_{max} - \int_0^{u_{x+}} r_{dx+}(u)du) + (\frac{1}{4}u_{max} - \int_0^{u_{x-}} r_{dx-}(u)du) \tag{9}$$

$$u_y = -(\frac{1}{4}u_{max} - \int_0^{u_{y+}} r_{dy+}(u)du) + (\frac{1}{4}u_{max} - \int_0^{u_{y-}} r_{dy-}(u)du) \tag{10}$$

Eqs. (9) and (10) can be rewritten as

$$u_x = -(\frac{1}{4}u_{max} - u_{x+}) + (\frac{1}{4}u_{max} - u_{x-}) \tag{11}$$

$$u_y = -(\frac{1}{4}u_{max} - u_{y+}) + (\frac{1}{4}u_{max} - u_{y-}) \tag{12}$$

Eqs. (7) and (8) can be rewritten as following

$$T_{s\alpha} = u_y PA_s\tilde{r}_d cos^2\alpha cos^2\gamma \tag{13}$$

$$T_{s\gamma} = -u_x PA_s\tilde{r}_d cos^2\alpha cos^2\gamma \tag{14}$$

For the absorption mode, Eq. (5) can be rewritten as the following:

$$\mathbf{F}_{SRP} = 2upA_s cos\alpha_s(u\mathbf{s} + 2(1-u)cos\alpha_s\mathbf{n}) \tag{15}$$

The acceleration exerted by solar radiation pressure can be written

$$\mathbf{a}_{SRP} = 2upA_s/m cos\alpha_s(u\mathbf{s} + 2(1-u)cos\alpha_s\mathbf{n}) \tag{16}$$

Define κ as the characteristic acceleration of the solar sail, Eq. (16) can be written as

$$\mathbf{a}_{SRP} = u\kappa cos\alpha_s(u\mathbf{s} + 2(1-u)cos\alpha_s\mathbf{n}) \tag{17}$$

For the MMS mission, two vectors expressed in the inertial frame can be written as

$$\mathbf{s} = [-cos\alpha_s, -sin\alpha_s, 0]^T \tag{18}$$

$$\mathbf{n} = [-cos\alpha cos(\gamma + \lambda_s), -cos\alpha sin(\gamma + \lambda_s), -sin\alpha]^T \tag{19}$$

The equation of motion of the sail in the inertial frame can be written as

$$\ddot{\mathbf{r}} = -\frac{\mu\mathbf{r}}{r^3} + \mathbf{a}_s \tag{20}$$

For the MMS mission, the rotation of the apse line of the orbit should be synchronous with the annual rotation of the sun.

$$\mathbf{a_s} = \frac{2}{3}\dot{\lambda}_s e\sqrt{\frac{\mu}{a(1-e^2)}}\mathbf{n} \tag{21}$$

where $\dot{\lambda}_s$ is the rate of the rotation of the sun. a and e are the semimajor axis and the eccentricity of the sailcraft's orbit. The characteristic acceleration of a solar sail κ is defined as the acceleration experienced by the sailcraft at the heliocentric distance of 1 AU. The sail normal is assumed to be along the sun line in the antisun direction.

$$\kappa = \frac{4}{-u+6}\dot{\lambda}_s \frac{e}{\sqrt{1-e^2}}\sqrt{\frac{\mu}{a}} \tag{22}$$

Assume $2PA_s/m = \kappa$ we can rewrite

$$T_{s\alpha} = \frac{1}{2}u_y \kappa m \tilde{r}_d cos^2\alpha cos^2\gamma \tag{23}$$

$$T_{s\gamma} = -\frac{1}{2}u_x \kappa m \tilde{r}_d cos^2\alpha cos^2\gamma \tag{24}$$

After carrying out algebraic manipulation and nondimensionalization, the nondimensionalized torque due to SRP is given by,

$$\hat{T}_{s\alpha} = \frac{1}{2}u_y \kappa m \tilde{r}_d cos^2\alpha cos^2\gamma \frac{\left(1-e^2\right)^3}{(1+e\cos f)^3} \tag{25}$$

$$\hat{T}_{s\gamma} = -\frac{1}{2}u_x \kappa m \tilde{r}_d cos^2\alpha cos^2\gamma \frac{\left(1-e^2\right)^3}{(1+e\cos f)^3} \tag{26}$$

4 Formation Flying

The small spinning spacecraft formation flying station keeping is controlled with Solar Radiation Pressure and RCDs. The equations of motion are the dynamics of a follower spacecraft relative to a leader spacecraft (Schaub and Junkin 2003). The motion of a follower spacecraft with respect to a leader spacecraft can be described with respect to a local vertical local horizontal coordinate frame attached to the leader. The position vector of the follower relative to the leader has components x, y and z, where x is the radial direction of the leader, z is the component in the direction of the leader's orbital angular momentum and y is the component along the direction completing the right handed orthogonal triad (assuming two body gravity).

The Spacecraft Formation Flying (SFF) equations of motion can be written as

$$\ddot{x} - 2\dot{f}(\dot{y} - y\frac{\dot{r}}{r}) - x\dot{f}^2 = -\mu\frac{x+r}{{r_d}^3} + \mu\frac{1}{r^2} + F_x \tag{27}$$

$$\ddot{y} + 2\dot{f}(\dot{x} - x\frac{\dot{r}}{r}) - y\dot{f}^2 = -\mu\frac{y}{{r_d}^3} + F_y \tag{28}$$

$$\ddot{z} = -\mu\frac{z}{r_d^3} + F_z \tag{29}$$

$r_d = \sqrt{(r+x)^2 + y^2 + z^2}$. $F = [F_x, F_y, F_z]^T$ are the control inputs provided by x, y, z solar sails.

The acceleration of x, y and z can be written as

$$F = R_z(-\lambda_s - f)(a_{sF} - a_{sL}) \tag{30}$$

where $F = R_z(-\lambda_s - f)[ax_{SRP}, ay_{SRP}, az_{SRP}]^T$. The acceleration of x, y and z can be written as

$$\begin{aligned} ax_{SRP} &= -(\kappa u)(cos\alpha)^2(cos\gamma)^2 + (\kappa(1-u))(cos\alpha)^2(cos\gamma)^2(-cos\alpha)cos(\gamma + \lambda_s) \\ ay_{SRP} &= -(\kappa u)cos\alpha cos\gamma sin\alpha sin\gamma + (\kappa(1-u))(cos\alpha)^2(cos\gamma)^2(-cos\alpha)sin(\gamma + \lambda_s) \\ az_{SRP} &= -(\kappa)(cos\alpha)^2(cos\gamma)^2 sin\alpha(1-u) \end{aligned} \tag{31}$$

The theoretical basis for developing the nonlinear control algorithms for the nonlinear non-affine mathematical model using solar radiation pressure is shown in this section (Boskovic et al. 2004; Gao et al. 2012). The methods to decompose the original non-affine system into an affine one in the control part and a non-affine part representing generalized modeling errors are the mean value theorem, the Taylor series expansion and the contraction mapping method (Labiod and Guerra 2007; Leu et al. 2005; Park et al. 2005a).

It is difficult to invert non-affine non-linearity to obtain the inverting control input. Fuzzy logic Systems or neural networks are used to approximate the desired feedback control input (Park et al. 2005b). First, a control method which can account for the non-affine SRP control inputs is developed. The concept behind this method is to differentiate the nonlinear state equations once so that the resultant augmented equations are linear in $\dot{U}$ and use this as the new control input.

Based on the Eqs. (27)–(31), the lower order states are defined as $\bar{x}_1 \in R^{6\times 1} = [\ x\ y\ z\ \dot{x}\ \dot{y}\ \dot{z}\ ]^T$, the higher order states $\bar{x}_2 \in R^{3\times 1} = [\ \ddot{x}\ \ddot{y}\ \ddot{z}\ ]^T$, the full state vector $X \in R^{9\times 1} = [\ \bar{x}_1\ \bar{x}_2\ ]^T$, and the reference trajectory $X^d \in R^{9\times 1} = [\ \bar{x}_1^d\ \bar{x}_2^d\ ]^T$. $\dot{\bar{x}}_1 \in R^{6\times 1} = [\ \dot{x}\ \dot{y}\ \dot{z}\ \ddot{x}\ \ddot{y}\ \ddot{z}\ ]^T$. $\dot{\bar{x}}_2 \in R^{3\times 1} = [\ \dddot{x}\ \dddot{y}\ \dddot{z}\ ]^T$.

Rewrite the system as

$$\begin{bmatrix} \dot{\bar{x}}_1 \\ \dot{\bar{x}}_2 \end{bmatrix} = \begin{bmatrix} \check{A}_{11} & \check{A}_{12} \\ \check{A}_{21} & \check{A}_{22} \end{bmatrix} \begin{bmatrix} \bar{x}_1 \\ \bar{x}_2 \end{bmatrix} + \begin{bmatrix} \check{B}[P(\bar{x}_2) + G(U) + F_{d1}] \\ \dot{P}(\bar{x}_2) + F_{d2} \end{bmatrix} + \begin{bmatrix} 0_{6\times 3} \\ \bar{R}_z J(U) \end{bmatrix} \dot{U} \tag{32}$$

where $P(\bar{x}_2) \in R^{3\times 1}$ is the nonlinear portion of the dynamics, $G(U) \in R^{3\times 1}$ is the SRP acceleration, $J(U) \in R^{3\times 1} = \partial G(U)/\partial U$, and $B = [\ 0_{3\times 3}\ I_{3\times 3}\ ]^T$. Also, the higher and lower order disturbances are $F_{d1} \in R^{6\times 1}$ and $F_{d2} \in R^{3\times 1}$ respectively. $\bar{R}_z = R_z(-\lambda_s - f)$.

The error dynamics of the system can be formulated using Eqs. (27)–(29) as

$$\begin{bmatrix} \dot{\bar{S}} \\ \ddot{\bar{S}} \end{bmatrix} = \begin{bmatrix} \check{A}_{11} & \check{A}_{12} \\ \check{A}_{21} & \check{A}_{22} \end{bmatrix} \begin{bmatrix} \bar{S} \\ \dot{\bar{S}} \end{bmatrix} + \begin{bmatrix} 0 \\ \bar{\check{A}}_{21}\bar{x_1}^d + \bar{\check{A}}_{22}\bar{x_2}^d \end{bmatrix} + \begin{bmatrix} \check{B}[P(\bar{x_2}) + G(U)] \\ \dot{P}(\bar{x_2}) \end{bmatrix} + \begin{bmatrix} 0_{6\times3} \\ \bar{R}_z J(U) \end{bmatrix} \dot{U} \tag{33}$$

where $\bar{\check{A}}_{21} = \check{A}_{21} - \check{A}^d_{21}$ and $\bar{\check{A}}_{22} = \check{A}_{22} - \check{A}^d_{22}$. The tracking error $e \in R^{6\times1} = X - X^d$ is composed of the lower order tracking error ($\bar{S} = \dot{e} + \bar{K}e$) and the higher order tracking error $\dot{\bar{S}}$.

To complete the control design, a description of possible disturbances is included. The disturbance is assumed to be unknown and bounded. The sliding manifold can be written

$$\bar{\sigma} \in R^{3\times1} = \dot{\bar{S}} + \hat{K}\bar{S} \tag{34}$$

where $\dot{\bar{\sigma}} = \ddot{\bar{S}} + \hat{K}\dot{\bar{S}}$.

The Jacobian for this case is

$$J(U) = \begin{bmatrix} J_{11} & J_{12} & J_{13} \\ J_{21} & J_{22} & J_{23} \\ J_{31} & J_{32} & J_{33} \end{bmatrix} \tag{35}$$

where

$$\begin{aligned}
J_{11} &= -\kappa cos^2\alpha cos^2\gamma + \kappa cos^3\alpha cos^2\gamma cos(\gamma + \lambda_s) \\
J_{12} &= 2\kappa u cos^2\alpha cos\gamma sin\gamma + \kappa(u-1)cos^3\alpha[-2cos\gamma sin\gamma cos(\gamma + \lambda_s) \\
&\quad -cos^2\gamma sin(\gamma + \lambda_s)] \\
J_{13} &= 2\kappa u cos\alpha sin\alpha cos^2\gamma + \kappa(1-u)3cos^2\alpha sin\alpha cos^2\gamma cos(\gamma + \lambda_s) \\
J_{21} &= \kappa cos\alpha sin\alpha cos\gamma sin\gamma + \kappa cos^3\alpha cos^2\gamma sin(\gamma + \lambda_s) \\
J_{22} &= -\kappa u cos\alpha sin\alpha[cos^2\gamma - sin^2\gamma] + \kappa(u-1)cos^3\alpha[-2cos\gamma sin\gamma sin(\gamma + \lambda_s) \\
&\quad +cos^2\gamma cos(\gamma + \lambda_s)] \\
J_{23} &= \kappa u sin^2\alpha cos\gamma sin\gamma + 3\kappa(u-1)cos^2\alpha sin\alpha cos^2\gamma sin(\gamma + \lambda_s) \\
J_{31} &= \kappa cos^2\alpha cos^2\gamma sin\alpha \\
J_{32} &= 2\kappa(1-u)cos^2\alpha cos^2\gamma sin\alpha sin\gamma \\
J_{33} &= \kappa(u-1)cos^2\gamma[cos^3\alpha - 2cos\alpha sin^2\alpha]
\end{aligned}$$

The error dynamics for this case are expressed as Eq. (33). The control laws were developed where all three control inputs $U = [\ u\ \alpha\ \gamma\ ]^T$ are available. In stead of using LQR, we use a nonlinear non-affine control design which is similar to the method used in Li et al. (2015).

The control law is

$$\dot{U} = \begin{bmatrix} \dot{u} \\ \dot{\alpha} \\ \dot{\gamma} \end{bmatrix} = [-\bar{R}_{\tilde{}}J(U)]^{-1}\left[\eta\bar{\sigma} + \hat{K}\dot{\bar{S}} + \check{A}_{21}\bar{x_1} + \check{A}_{22}\bar{x_2} + \dot{P}(\bar{x}_2) - \dot{\bar{x}}_2^{\ d}\right] \quad (36)$$

5 Attitude Control Using Reflectivity Control Devices

The sail with attached RCDs is assumed to be a rigid sail with negligible flexibility relative to the sail loading. We assume the spin axis is roll ($\dot{\phi} = \phi_0$ and $\phi = \phi_0 t$). For an elliptic orbit considering gravity gradient torques, the governing nonlinear coupled differential equations of motion of the system (pitch α and yaw γ) are obtained as Varma (2011)

$$\begin{bmatrix} \ddot{\alpha} \\ \ddot{\gamma} \end{bmatrix} = \begin{bmatrix} M_{11} & M_{12} \\ M_{21} & M_{22} \end{bmatrix}\begin{bmatrix} F_\alpha \\ F_\gamma \end{bmatrix} + \begin{bmatrix} M_{11} & M_{12} \\ M_{21} & M_{22} \end{bmatrix}\begin{bmatrix} \widehat{T_{s\alpha}} \\ \widehat{T_{s\gamma}} \end{bmatrix} \quad (37)$$

where $M_{11} = \dfrac{\sin^2\gamma + k_{yz}\cos^2\gamma}{k_{yz}\cos^2\phi}$. $M_{12} = M_{11}$. $M_{21} = M_{12}$. $M_{22} = M_{11}\sin^2\phi + \dfrac{1}{k_{xz}}$.

The dimensionless parameters are defined as $k_{xz} = \frac{I_x}{I_z} = \frac{1-k_1}{1-k_1k_2}$. $k_{yz} = \frac{I_y}{I_z} = \frac{1-k_2}{1-k_1k_2}$. $k_1 = \frac{I_z-I_x}{I_y}$. $k_2 = \frac{I_z-I_y}{I_x}$.

F_α, and F_γ are the nonlinear terms given by

$$\begin{aligned} F_\alpha &= N_\alpha\cos\phi\cos\gamma + N_\phi\cos\phi\sin\gamma - N\gamma\sin\gamma \\ F_\gamma &= N_\gamma \end{aligned} \quad (38)$$

The coefficients N_α, and N_γ in the preceding Eq. (38) are given by,

$$\begin{aligned} N_\alpha = {} & [(1 - k_{xz} + k_{yz})(\dot{f} + \dot{\alpha})\dot{\phi}\sin\phi\ \cos\gamma] - (k_{xz} - k_{yz})(\dot{f} + \dot{\alpha})^2\sin\phi \\ & \cos\phi\sin\gamma + (1 + k_{xz} - k_{yz})[(\dot{f} + \dot{\alpha})(\dot{\gamma} + \dot{\lambda}_s)\cos\phi\sin\gamma + \dot{\phi}(\dot{\gamma} + \dot{\lambda}_s)\cos\gamma] \\ & - 3(k_{xz} - k_{yz})(\cos\alpha\sin\phi\sin\gamma - \sin\alpha\cos\gamma)\cos\alpha\cos\phi \end{aligned} \quad (39)$$

$$N_\phi = [(1 - k_{xz} + k_{yz})(\dot{f} + \dot{\alpha})(\dot{\gamma} + \dot{\lambda}_s)\sin\phi\ \sin\gamma] - (1 - k_{xz})(\dot{f} + \dot{\alpha})^2\sin\phi \quad (40)$$

$$\cos\phi\cos\gamma + (1 - k_{xz} - k_{yz})[(\dot{f} + \dot{\alpha})(\dot{\gamma} + \dot{\lambda}_s)\cos\phi\cos\gamma - \dot{\phi}(\dot{\gamma} + \dot{\lambda}_s)\sin\gamma] + 3(1 - k_{xz})(\cos\alpha\sin\phi\cos\gamma + \sin\alpha\sin\gamma)\cos\alpha\cos\phi$$

$$N_\gamma = [k_{xz} - (1 - k_{yz})\cos 2\gamma](\dot{f} + \dot{\alpha})\dot{\phi}\cos\phi - (1 - k_{yz})[(\dot{f} + \dot{\alpha})^2\cos^2\phi - \dot{\phi}^2]\sin\gamma\cos\gamma + 3(1 - k_{yz})(\cos\alpha\sin\phi\cos\gamma + \sin\alpha\sin\gamma)(\cos\alpha\sin\phi\sin\gamma - \sin\alpha\cos\gamma) \quad (41)$$

In this proposed system configuration, the control torques $\hat{T}_{s\alpha}$ and $\hat{T}_{s\gamma}$ are obtained using solar radiation pressure as given by Eqs. (25) and (26). An adaptive control law based on sliding mode technique is designed. Parameter uncertainties are associated with the SRP stabilized system, and these uncertainties can cause numerous problems in the control tasks and lead to inaccuracy and instability of the control system. Hence an adaptive control methodology is proposed to achieve consistent stable performance. The control inputs are RCDs on the solar sail. To facilitate the control law design, the state space representation of the system can be formulated as follows. Using Eqs. (25), (26) and (37), we can rewrite the equations for the controller design. Define $x_1 = [\alpha, \gamma]^T$ and $x_2 = [\dot{\alpha}, \dot{\gamma}]^T$

$$\begin{bmatrix} x_1{}' \\ x_2{}' \end{bmatrix} = \begin{bmatrix} \bar{A}_{11} & \bar{A}_{12} \\ \bar{A}_{21} & \bar{A}_{22} \end{bmatrix} \left\{ \begin{bmatrix} F_1 \\ F_2 \end{bmatrix} + \begin{bmatrix} 0 \\ U_a \end{bmatrix} + \begin{bmatrix} d_1 \\ d_2 \end{bmatrix} \right\} \quad (42)$$

$U_a \in R^{2\times1} = [\hat{T}_{s\alpha}, \hat{T}_{s\gamma}]^T$, d_1 and d_2 represents the external disturbances, $F_1 = [0, 0]^T$ and $F_2 = [F_\alpha, F_\gamma]^T$. $\bar{A}_{11} = \begin{bmatrix} 0 & 0 \\ 0 & 0 \end{bmatrix}$, $\bar{A}_{12} = \begin{bmatrix} 1 & 0 \\ 0 & 1 \end{bmatrix}$, $\bar{A}_{21} = \begin{bmatrix} 0 & 0 \\ 0 & 0 \end{bmatrix}$ and $\bar{A}_{22} = \begin{bmatrix} M_{11} & M_{12} \\ M_{21} & M_{22} \end{bmatrix}$ where

$$U_a = \frac{1}{2}\kappa m\tilde{r}_d cos^2\alpha cos^2\gamma \frac{(1 - e^2)^3}{(1 + e\cos f)^3} \begin{bmatrix} 0 & 1 \\ -1 & 0 \end{bmatrix} \begin{bmatrix} u_x \\ u_y \end{bmatrix} \quad (43)$$

The torque equations are then converted into dimensionless form by making use of the term

$$C_{srp} = \frac{1}{2}\kappa m\tilde{r}_d cos^2\alpha cos^2\gamma \frac{(1 - e^2)^3}{(1 + e\cos f)^3} \quad (44)$$

$$U_a = C_{srp} \begin{bmatrix} 0 & 1 \\ -1 & 0 \end{bmatrix} \begin{bmatrix} u_x \\ u_y \end{bmatrix} \quad (45)$$

$$x'' = F(x, x') + \begin{bmatrix} \bar{A}_{11} & \bar{A}_{12} \\ \bar{A}_{21} & \bar{A}_{22} \end{bmatrix} \begin{bmatrix} 0 & 0 \\ 0 & 0 \\ 0 & C_{srp} \\ -C_{srp} & 0 \end{bmatrix} \begin{bmatrix} 0 \\ 0 \\ u_x \\ u_y \end{bmatrix} \tag{46}$$

where $x = [x_1, x_2]^T$, $F(x, x')$, $D \in R^4$, and $U_a = [\hat{T}_{s\alpha}, \hat{T}_{s\gamma}]^T$. $F(x, x')$ represents the vector containing all the nonlinear terms including the Coriolis and centrifugal contributions and is described in Eq. (38).

Next, we design the linear sliding manifold which guarantees the desired dynamic behavior for the nominal system in the presence of uncertainties and disturbances, and this is followed by design of the sliding mode control law such that the system states are driven to this sliding plane and the system remains stable.

For the proposed system given by Eq. (45), u_x and u_y are the control inputs. Hence a nonlinear control algorithm based on the sliding mode is developed. These control inputs are then used in the system dynamics to calculate the torque due to SRP. First a sliding surface with order less than the system dynamics is introduced, it is given as

$$\sigma = \tilde{x}' + K_a \tilde{x} \tag{47}$$

where $\tilde{x}' = x' - x'_d$, and $\tilde{x} = x - x_d$ are the state error variables and K_a are positive constants.

The control law is given by,

$$[u_x, \ u_y] = -\alpha_0 \sigma - (\rho + 1)\eta \left[\frac{\sigma}{|\sigma| + \delta} \right] \tag{48}$$

where the adaptive control law is

$$\dot{\rho} = -b_1 \rho + b_2 \eta \|\sigma\| \tag{49}$$

where $\alpha_0, \ b_1, \ b_2 > 0$ are scalar constants,

6 Performance Evaluation

The formation flying dynamics Eqs. (27)–(29) with the nonaffine control law Eq. (36) provide the simulation results for formation keeping. The performance of the proposed RCD is evaluated by numerically simulating the set of governing equations of motion given by Eq. (37) along with proposed adaptive control law given by Eqs. (48) and (49).

The spacecraft orbit elements in the initial epoch is $a_0 = 127{,}000$ km, $e_0 = 0.5$, $i_0 = 0$ deg, $\Omega_0 = 0$ deg, $\omega_0 = 0$ deg, $f_0 = 0$ deg. The reference formation

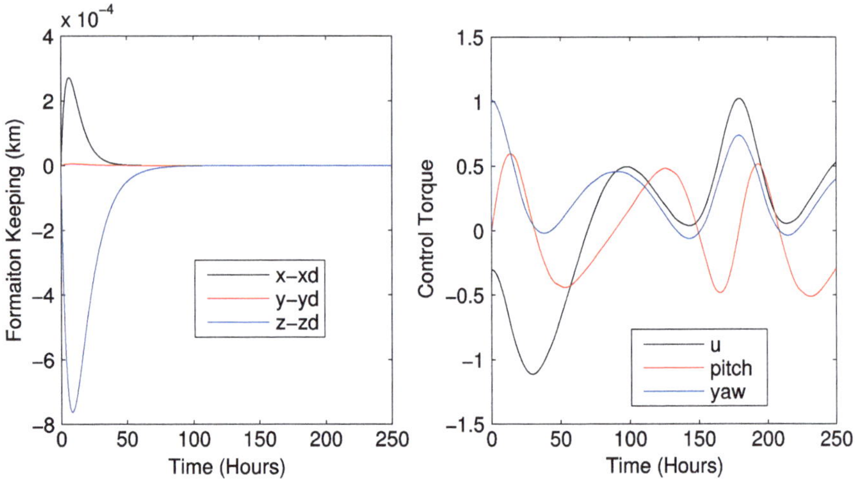

Fig. 4 Controlled relative orbit errors and relative reflectivity and relative attitude angles

trajectories are time varied *sin*/*cos* functions. The max of u is 1.2. The roll axis spin rate is 0.5 rpm. The units of pitch and yaw are degrees. A small spacecraft with RCD and SRP as a MMS mission is considered here and hence the formation flying control was using a non-affine controller. Pitch and yaw angle was also slewed by RCDs instead of motors control for formation keeping.

Figure 4 illustrates the spacecraft formation keeping. It can be observed that the change of reflectivity of the sail, pitch and yaw angle rotation successfully keeping the spacecraft in formation flying. The angular velocity of the spacecraft about the roll axis converges to a value equal to 0.5 rpm. The pitch and yaw axis of the sail rotate regularly to keep the y and z axis in LVLH frame. The RCDs are used to actuating the pitch and yaw rotation.

7 Conclusions

The use of solar radiation pressure to achieve controlled small spacecraft formation flying is studied in this paper. A spacecraft system configuration of a $12U$ CubeSat bus and a circular sail with reflectivity control devices is considered with attitude manoeuvring and stabilization accomplished by the variation of reflectivity from RCDs. Explicit control torque using RCDs is available only along the spacecraft pitch and yaw axes, but follows the spinning roll axis motion of the sail. The spin axis actuator is a reaction wheel. The performance of the control strategy in formation keeping is examined in detail using a high-fidelity nonlinear system model. Numerical simulation is carried out and the controller demonstrates good formation keeping performance. These results confirm that the proposed solar sail small spacecraft can achieve formation flying control and therefore will be a good candidate for use in future space missions.

References

Borggrafe, A., Heiligers, J., Ceriotti, M., McInnes, C.R.: Distributed reflectivity solar sails for extended mission applications. In: Proceedings of 3th International Symposium on Solar Sailing, Glasgow (2013)

Boskovic, J.D., Chen, L.J., Mehra, R.K.: Adaptive control design for nonaffine models arising in flight control. J. Guid. Control Dyn. **27**(2), 209–217 (2004)

Cay, Y., Orger, N.C., Kaymaz, Z.: Design of a solar sail driven cubesat to investigate the polar magnetospheric regions: POLARBEESAIL. In: 8th Ankara International Aerospace Conference, Ankara (2015)

Funase, R., Shirasawa, Y., Mimasu, Y., Mori, O., Tsuda, Y., Saiki, T., Kawaguchi, J.I.R.: On-orbit verification of fuel-free attitude control system for spining solar sail utilizing solar radiation pressure. Adv. Space Res. **48**, 1740–1746 (2011)

Gao, D.X., Lu, D.M., Sun, Z.Q.: Fuzzy adaptive control for a class of non-affine systems via time scale separation. In: Proceedings of the World Congress on Engineering and Computer Science 2012, Vol II WCECS 2012, San Francisco, 24–26 October 2012

Guzman, J.J., Edery, A.: Mission design for the MMS tetrahedron formation. In: Proceedings of IEEE Aerospace Conference (2004)

Joshi, V.K., Kumar, K.: New solar attitude control approach for satellites in elliptic orbits. J. Guid. Control **3**(1), 42–47 (1979)

Kumar, K.D., Bang, H.C., Tahk, M.J.: New attitude control approach for satellites in elliptic orbits using solar radiation pressure. Acta Astronaut. **59**, 462–473 (2006)

Labiod, S., Guerra, T.M.: Adaptive fuzzy control of a class of SISO nonaffine nonlinear systems. Fuzzy Sets Syst. **158**, 1126–1137 (2007)

Lappas, V., Vie, B., McInnes, C., Tarabini, L., Gomes, L., Wallace, K.: Microsolar sails for earth magnetotail monitoring. J. Spacecr. Rocket. **44**(4), 401–411 (2007)

Lapps, V., Mengali, G., Quarta, A.A., Gil-Fernandez, J., Schmidt, T., Wie, B.: Practical systems design for an earth-magnetotail-monitoring solar sail mission. J. Spacecr. Rocket. **46**(2), 381–393 (2009)

Leu, Y.G., Wang, W.Y., Lee, T.T.: Observer based direct adaptive fuzzy neural control for nonaffine nonlinear systems. IEEE Trans. Neural Netw. **5**(4), 853–861 (2005)

Li, J., Greenland, S., Clark, C., Post, M.: Microsatellites formation flying using pulsed plasma thruster and solar sailing at earth moon L4. In: 8th International Workshop on Satellite Constellations and Formation Flying, Delft, 8–10 June 2015

Macdonald, M., McInnes, C., Alexander, D., Sandman, A.: GEOSAIL: exploring the magnetosphere using a low cost solar sail. In: Proceeding of Fifth IAA International Conference on Low Cost Planetary Missions. ESTEC, Noordwijk (2003)

Macdonald, M., Hughes, G.W., McInnes, C., Lyngvi, A., Falkner, P., Atzei, A.: GeoSail: an elegant solar sail demonstration mission. J. Spacecr. Rocket. **44**(4), 784–796 (2007)

Mu, J.S., Gong, S.P., Li, J.F.: Coupled control of reflectivity modulated solar sail for geosail formation flying. J. Guid. Control Dyn. **38**(4), 740–751 (2015)

NASA: Magnetospheric Multiscale (MMS) Mission. http://mms.gsfc.nasa.gov (2015)

Norton, C.D., Pellegrion, S., Johnson, M., Angelopoulous, V., Arya, M., Baker, J., Rogez, J.C., Duncan, C., Erwin, D., Freeman, L., Gamble, K.B., Hyon, J., Imken, T., Jensen-Clern, R., Jones, L., Klesh, A., Lee, J., Liewer, P., Lightsey, G., Lisano, M., Lo, M., Mueller, J., Purphy, N., Murray, S., Nicholson, P., Suari, J.P., Shaffer, C., Spangalo, S., Springmann, J., Staehle, R., Steeves, J., Unwin, S., Vourlidas, A., Willams, A.: Small Satellites: A Revolution in Space

Science, Final Report. Keck Institute for Space Studies California Institute of Technology, Pasadena (2014)
Park, J.-H., Hun, S.H., Kim, S.-H., Seo, S.J., Park, G.: Direct adaptive controller for nonaffine nonlinear systems using self structuring networks. IEEE Trans. Neural Netw. **16**(2), 414–422 (2005)
Park, J.-H., Park, G.-T., Kim, S.-H., Moon, C.-J.: Direct adaptive self structuring fuzzy controller for nonaffine nonlinear system. Fuzzy Sets Syst. **153**, 429–445 (2005)
Schaub, H., Junkin, J.L.: Analytical Mechanics of Space Systems. AIAA, Reston (2003)
Sohn, R.L.: Attitude stabilization by means of solar radiation pressure. ARS J. **29**(5), 371–373 (1959)
Tsuda, Y., Saiki, T., Minzasu, Y.: Realization of minimal attitude determination system for small spinner spacecraft IKAROS. In: Proceedings of Small Satellite Conference, SSC13-IV-4, Utah (2013)
Varma, S.: Control of satellites using environmental forces: aerodynamics drag solar radiation pressure. PhD Thesis, Ryerson University (2011)

A Note on Dynamics About the Coherent Sun–Earth–Moon Collinear Libration Points

Zubin P. Olikara, Gerard Gómez, and Josep J. Masdemont

Abstract Orbits about the Sun–Earth libration points are perturbed by the Moon, and orbits about the Earth–Moon libration points are perturbed by the Sun. We study both situations in the framework of a single, coherent model, the Hill restricted four-body problem. This model is presented in a Sun–Earth rotating reference frame to complement its earlier Earth–Moon frame formulation. We provide an overview of the planar quasi-periodic orbits that originate from the L_1 and L_2 Lyapunov periodic orbits in Hill's problem and the circular restricted three-body problem. The role of resonances is also discussed.

1 Introduction

Recent spacecraft have orbited the L_1 and L_2 libration points in the Sun–Earth and Earth–Moon systems. The libration points are equilibrium solutions of the circular restricted three-body problem (CR3BP) or Hill's problem, which is closely related. Dynamical systems techniques enable the construction of additional invariant solutions such as periodic and quasi-periodic orbits and associated stable and unstable manifolds. This geometric perspective has proven useful for trajectory design (Gómez et al. 2001; Howell et al. 1997).

Spacecraft operating in the vicinity of a libration point are subject to forces that are not present in a simple model. This includes the gravitational attraction of additional bodies and solar radiation pressure. Furthermore, the primary bodies (the Sun and Earth, or the Earth and Moon) do not move in truly circular orbits about

Z.P. Olikara (✉) • G. Gómez
Institut d'Estudis Espacials de Catalunya and Universitat de Barcelona, Gran Via 585, 08007 Barcelona, Spain
e-mail: zubin@maia.ub.es; gerard@maia.ub.es

J.J. Masdemont
Institut d'Estudis Espacials de Catalunya and Universitat Politècnica de Catalunya, Diagonal 647, 08028 Barcelona, Spain
e-mail: josep.masdemont@upc.edu

G. Gómez, J.J. Masdemont (eds.), *Astrodynamics Network AstroNet-II*,
Astrophysics and Space Science Proceedings 44,
DOI 10.1007/978-3-319-23986-6_13

their barycenter. As a consequence, invariant solutions no longer exist, though a "shadow" of them may persist.

Intermediate models between the CR3BP and reality allow us to consider the most significant perturbations while neglecting lesser ones. The primary perturbation on a spacecraft near an Earth–Moon libration point is from the Sun. Similarly, Sun–Earth libration point orbits can be significantly perturbed by the Moon. Both of these situations are addressed by considering a restricted four-body model consisting of the Sun, Earth, Moon, and spacecraft. Notably, in such a model invariant solutions can still exist. The difference between modeling the Sun–Earth and Earth–Moon libration points in a restricted four-body problem is simply the choice of coordinates; the dynamics are unchanged between the reference frames.

The simplest restricted four-body problem is the bicircular model (Simó et al. 1995). In this model, we assume that the Earth–Moon barycenter travels in a circular orbit about the Sun, and the Earth and Moon travel in circular orbits about their barycenter. This assumption is not consistent, however, with Newtonian motion since the three primary bodies do not follow a solution to the general three-body problem.

Another option is to use a coherent restricted four-body model. It is possible to find a periodic solution to the general three-body problem such that the bodies' masses match the Sun, Earth, and Moon, and their relative configuration repeats each synodic month (about 29.53 days). The motion of the fourth body with infinitesimal mass has been studied in earlier works (Andreu 1999; Gúzman 2001) and is sometimes referred to as the quasi-bicircular model. An alternative coherent approach is to construct the relative motion of the Earth and Moon as a periodic solution to Hill's problem with the Sun as the much more massive primary. Then when we consider the motion of the fourth body, we have the Hill restricted four-body problem (HR4BP) (Mohn and Kevorkian 1967; Scheeres 1998). In either situation we assume that the Earth and Moon lie in the ecliptic plane, while in reality their relative orbit has an average inclination of 5.14°. Also note that while these models are coherent, we do not match the eccentricities of Earth's orbit about the Sun (0.0167) or the Moon's orbit about the Earth (0.0549).

In this document we consider motion near the L_1 and L_2 collinear libration points of the HR4BP in the Sun–Earth and Earth–Moon reference frames. Note that since the Sun is about 1 AU from the Earth and Moon, the magnitude of the Sun's perturbation is approximately constant in the vicinity of the Earth–Moon libration points. Within families of orbits about the Sun–Earth L_1 and L_2 libration points, however, the minimum distance to the Moon varies significantly, and the Moon's influence is much greater for larger amplitude orbits.

Unlike Hill's problem or the CR3BP, which are autonomous systems, time appears periodically in the HR4BP equations of motion and repeats each synodic month. As a consequence, invariant objects from the autonomous system generally turn into objects one dimension higher (Jorba and Villanueva 1997). The equilibrium points become periodic orbits, and most periodic orbits become quasi-periodic orbits lying on two-dimensional invariant tori. Two-dimensional invariant tori (such as Lissajous and quasi-halo orbits) generally become three-dimensional invariant tori in the HR4BP, but these orbits are not currently considered.

The results presented in this work are obtained numerically. The advantage of a numerical approach over a semi-analytic one is that we are not limited by the convergence radius of an expansion about a reference solution. We do, however, need to compute objects one at a time, and separate computation of resonances is required. Numerical and semi-analytic results complement each other well. For instance, we can numerically verify some of the Earth–Moon results obtained semi-analytically by Andreu (1999) in a similar model and expand them to a larger region.

The current document provides a brief summary of the main results. A more complete analysis, particularly of the spatial motion, is intended for a later paper.

2 HR4BP in Sun–Earth Frame

The Hill restricted four-body problem is developed by Scheeres (1998) and constructed in a rotating reference frame of the smaller two primaries, in our case the Earth and the Moon. The motion of these primaries is the classical Hill variation orbit that can be computed to any desired accuracy in powers of a period parameter m. We use the parameter value $m = 0.0808$ that relates the period of the Earth–Moon barycenter about the Sun to the period of the Earth and Moon about their barycenter. The model depends on a second parameter μ corresponding to the mass ratio of the smaller primaries, $\mu = 0.01215$ for the Earth and the Moon.

The system is constructed to be 2π periodic, which corresponds to the time period after which the primaries' configuration repeats (a synodic month for the Sun–Earth–Moon system). The origin is located at the Earth–Moon barycenter. A reference frame that rotates at a constant rate is selected along with a length normalization by $a_0(m) = m^{2/3}(1 - \frac{2}{3}m + O(m^2))$, the average distance between the smaller primaries along Hill's variation orbit. Consequently, the Earth and Moon move slightly in their reference frame. Note that this is simply a coordinate choice and does not effect the dynamics; equivalently, the Earth and Moon positions could be fixed as in Mohn and Kevorkian (1967).

A convenient aspect of the HR4BP is that it reduces to the Earth–Moon CR3BP when $m = 0$, and to the Sun–Earth Hill's problem when $\mu = 0$. Notably, there is a natural route to perform continuation from the autonomous limiting cases to the Sun–Earth–Moon HR4BP such that the model is coherent at every step. The connection between the CR3BP and HR4BP is clear when the equations of motion are formulated in an Earth–Moon frame. The connection between Hill's problem and the HR4BP is most apparent if we formulate the equations in a Sun–Earth frame. We will use the same origin, length, and time scaling as the Earth–Moon frame, and refer the reader to Scheeres (1998) for Hill's expansion of the Earth and Moon motion.

To define the Sun–Earth reference frame, let the x-axis point from the Sun to the Earth–Moon barycenter, and the z-axis point in the direction of the primaries' angular momentum. The y-axis completes the right-handed frame. The Sun–Earth

HR4BP equations of motion are

$$\begin{aligned}\ddot{x} - 2m\dot{y} &= U_x \\ \ddot{y} + 2m\dot{x} &= U_y \\ \ddot{z} &= U_z,\end{aligned}$$

which depend on the partial derivatives of the force potential-like function

$$U(\mathbf{r}, t; m, \mu) = \frac{1}{2}m^2(3x^2 - z^2) + \frac{m^2}{a_0(m)^3}\left(\frac{1-\mu}{r_1(\mathbf{r}, t; m, \mu)} + \frac{\mu}{r_2(\mathbf{r}, t; m, \mu)}\right).$$

The lengths r_1 and r_2 are the distances of the fourth body located at $\mathbf{r} := (x, y, z)$ to the Earth and Moon. When we set $\mu = 0$, the Earth is located at the origin, length $r_1 = \|\mathbf{r}\|$, and we recover Hill's problem up to a constant length and time scaling.

The HR4BP in a Sun–Earth frame has several symmetries. These are similar to Hill's problem except time plays an important role since it orients the primaries. If we define $t = 0$ to be the time when the Sun, Earth, and Moon are in a line, the symmetries can be represented as follows:

$$\begin{aligned}(x,\ y,\ z,\ v_x,\ v_y,\ v_z,\ t) &\mapsto (-x,\ \ y,\ \ z,\ \ v_x, -v_y, -v_z,\ \pi - t) \\ (x,\ y,\ z,\ v_x,\ v_y,\ v_z,\ t) &\mapsto (\ \ x, -y,\ \ z, -v_x,\ \ v_y, -v_z,\ \ -t) \\ (x,\ y,\ z,\ v_x,\ v_y,\ v_z,\ t) &\mapsto (\ \ x,\ \ y, -z,\ \ v_x,\ \ v_y, -v_z,\ \ +t)\end{aligned}$$

where (v_x, v_y, v_z) are the velocities associated with the coordinates (x, y, z). Additional solutions can be generated by composing the symmetries.

The Sun–Earth–Moon HR4BP is intended for studying the motion in the vicinity of the Sun–Earth L_1 and L_2 libration points and the region between them (including motion near the Earth and Moon and their libration points). When μ is nonzero, the influence of the Moon is included and the Sun–Earth equilibrium points located on the x-axis at $x = \pm 3^{-1/3}/a_0(m) = \pm 3.91$ become small 2π-periodic orbits serving as dynamical substitutes.

In general, the planar Lyapunov orbits become two-dimensional quasi-periodic tori with an internal frequency associated with their period in Hill's problem, and an external frequency associated with the Moon's synodic period. Close to the Sun–Earth L_1 and L_2 libration points, the internal period is slightly less than 6 synodic months. As we continue along the planar orbit family, the period increases, and we cross a 1:6 resonance followed by 3:19, 2:13, and 3:20 resonances. Since the orbits near the libration point do not pass close to the Moon, the early resonances do not perturb the dynamics significantly. However, larger orbits pass nearer the Moon, and the family of quasi-periodic orbits appears to break down approaching the 1:7 resonance. One of the last computed orbits before the 1:7 resonance is shown in Fig. 1. Due to symmetry, the L_1 and L_2 orbits are identical up to a rotation about the origin.

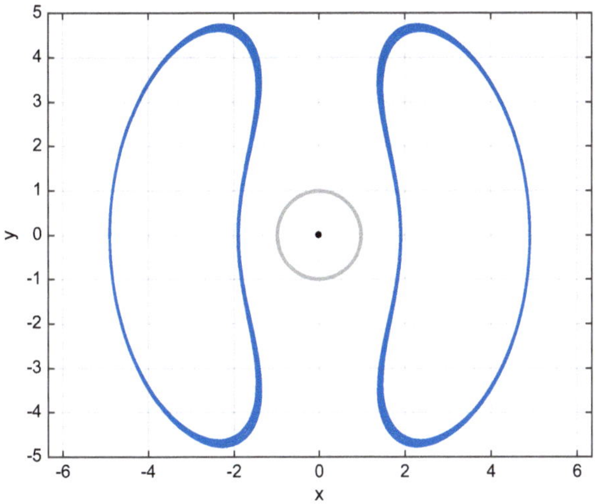

Fig. 1 Sun–Earth L_1 and L_2 planar quasi-periodic orbits near 1:7 resonance with Moon

3 Motion in Earth–Moon Frame

We now consider the HR4BP dynamics about the L_1 and L_2 libration points in an Earth–Moon frame (see Scheeres 1998 for the equations of motion). Unlike the Sun–Earth case where the Moon's influence increases as we move along the planar family of libration point orbits, the magnitude of the Sun's perturbation near the Earth–Moon libration points is approximately constant.

3.1 *Earth–Moon L_1 Planar Orbits*

The first step in analyzing the planar vicinity of the Earth–Moon L_1 point in the HR4BP is to find a substitute for the libration point itself. We initialize the computation from the Earth–Moon L_1 point in the CR3BP ($m = 0$). For nonzero m, the equations of motion are 2π-periodic,[1] and the equilibrium point becomes a periodic orbit of the same period. If the orbit is continued until $m = 0.0808$, the dynamical equivalent including the Sun's perturbation is obtained. This continuation process is illustrated by branch A in Fig. 2, where the orbit's x-component when $t \,(\mathrm{mod}\, 2\pi) = 0$, i.e., the Sun, Earth, and Moon are aligned, is shown as m is varied. The substitute has stability center $\times$ saddle.

[1]The Earth–Moon HR4BP equations of motion in fact have a minimal period π due to symmetry, but this is not essential for the discussion.

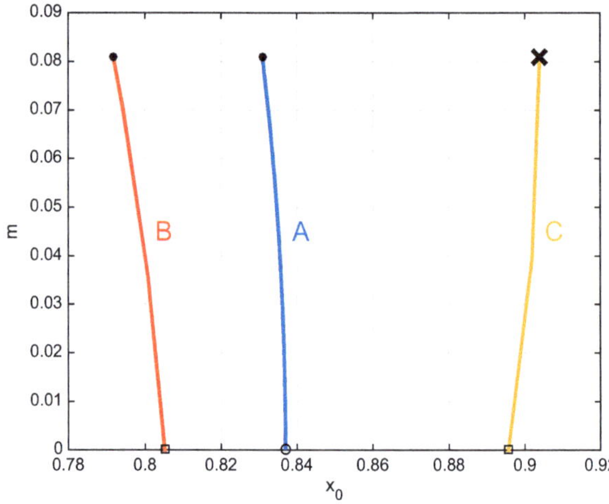

Fig. 2 Bifurcation diagram of Earth–Moon L_1 dynamical substitute periodic orbits

The planar L_1 Lyapunov orbits in the Earth–Moon CR3BP begin with a period of about $T = 2.69$ and increase along the family. Within this family there is a periodic orbit of period π. If we continue this orbit to nonzero m, it splits into two orbits of period 2π, one with stability center $\times$ saddle and the other with stability saddle $\times$ saddle. The continuation of these orbits is shown by branches B and C, respectively, in Fig. 2. While initial orbit is identical when $m = 0$, their phasing differs by half a revolution, so the initial x-component in the figure is on either side of the CR3BP L_1 libration point.

We expect most of the planar periodic orbits in the Earth–Moon CR3BP to become two-dimensional tori in the HR4BP. One frequency $\omega_0 = 2\pi/T$ is associated with the period T that varies along the family of CR3BP planar Lyapunov orbits. The other frequency $\omega_1 = 1$ matches the periodic influence of the Sun. Starting from the L_1 dynamical substitute (branch A periodic orbit at $m = 0.0808$), we initialize a family of Sun–Earth–Moon two-dimensional tori along which the frequency ω_0 varies. The family is continued (with fixed m) until a dynamical boundary is reached. The family of tori is illustrated in Fig. 3 using a stroboscopic map $t \pmod{2\pi} = 0$ and corresponds to the inner curves. Note that the axes are scaled independently to emphasize the detail.

The boundary is associated with the 2:1 resonance when $2\pi/\omega_0$ is half a synodic month. We can understand the structure in this resonance region by returning to the two additional periodic orbits we have computed. The orbit on branch B, which has an elliptic (center) component, has a small region of approximately resonant tori around it shown in Fig. 3. This family is bounded on the planar center manifold by the stable and unstable manifolds of the hyperbolic (saddle) periodic orbit on branch C. These manifolds will not exactly coincide, so there will be a small region of chaos on the center manifold in its vicinity.

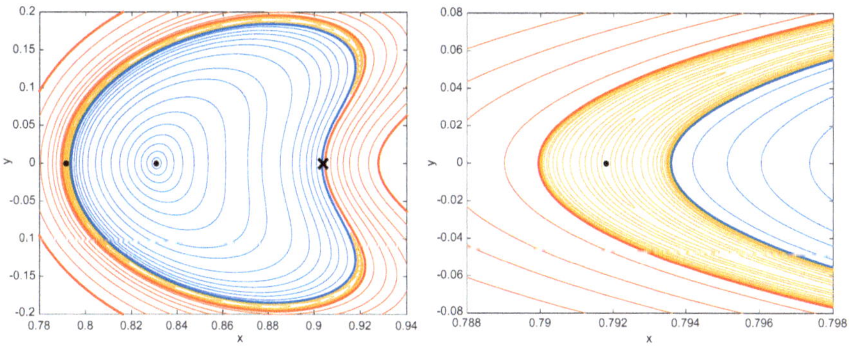

Fig. 3 Stroboscopic map of Earth–Moon L_1 planar quasi-periodic orbits (expanded view of 2:1 resonance region on *right*)

Finally, beyond the resonance we have more quasi-periodic orbits replacing the planar Lyapunov orbits with larger amplitude. These are the outer curves in Fig. 3. The family is continued until a 5:3 resonance is reached at which there are more resonant "islands" that we do not currently compute. Note that by using a numerical approach we are able to extend the family beyond what can be conveniently obtained using center manifold reduction about the Earth–Moon L_1 point. This allows us to compute the 2:1 resonance, which may otherwise be outside the region of convergence of a semi-analytic method (Andreu 1999).

3.2 *Earth–Moon L_2 Planar Orbits*

The situation for the Earth–Moon L_2 point differs in some significant ways from the L_1 case. Here the CR3BP planar Lyapunov orbits emanating from the libration point have a period T of about 3.37, which corresponds to a frequency ω_0 of about 1.86. However, this frequency crosses 2 as we perform continuation in m of the dynamical substitute.[2] The crossing leads to interesting bifurcation behavior of the L_2 dynamical substitute itself.

The bifurcation diagram of the 2π periodic orbit replacing the Earth–Moon L_2 point is shown in Fig. 4. At first, it appears that there is a pitchfork-like bifurcation that occurs along branch A, which originates from the CR3BP equilibrium point. A close inspection, however, shows that the bifurcation is perturbed (at least when Hill's variational orbit is expanded to order m^6), and the branch only goes to the periodic orbit on the right. Thus, the closest dynamical substitute to the L_2 libration

[2]As m varies, the dimensional time unit is scaled by $(1+m)$. When $m = 0$ (Earth–Moon CR3BP), 2π time units corresponds to a sidereal month (27.32 days). When $m = 0.0808$ (Sun–Earth–Moon HR4BP), 2π time units is a synodic month (29.53 days). See Scheeres (1998) for more details.

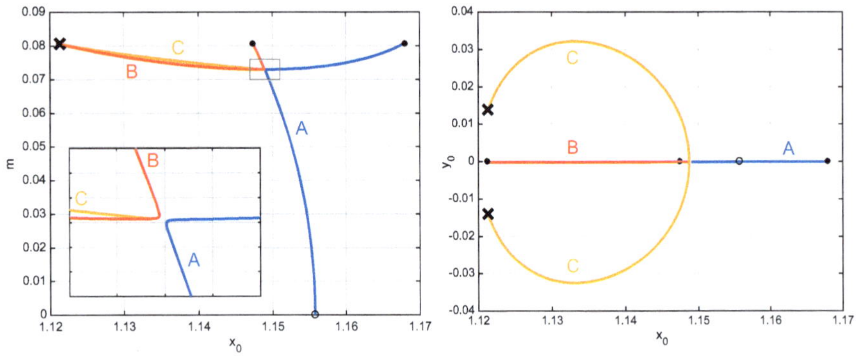

Fig. 4 Bifurcation diagram of Earth–Moon L_2 dynamical substitute periodic orbits

point does not belong to branch A. It belongs to branch B, which ends in another periodic orbit in the Sun–Earth–Moon HR4BP after passing a turning point in m. All three orbits have stability center × saddle. Near the turning point, branch B undergoes an additional bifurcation that gives rise to a periodic orbit family C that is symmetric across the x-axis (though the individual orbits themselves are not). This branch is most apparent on the right of Fig. 4 where the periodic orbit coordinates (x, y) at time $t \pmod{2\pi} = 0$ are plotted as m varies. The two periodic orbits in the Sun–Earth–Moon HR4BP along branch C have stability saddle × saddle.

The bifurcation behavior of the Earth–Moon L_2 dynamical substitute may clarify the observations by Andreu in the (non-coherent) bicircular four-body problem. The substitute computed by direct continuation is a fairly large orbit. It may be possible that a periodic orbit similar to the middle HR4BP Earth–Moon L_2 dynamical substitute exists on a disconnected branch. The branches may not be as close to intersecting as in Fig. 4 due to the lack of coherence.

The five periodic orbits obtained allow us to understand the geometry of the 2:1 resonance that is passed along the family of two-dimensional quasi-periodic orbits. As with the L_1 case, we initialize the family of tori near the middle dynamical substitute. This is shown in the inner region of Fig. 5, which is a stroboscopic map $t \pmod{2\pi} = 0$ of the invariant tori. The continuation stops when we approach the 2:1 resonance, $\omega_0 \approx 2\omega_1$. Here we have small families of nearly resonant tori that emanate from the two elliptic periodic orbits on the center manifold. The ones on the left arc particularly small. Beyond the resonance region, there are additional invariant tori that replace the larger Earth–Moon L_2 planar Lyapunov orbits. The inner, two resonant, and outer families of tori are all separated on the center manifold by the invariant manifolds of the two hyperbolic fixed points denoted by crosses. This structure appears to match that of the quasi-bicircular problem computed semi-analytically in Andreu (1999).

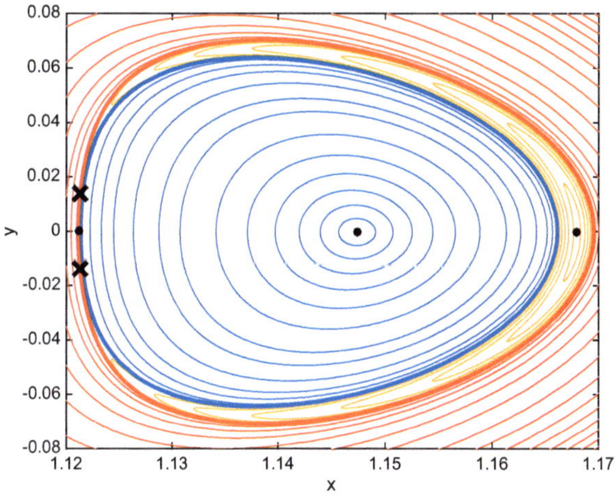

Fig. 5 Stroboscopic map of Earth–Moon L_2 planar quasi-periodic orbits (including 2:1 resonance region)

4 Concluding Remarks

In this note, we consider motion in the Sun–Earth–Moon HR4BP. We first formulate the equations of motion in a Sun–Earth frame and present the symmetry properties of these coordinates. We discuss the role of planar resonances and note the possible breakdown of the L_1 and L_2 orbit families as the internal period approaches 7 synodic months.

Resonances also play a role in the HR4BP for the Earth–Moon L_1 and L_2 families of planar quasi-periodic orbits. For the Earth–Moon L_1 point, the 2:1 resonance appears at a y-amplitude of approximately 0.18 and can be understood by considering π-periodic orbits originating in the CR3BP. For the Earth–Moon L_2 point, the 2:1 resonance appears closer to the libration point at an amplitude of approximately 0.07. The resonance structure can be understood by studying bifurcations of the 2π-periodic orbit substituting the libration point.

A potential next step would be to consider spatial HR4BP solutions, specifically non-planar quasi-periodic orbits lying on either two- or three-dimensional invariant tori. For the Earth–Moon L_2 point, the halo bifurcation appears close to the planar 2:1 resonance, so the interaction with the out-of-plane resonance may have interesting behavior. An additional avenue for future work would be to study the effect of the Sun on the motion about the Earth–Moon L_3 point.

Acknowledgements This work was supported by the Marie Curie Initial Training Network PITN-GA-2011-289240, AstroNet-II.

References

Andreu, M.A.: The quasi-bicircular problem. PhD thesis, Univ de Barcelona (1999)

Gómez, G., Llibre, J., Martínez, R., Simó, C.: Dynamics and Mission Design Near Libration Points I: Fundamentals: The Case of Collinear Libration Points. World Scientific, Singapore (2001)

Gúzman, J.: Spacecraft trajectory design in the context of a coherent restricted four-body problem. PhD thesis, Purdue University (2001)

Howell, K.C., Barden, B.T., Lo, M.W.: Application of dynamical systems theory to trajectory design for a libration point mission. J. Astronaut. Sci. **45**(2), 161–178 (1997)

Jorba, À., Villanueva, J.: On the persistence of lower dimensional invariant tori under quasi-periodic perturbations. J. Nonlinear Sci. (1997). doi:10.1007/s003329900036

Mohn, L., Kevorkian, J.: Some limiting cases of the restricted four-body problem. Astron. J. **72**(8), 959–963 (1967)

Scheeres, D.J.: The restricted Hill four-body problem with applications to the Earth–Moon–Sun system. Celest. Mech. Dyn. Astr. (1998). doi:10.1023/a:1026498608950

Simó, C., Gómez, G., Jorba, À., Masdemont, J.: The bicircular model near the triangular libration points of the RTBP. In: Roy, A.E., Steves, B.A. (eds.) From Newton to Chaos, pp. 343–370. Springer, New York (1995)

The Trojan Problem from a Hamiltonian Perturbative Perspective

Rocío Isabel Páez, Ugo Locatelli, and Christos Efthymiopoulos

Abstract The study of the Trojan problem (i.e. the motion in the vicinity of the equilateral Lagrangian points L_4 or L_5) has a long history in the literature. Starting from a representation of the Elliptic Restricted 3-Body Problem in terms of modified Delaunay variables, we propose a sequence of canonical transformations leading to a Hamiltonian decomposition in the three degrees of freedom (fast, synodic and secular). From such a decomposition, we introduce a model called the 'basic Hamiltonian' H_b, corresponding to the part of the Hamiltonian independent of the secular angle. Averaging over the fast angle, the $\langle H_b \rangle$ turns to be an integrable Hamiltonian, yet depending on the value of the primary's eccentricity e'. This allows to formally define action-angle variables for the synodic degree of freedom, even when $e' \neq 0$. In addition, we introduce a method for locating the position of secondary resonances between the synodic libration frequency and the fast frequency, based on the use of the normalized $\langle H_b \rangle$. We show that the combination of a suitable normalization scheme and the representation by the H_b is efficient enough so as to allow to accurately locate secondary resonances as well as higher order resonances involving also the very slow secular frequencies.

1 Introduction

In recent years, the equilateral Lagrangian points L_4 and L_5 have become the subject of several mission proposals, as a privileged position for solar observatories (see Gopalswamya et al. 2011 and references therein). While these studies show the

R.I. Páez (✉) • U. Locatelli
Dipartimento di Matematica, Universitá di Roma "Tor Vergata", Via della Ricerca Scientifica 1, 00133 Roma, Italy
e-mail: paez@mat.uniroma2.it; locatell@mat.uniroma2.it

C. Efthymiopoulos
RCAAM (Research Center for Astronomy and Applied Mathematics), Academy of Athens, Soranou Efesiou 4, GR-11527 Athens, Greece
e-mail: cefthim@academyofathens.gr

G. Gómez, J.J. Masdemont (eds.), *Astrodynamics Network AstroNet-II*, Astrophysics and Space Science Proceedings 44, DOI 10.1007/978-3-319-23986-6_14

high feasibility of a mission towards the equilateral points, a deep understanding of the Trojan dynamics (i.e. the motion in the vicinity of the equilateral Lagrangian points) is mandatory for the success of such missions.

From a theoretical point of view, many studies faced the problem of the Trojan motion by using the circular approximation of the Restricted 3-Body Problem (CR3BP). In this framework, the linear stability around L_4 or L_5 is guaranteed for values of the mass parameter $\mu \leq \mu_R = \frac{27-\sqrt{621}}{54} \approx 0.0385$, known as the Routh criterion (Gascheau 1843). The linearized orbits, in the vicinity of L_4 and L_5, are given by the composition of two different oscillatory motions, with frequencies $\omega_f = 1 - \frac{27}{8}\mu + \mathcal{O}(\mu^2)$ and $\omega_s = \sqrt{\frac{27}{4}\mu} + \mathcal{O}(\mu)$, where the subscripts f and s stand for 'fast' and 'synodic' respectively. These two oscillations have very different timescales. On one hand, ω_f is approximately equal to 1, i.e. it gives an oscillation of period similar to the mean motion of the primary. On the other hand, ω_s is proportional to the square root of the mass parameter μ, which is a small parameter itself. Thus, the motion of the test particle can be decomposed in two different contributions (Murray and Dermott 1999): the slow motion, associated to a *guiding center* motion around the position of equilibrium, with period $2\pi/\omega_s$ (known as *synodic libration*), and the fast one, attributed to the short period motion of the particle around the guiding center.

Besides the 1:1 mean motion resonance between the primary and the Trojan object, there may exist secondary resonances, corresponding to commensurabilities between the fast and the synodic frequency of the type $\omega_f \approx n\omega_s$, with n integer. These resonances generate periodic orbits forming n epicyclic oscillations (e.g. loops) while they accomplish one full synodic libration. The presence of the secondary resonances affects the quasiperiodic orbital solutions in terms of the two main frequencies ω_f and ω_s (e.g. Deprit and Delie 1965), as it gives rise to so-called critical terms, i.e., terms depending on a resonant combination of the angles. These resonant terms in the series require a special treatment (e.g. Garfinkel 1977). On the other hand, in the Trojan problem it can be shown that their effect on the slow (secular) motions is rather limited (Namouni 1999). Let us mention, finally, that the use of averaging techniques allows to simplify the study of the synodic librations by finding a simplified form of the equation of motion for the so-called critical argument $\tau = \lambda - \lambda'$, with λ, λ' the mean longitude of the test particle and the primary, respectively.

Whereas the CR3BP may be a good first model for developing the theory of Trojan orbits, it clearly does not suffice to represent more realistic problems. As a natural extension, there exist several approximations to the analytical solution of the Trojan problem in the framework of the Elliptic Restricted 3-Body problem (ER3BP). This generalization brings new interesting features to the formulation. While most works on the CR3BP consider two time scales (associated to ω_f and ω_s), in the ER3BP *three* times scales are necessary (Érdi 1977), associated to the fast, synodic and secular frequency. From the physical point of view, these three scales are associated to the epicyclic oscillation (fast, $\mathcal{O}(1)$), the libration around the libration center (synodic, $\mathcal{O}(\sqrt{\mu})$) and the slow precession of the perihelion of the orbit of the Trojan body (secular, $\mathcal{O}(\mu)$) (Érdi 1978).

In the present work, we obtain a Hamiltonian decomposition individualizing the three time scales, starting from the ER3BP. From this decomposition, we introduce a model called the 'basic Hamiltonian' H_b, representing only the fast and synodic d.o.f. of the elliptic problem. Averaging over the fast angle, the $\langle H_b \rangle$ turns to be an integrable Hamiltonian, yet depending on the value of the primary's eccentricity e'. From the latter, we can define action-angle variables for the synodic degree of freedom, even when $e' \neq 0$. We introduce a method, based on the use of the normalized $\langle H_b \rangle$, for locating the position of the secondary resonances between the synodic libration frequency and the fast frequency. We show that the H_b normalized under a suitable scheme is efficient enough to accurately locate both secondary resonances and higher order resonances involving also the secular frequency.

2 The Basic Hamiltonian H_b

We start the construction of the H_b from the Hamiltonian corresponding to the planar ER3BP.

$$H_{ell} = \frac{\|\mathbf{p}\|^2}{2} - \frac{1}{\|\mathbf{r}\|} - \mathcal{G}m' \left(\frac{1}{\Delta} - \frac{1}{\|\mathbf{r}\|} - \frac{\mathbf{r} \cdot \mathbf{r}'}{\|\mathbf{r}'\|^3} \right) , \tag{1}$$

where $\mathbf{r}'$ and $\mathbf{r}$ are the heliocentric position vectors for the planet and for the massless body, respectively, $\Delta = \|\mathbf{r} - \mathbf{r}'\|$, $\mathbf{p} = \dot{\mathbf{r}}$ and $\mathcal{G}m' = \mu$. We introduce modified Delaunay variables (x, y, λ, ϖ), independent of the mass parameter μ (Brown and Shook 1933; Morais 2001), given by

$$x = \sqrt{a} - 1 , \quad \lambda , \quad y = \sqrt{a}\left(\sqrt{1 - e^2} - 1\right) , \quad \varpi , \tag{2}$$

where λ, ϖ, a and e are the mean longitude, longitude of the pericenter, major semi-axis and eccentricity of the orbit of the Trojan body (primed symbols correspond to the primary). The Hamiltonian (1) in the new variables reads

$$H_{ell} = -\frac{1}{2(1+x)^2} + I - \mu R(x, y, \lambda, \varpi, \lambda'; \varpi', e') \tag{3}$$

where we introduce a 'dummy' action variable I conjugate to λ', and $\lambda' = nt$. The present expression of the Hamiltonian corresponds to an autonomous system of 3 d.o.f.

For the study of the Trojan dynamics, we define two new angles. The angle $\tau = \lambda - \lambda'$ is the resonant angle corresponding to the 1:1 MMR resonance, with value $\tau = \pi/3$ at the Lagrangian point L_4. The angle $\delta\varpi = \varpi - \varpi'$ expresses the relative position of the pericenter of the Trojan body from the pericenter of the planet. We introduce these new angles through a generating function S_2 depending on the old angles $(\lambda, \lambda', \varpi)$ and the new actions (X_1, X_2, X_3),

$$S_1 = (\lambda - \lambda')X_1 + \lambda' X_2 + (\varpi - \varpi')X_3 , \tag{4}$$

yielding the following transformation rules

$$\tau = \lambda - \lambda' , \quad \tau_2 = \lambda' , \quad \delta\varpi = \varpi - \varpi' , \quad x = X_1 , \quad I = X_2 - X_1 , \quad y = X_3 . \tag{5}$$

We keep the old notation for all variables involved in an identity transformation ($X_1 = x$, $\tau_2 = \lambda'$, $X_3 = y$). The Hamiltonian then reads:

$$H_{ell} = -\frac{1}{2(1+x)^2} - x + X_2 - \mu R(x, y, \tau, \delta\varpi, \lambda'; e', \varpi') . \tag{6}$$

This expression can be recast under the form

$$H_{ell} = \langle H \rangle + H_1 \tag{7}$$

where

$$\langle H \rangle = -\frac{1}{2(1+x)^2} - x + X_2 - \mu \langle R \rangle (\tau, \delta\varpi, x, y; e', \varpi') \tag{8}$$

and

$$H_1 = -\mu \tilde{R}(\tau, \delta\varpi, x, y, \lambda'; e', \varpi') ,$$

with

$$\langle R \rangle = \frac{1}{2\pi} \int_0^{2\pi} R \mathrm{d}\lambda', \quad \tilde{R} = R - \langle R \rangle \ .$$

The action X_2 is an integral of motion under the Hamiltonian flow of $\langle H_{ell} \rangle$. Thus, $\langle H_{ell} \rangle$ represents a system of two d.o.f. We call *position of the forced equilibrium* $(\tau_0, \delta\varpi_0, x_0, y_0)$ the solution of the system of equations

$$\dot{\tau} = \frac{\partial \langle H \rangle}{\partial x} = 0, \ \ \delta\dot{\varpi} = \frac{\partial \langle H \rangle}{\partial y} = 0, \ \ \dot{x} = -\frac{\partial \langle H \rangle}{\partial \tau} = 0, \ \ \dot{y} = -\frac{\partial \langle H \rangle}{\partial \delta\varpi} = 0 \ . \tag{9}$$

We find

$$(\tau_0, \delta\varpi_0, x_0, y_0) = \left(\pi/3, \pi/3, 0, \sqrt{1 - e'^2} - 1\right) \ . \tag{10}$$

Let us note that the equilibrium point given by (10) does not represent a fixed point in the synodic frame of reference, as in the circular case, but a short-period epicyclic loop around L_4, corresponding to a fixed ellipse of eccentricity $e = e'$ in the inertial frame.

We now introduce local action-angle variables around the point of forced equilibrium. To this end, we consider the 'shift of center' canonical transformation given by[1]:

$$v = x - x_0, \ \ u = \tau - \tau_0, \ \ Y = -(W^2 + V^2)/2, \ \ \phi = \arctan(V, W) \tag{11}$$

where

$$V = \sqrt{-2y} \sin \delta\varpi - \sqrt{-2y_0} \sin \delta\varpi_0, \quad W = \sqrt{-2y} \cos \delta\varpi - \sqrt{-2y_0} \cos \delta\varpi_0 \ ,$$

where Y is defined negative so as to keep the canonical structure with respect to ϕ. Re-organising terms, the Hamiltonian (6) takes the form:

$$\begin{aligned} H_{ell} = & -\frac{1}{2(1+v)^2} - v + X_2 - \mu \left(\mathscr{F}^{(0)}(u, \lambda' - \phi, v, Y; e', \varpi') \right. \\ & \left. + \mathscr{F}^{(1)}(u, \phi, \lambda', v, Y; e', \varpi') \right) \end{aligned} \tag{12}$$

where $\mathscr{F}^{(0)}$ contains terms depending on the angles λ' and ϕ only through the difference $\lambda' - \phi$, and $\mathscr{F}^{(1)}$ contains terms dependent on non-zero powers of e'.

[1] We symbolize with arctan (a, b) the function $\tan^{-1}(a/b) : \mathbb{R}^2 \to \mathbb{T}^1$, of two variables, that maps the value of the arctangent to the corresponding quadrant in the coordinate system with b as the abscissa and a as the ordinate.

The part of the Hamiltonian corresponding to $\mathscr{F}^{(0)}$ can be formally reduced to a system of 2 d.o.f. through the generating function

$$S_2(u, \lambda', \phi, Y_u, Y_s, Y_p) = uY_u + (\lambda' - \phi)Y_f + \phi Y_p \tag{13}$$

yielding

$$\phi_u = u\ , \quad \phi_f = \lambda' - \phi\ , \quad \phi_p = \phi\ , \quad v = Y_u\ , \quad X_2 = Y_f\ , \quad Y = Y_p - Y_f\ . \tag{14}$$

The subscripts 'f' and 'p' stand for 'fast' and 'proper' respectively. As before, we keep the old notation for the variables transforming by the identity $\phi_u = u, \phi_p = \phi$, and $Y_u = v$, except for the action $Y_f \equiv X_2$. The Hamiltonian (12) in the new canonical variables reads

$$\begin{aligned} H_{ell} = & -\frac{1}{2(1+v)^2} - v + Y_f - \mu\mathscr{F}^{(0)}(v, Y_p - Y_f, u, \phi_f; e', \varpi') \\ & - \mu\mathscr{F}^{(1)}(v, Y_p - Y_f, u, \phi_f, \phi; e', \varpi')\ . \end{aligned} \tag{15}$$

Collecting terms linear in $(Y_p - Y_f)$, we find:

$$\omega_f \equiv \dot{\phi}_f = \frac{\partial H_{ell}}{\partial Y_f} = 1{-}27\mu/8{+}\mathscr{O}(\mu^2)\ldots,\quad g \equiv \dot{\phi} = \frac{\partial H_{ell}}{\partial Y_p} = 27\mu/8{+}\mathscr{O}(\mu^2)\ldots \tag{16}$$

We identify ω_f and g as the short-period and secular frequencies, respectively, of the Trojan body. Therefore, the set of variables constructed in (13) allows to separate the three time-scales by the corresponding 3 d.o.f. in the Hamiltonian, and it allows to consider various 'levels' of perturbation. We call *basic model* the one of Hamiltonian

$$H_b = -\frac{1}{2(1+v)^2} - v + Y_f - \mu\mathscr{F}^{(0)}(v, Y_p - Y_f, u, \phi_f; e', \varpi')\ . \tag{17}$$

The total Hamiltonian takes the form $H_{ell} = H_b + H_{sec}$, where

$$H_{sec} = -\mu\mathscr{F}^{(1)}(v, Y_p - Y_f, u, \phi_f, \phi; e', \varpi') \tag{18}$$

contains terms of at least order $\mathscr{O}(e'\mu)$. Figure 1 summarizes the physical meaning of the action-angle variables $(\phi_f, u, \phi, Y_f, v, Y_p)$.

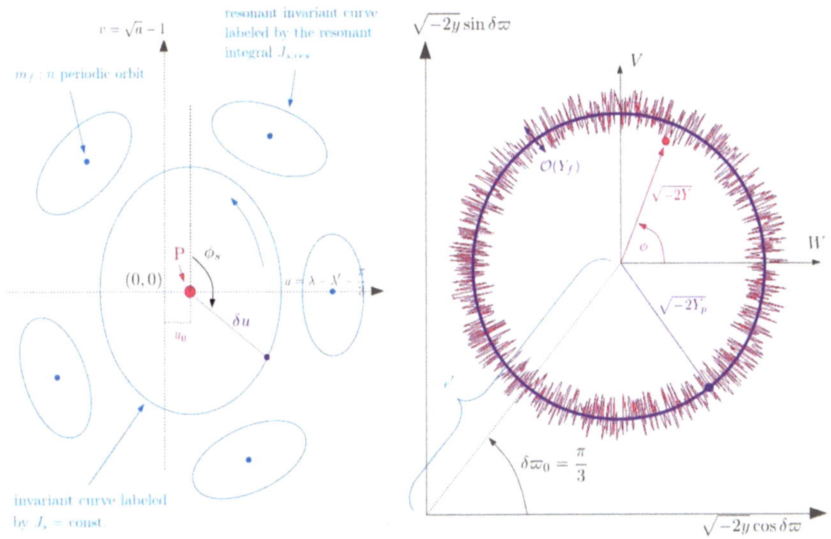

Fig. 1 Schematic representation of the physical meaning of the action-angle variables used for the H_{ell} in Eq. (15). The plane (u, v) corresponds to the 'synodic' motion of the Trojan body. Under the Hamiltonian H_b, the phase portrait can be represented by a Poincaré section corresponding to every time when the angle ϕ_f accomplishes a full cycle. The *left panel* shows schematically the form of the projection of this section on the plane (u, v). The central point P represents a stable fixed point corresponding to the short-period periodic orbit around L4. The orbit has frequency ω_f, while its amplitude increases monotonically with Y_f. The forced equilibrium corresponds to $u_0 = 0$, $Y_f = 0$. The point P, however, has in general a shift to positive values $u_0 > 0$ for proper eccentricities larger than zero [see later Eq. (24)]. Far from resonances, the invariant curves around P are labeled by a constant action variable J_s, and its associated angle (phase of the oscillation) ϕ_s. Resonances, and their island chains correspond to rational relations between the frequencies ω_f and ω_s. Within the resonant islands, J_s is no longer preserved, but we have, instead, the preservation of a resonant integral $J_{s,res}$. The plane (W, V) (*right panel*) depicts the evolution of the Trojan body's eccentricity vector under the Hamiltonian H_b. The motion of the endpoint of the eccentricity vector can be decomposed to a circulation around the forced equilibrium, with angular frequency g, and a fast (of frequency ω_f) 'in-and-out' oscillation with respect to a circle of radius e_p, of amplitude which is of order $\mathcal{O}(Y_f)$. Under H_b alone, the quantities Y_p, J_s, or $Y_p, J_{s,res}$ are quasi-integrals for all the regular orbits. Furthermore, all extra terms with respect to H_b in the Hamiltonian (6) depend on the slow angles ϕ. Thus, all these terms can only slowly modulate the dynamics under H_b, and this modulation can produce a long-term drift of the values of (Y_p, J_s), or $Y_p, J_{s,res}$

In a first approximation, the quasi-integral of the proper eccentricity (Érdi 1996) can be defined as

$$e_{p,0} = \sqrt{V^2 + W^2} = \sqrt{-2Y} \ . \tag{19}$$

However, Y is subject to fast variations due to its dependence on Y_f $\omega_f \sim \mathcal{O}(1)$. The time variations of Y_f become particularly important when one of the following two conditions holds: i) $e' < \mu$, or ii) the orbit of the Trojan body is subject to a low-order resonance. On the other hand, since ϕ is ignorable, Y_p remains an exact integral of the Hamiltonian (17) even in the cases (i) or (ii). We thus adopt the following definition of the proper eccentricity:

$$e_p = \sqrt{-2Y_p} \ . \tag{20}$$

Since $\mathscr{F}^{(1)}$ is at least $\mathscr{O}(e')$, $\dot{Y}_p = \mathscr{O}(e')$ under the full Hamiltonian (15). Thus, e_p remains a good quasi-integral for not very high values of the primary's eccentricity (Páez and Efthymiopoulos 2015).

A second averaging over the fast angle ϕ_f yields the Hamiltonian

$$\overline{H_b}(u, v; Y_f, Y_p, e', \varpi') = -\frac{1}{2(1+v)^2} - v + Y_f - \mu\overline{\mathscr{F}^{(0)}}(u, v, Y_p - Y_f; e', \varpi') \tag{21}$$

with

$$\overline{\mathscr{F}^{(0)}} = \frac{1}{2\pi}\int_0^{2\pi} \mathscr{F}^{(0)} \mathrm{d}\phi_f \ .$$

The Hamiltonian $\overline{H_b}(u, v; Y_f, Y_p, e', \varpi)$ represents a system of one degree of freedom, all three quantities Y_f, Y_p, e' serving now as parameters, i.e. constants of motion under the dynamics of $\overline{H_b}$. The Hamiltonian $\overline{H_b}$ describes the synodic (guiding-center) motions of the Trojan body, with the additional point that, since it depends on e', it *does not* correspond to the averaged (over fast angles) Hamiltonian of the circular RTBP. Thus, it allows to find an integrable approximation to synodic motions even when $e' \neq 0$.

The equilibrium point (u_0, v_0) given by

$$\frac{\partial\overline{\mathscr{F}^{(0)}}}{\partial u} = \frac{\partial\overline{\mathscr{F}^{(0)}}}{\partial v} = 0$$

corresponds to a short-period periodic orbit of the Hamiltonian H_b around the forced equilibrium point. We define the action variable

$$J_s = \frac{1}{2\pi}\int_C (v - v_0)\mathrm{d}(u - u_0) \tag{22}$$

where the integration is over a closed invariant curve C around (u_0, v_0) and 's' stands for 'synodic' (see Fig. 1). The angular variable ϕ_s, conjugate to J_s, evolves in time according to the synodic frequency ω_s given by [see Eq. (28)]

$$\omega_s = \dot{\phi}_s = -\sqrt{\frac{27\mu}{4}} + \dots \tag{23}$$

Some manipulation of Eq. (21) allows to find a first order approximation to the values of the frequencies ω_s and g. We deduce the shift in position, with respect to L4, of the fixed point of $\overline{\mathscr{F}^{(0)}}$, corresponding to the short-period orbit around L4 (Namouni and Murray 2000). The shift is given by $u_0 = \tau_0 - \pi/3$, where τ_0 is the solution of $\partial\overline{\mathscr{F}^{(0)}}/\partial\tau = 0$. We find:

$$u_0 = \frac{29\sqrt{3}}{24} e_{p,0}^2 + \dots \tag{24}$$

where the error is of order 4 in the eccentricities $e_{p,0}$, e'.

We introduce the following canonical transformation to analyze the motion around the position of the periodic orbit given by u_0, v_0

$$S_3(\phi_f, u, \phi, V, J_f, J_p) = (u - u_0)V + \phi_f J_f + \phi J_p\,, \tag{25}$$

where, in terms of the new actions, we have $u_0 = \frac{29\sqrt{3}}{12}(J_f - J_p)$, yielding

$$v = V\,,\ Y_f = J_f\,,\quad Y_p = J_p\,,\quad \delta u = u - u_0\,,\quad q_f = \phi_f - \frac{29\sqrt{3}}{12} V \quad q_p = \phi + \frac{29\sqrt{3}}{12} V\,.$$

Since $\overline{H}_b$ in Eq. (21) does not explicitly depend on the angles q_p and q_f, the conjugated actions $J_p = Y_p$ and $J_f = Y_f$ remain integrals of motion. We keep the notation for Y_f, Y_p and v. Taylor-expanding $\overline{H}_b$, around u_0 up to terms of order $\mathscr{O}(\delta u^2)$, we find (up to terms of first order in μ and second order in the eccentricities):

$$\begin{aligned}\overline{H}_{b,ell} &= -\frac{1}{2} + Y_f - \mu\left(\frac{27}{8} + \dots\right)\frac{e_{p,0}^2}{2} - \frac{3}{2}x^2 + \dots \\ &\quad - \mu\left(\frac{9}{8} + \frac{63e'^2}{16} + \frac{129e_{p,0}^2}{64} + \dots\right)\delta u^2 + \dots\end{aligned} \tag{26}$$

where $\frac{e_{p,0}^2}{2} = Y_f - Y_p$. Since Y_f is of order $\mathscr{O}(\mu)$, up to terms linear in μ the part

$$H_{syn} = -\frac{3}{2}v^2 - \mu\left(\frac{9}{8} + \frac{63e'^2}{16} + \frac{129e_p^2}{64} + \dots\right)\delta u^2 \tag{27}$$

defines a harmonic oscillator for the synodic degree of freedom. The corresponding synodic frequency is

$$\omega_s = -\sqrt{6\mu\left(\frac{9}{8} + \frac{63e'^2}{16} + \frac{129e_p^2}{64} + \dots\right)}\,. \tag{28}$$

On the other hand, the secular frequency is given by $g = \partial\overline{H}_b/\partial Y_p$. Assuming a harmonic solution $\delta u = \delta u_0 \cos(\omega_s t + \phi_{0s})$, and averaging over the synodic period $\langle \delta u^2 \rangle = \delta u_0^2/2$, we find

$$g = \mu \left(\frac{27}{8} + \frac{129}{64} \delta u_0^2 + \ldots \right) \ , \tag{29}$$

completing the estimation of the frequencies.

2.1 Secondary Resonances in the ER3BP

The Trojan domain describes itself a resonant regime, defined by the 1:1 commensurability of the mean motions of the Trojan body and the primary. In addition, within this domain, we can find *secondary resonances* of the form

$$m_f \omega_f + m_s \omega_s + mg = 0 \tag{30}$$

with m_f, m_s, m integers. The most important resonances are those involving low order conmensurabilities between ω_f and ω_s, which exist also in the pCR3BP ($e' = 0$). They are of the form

$$\omega_f + n\omega_s = 0 \tag{31}$$

with $n = m_s$. We refer to (31), as the '1:n' resonance. For $m_f = 1$ and μ in the range $0.001 \leq \mu \leq 0.01$, n is in the range $4 \leq n \leq 12$. In the frequency space (ω_f, ω_s, g), the relations (31) represent planes normal to the plane (ω_f, ω_s) which intersect each other along the g–axis. We refer to the resonances with $m \neq 0$ as 'transverse', since they intersect tranversally such planes. In the numerical examples below, we use the notation (m_f, m_s, m), for the integers of the resonant condition (30).

Figure 2 presents stability maps produced by the computation of the chaotic indicator Ψ FLI (Froeschlé et al. 2000), in the space of proper elements $e_{p,0}$ [given in Eq. (19)] and $\Delta u = u - u_0$, with u_0 given in Eq. (24), (for a conspicuous discussion of the initial conditions, see Páez and Efthymiopoulos 2015). In color scale, we differentiate chaotic orbits (yellow) from regular orbits (dark purple). For the different combinations of μ and e' considered, we can distinguish the structures of the resonances 1:m_s and (1,m_s,m), for $m_s = 6, 7, 8, 10$ and $m = \pm 1, \pm 2, 3$. A validation of the resonant nature of the orbits within these structures is done by means of Frequency Analysis (Laskar 2004).

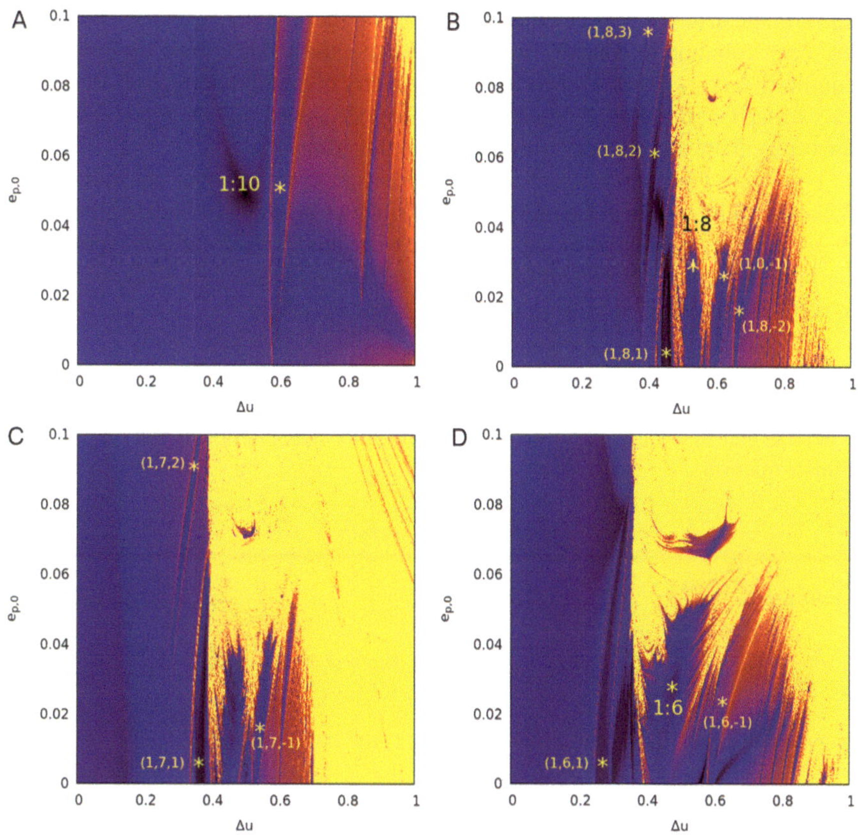

Fig. 2 Stability maps in terms of the FLI chaotic indicator. *Light colors* (*yellow*) correspond to chaotic orbits, *while dark colors* (*dark purple*) correspond to regular orbits. The maps show different resonances appearing for different combinations of physical parameters: $\mu = 0.0016$ and $e' = 0$ (**a**), $\mu = 0.0024$ and $e' = 0.06$ (**b**), $\mu = 0.0031$ and $e' = 0.04$ (**c**), $\mu = 0.0016$ and $e' = 0.02$ (**d**)

3 The Normalized Basic Hamiltonian $\langle H_b \rangle$

As already emphasized, the Trojan motion in the ER3BP has three well separated temporal scales. The most basic form of Hamiltonian normalization stems from averaging the Hamiltonian over the fast angles. Independently of the formalism used, what remains after such averaging gives the synodic motion around the libration center. However, the Hamiltonian of the ER3BP has a real singularity corresponding to close encounters of the massless body with the primary m'. This singularity corresponds to $a = a'$, $\tau = \lambda - \lambda' = 0$ and it is inherited by the H_b. The key remark is that any *polynomial* series expansion of the equations of motion (or the Hamiltonian) with respect to τ around a fixed value is convergent in a disk of radius equal to the distance between the fixed point and the singularity. In the literature, it has been common to consider such polynomial expansions around the position of equilibrium ($\tau_{L_4,L_5} = \pm\frac{\pi}{3}$). Due to the asymmetry of the librations (Érdi 1996), it is easy to see that the above limited convergence affects severely the representation of the orbits mainly in the opposite direction to the primary.

Following Páez and Locatelli (2015), we average the H_b over the short period *without* making expansions affected by the singular behavior of the Hamiltonian at $|\tau - \pi/3| = \pi/3$. We start by expressing the basic model H_b in variables appropriate for introducing the normalization scheme of Páez and Locatelli (2015). Hence, the synodic degree of freedom is re-expressed by the variables

$$x = v + x_0, \quad \tau = u + \tau_0 \ , \tag{32}$$

where x_0 and τ_0 are given in Eq. (10). We introduce new canonical pairs, though the transformation

$$\mathscr{S}_3 = (Y_f - Y_p)\theta_1 + Y_p\theta_2 + x\theta_3 \ , \tag{33}$$

yielding

$$\mathscr{Y} = Y_f - Y_p \ , \quad Y_p = Y_2 \ , \quad x = Y_3 \ , \quad \theta_1 = \phi_f \ , \quad \theta - \theta_1 = \phi \ , \quad \theta_3 = \tau \ . \tag{34}$$

Keeping the previous notation for Y_p, x, ϕ_f, τ, the basic model H_b reads

$$H_b = -\frac{1}{2(1+x)^2} - x + \mathscr{Y} + Y_p - \mu\mathscr{F}^{(0)}(x, \mathscr{Y}, \tau, \phi_f; e') \ . \tag{35}$$

In terms of these variables, the dependence of H_b on τ is of the form $\frac{\cos^{k_1}\tau}{(2-2\cos\tau)^{j/2}}$ or $\frac{\cos^{k_2}\tau}{(2-2\cos\tau)^{j/2}}$, $j = 2n - 1$ with k_1, k_2 and n integers. Additionally, we express the Hamiltonian in terms of modified Poincaré variables,

$$x \ , \quad \tau \ , \quad \xi = \sqrt{2\mathscr{Y}}\cos\phi_f \ , \quad \eta = \sqrt{2\mathscr{Y}}\sin\phi_f \ . \tag{36}$$

The new expression for the Hamiltonian reads

$$H_b(x, \tau, \xi, \eta, Y_p) = -\frac{1}{2(1+x)^2} - x + Y_p + \frac{\xi^2 + \eta^2}{2} - \mu\mathscr{F}^{(0)}(\tau, x, \xi, \eta; Y_p, e') \ . \tag{37}$$

Finally, we expand the Hamiltonian in terms of every variable except τ, obtaining

$$
\begin{aligned}
H_b(x,\tau,\xi,\eta,Y_p) = & -x + \sum_{i=0}^{\infty}(-1)^{i-1}(i+1)\frac{x^i}{2} + \frac{\xi^2+\eta^2}{2} + Y_p \\
& + \mu \sum_{\substack{m_1,m_2,m_3 \\ k_1,k_2,k_3,j}} a_{m_1,m_2,m_3,k_1,k_2,j}\, e'^{k_3} x^{m_1}\, \xi^{m_2}\, \eta^{m_3}\, \cos^{k_1}(\tau)\, \sin^{k_2}(\tau)\, \beta^j(\tau)\ ,
\end{aligned}
\tag{38}
$$

where the $a_{m_1,m_2,m_3,k_1,k_2,j}$ are constant coefficients and $\beta(\tau) = \frac{1}{\sqrt{2-2\cos\tau}}$. The Hamiltonian H_b in (38) corresponds to the 'zero-th' step in the normalizing scheme, i.e., before any normalization. This is denoted as $H^{(1,0)}$.

3.1 Normalization Scheme

The normalizing algorithm defines a sequence of Hamiltonians by an iterative procedure. Let us first introduce the following definition

Definition 1 A generic function $g = g(x,\xi,\tau,\eta)$ belongs to the class $\mathscr{P}_{l,s}$, if its expansion is of the type:

$$
\sum_{2m_1+m_2+m_3=l} \ \sum_{\substack{k_1+k_2\le l+4s-3 \\ j\le 2l+7s-6}} c_{m_1,m_2,m_3,k_1,k_2,j}\, \mu^s\, \rho^{m_1} \xi^{m_2} \eta^{m_3} (\cos\tau)^{k_1} (\sin\tau)^{k_2} \left(\beta(\tau)\right)^j ,
$$

where the real coefficients $c_{m_1,m_2,m_3,k_1,k_2,j}$ gather also the dependence on the primary's eccentricity e'.

At a generic normalizing step (r_1,r_2), the expansion of the Hamiltonian is given by

$$
\begin{aligned}
H^{(r_1,r_2)}(x,\xi,\tau,\eta,Y_p) = & \ Y_p + \frac{\xi^2+\eta^2}{2} + \sum_{l\ge 4} Z_l^{(0)}\left(x,(\xi^2+\eta^2)/2\right) \\
& + \sum_{s=1}^{r_1-1}\left(\sum_{l=0}^{R_2}\mu^s Z_l^{(s)}\left(x,(\xi^2+\eta^2)/2,\tau\right) + \sum_{l>R_2}\mu^{r_1} f_l^{(r_1,r_2-1;s)}(x,\xi,\eta,\tau)\right) \\
& + \sum_{l=0}^{r_2}\mu^{r_1} Z_l^{(r_1)}(x,(\xi^2+\eta^2)/2,\tau) + \sum_{l\ge r_2+1}\mu^{r_1} f_l^{(r_1,r_2-1;r_1)}(x,\tau,\xi,\eta) \\
& + \sum_{s>r_1}\sum_{l\ge 0}\mu^s f_l^{(r_1,r_2-1;s)}(x,\tau,\xi,\eta)\ .
\end{aligned}
\tag{39}
$$

All the terms $Z_l^{(s)}$ and $f_l^{(r_1,r_2;s)}$ appearing in (39) are produced by expansions including a *finite* number of monomials of the type given by the class $\mathscr{P}_{l,s}$. More specifically $Z_l^{(0)} \in \mathscr{P}_{l,0}\ \forall\ l \geq 4$, $Z_l^{(s)} \in \mathscr{P}_{l,s}\ \forall\ 0 \leq l \leq R_2\,,\ 1 \leq s < r_1\,,$ $Z_l^{(r_1)} \in \mathscr{P}_{l,r_1}\ \forall\ 0 \leq l < r_2\,, f_l^{(r_1,r_2-1;r_1)} \in \mathscr{P}_{l,r_1}\ \forall\ l \geq r_2\,, f_l^{(r_1,r_2-1;s)} \in \mathscr{P}_{l,s}\ \forall\ l > R_2\,,\ 1 \leq s < r_1$ and $\forall\ l \geq 0,\ s > r_1$. We can distinguish the terms in normal form Z (i.e. the terms depending on ξ and η exclusively through $(\xi^2+\eta^2)/2$), of order up to r_1 and r_2, from those that still keep a generic dependence on these variables.

The (r_1, r_2)th step of the algorithm formally defines the latter Hamiltonian $H^{(r_1,r_2)}$ by

$$H^{(r_1,r_2)} = \exp\left(\mathscr{L}_{\mu^{r_1}\chi_{r_2}^{(r_1)}}\right) H^{(r_1,r_2-1)}\ , \tag{40}$$

where the Lie series operator is

$$\exp \mathscr{L}_\chi \cdot = \mathbb{I}\ \cdot + (\mathscr{L}_\chi \cdot) + \frac{1}{2}(\mathscr{L}_\chi^2 \cdot) + \ldots \tag{41}$$

and

$$\mathscr{L}_\chi \equiv \{\,\cdot\,, \chi\}, \tag{42}$$

is the Poisson bracket with χ. The generating function $\mu^{r_1}\chi_{r_2}^{(r_1)}$ is determined by solving the following homological equation with respect to the unknown $\chi_{r_2}^{(r_1)} = \chi_{r_2}^{(r_1)}(x, \xi, \tau, \eta)$:

$$\mathscr{L}_{\mu^{r_1}\chi_{r_2}^{(r_1)}} Z_2^{(0)} + f_{r_2}^{(r_1,r_2-1;r_1)} = Z_{r_2}^{(r_1)}\ , \tag{43}$$

where $Z_{r_2}^{(r_1)}$ is the new term in the normal form, and $Z_2^{(0)}$ represents the kernel of the homological equation. By construction, the Hamiltonian produced at ever step inherits the structure presented in (39). From the latter, we point out that the splitting of the Hamiltonian in sub-functions of the form $\mathscr{P}_{l,s}$, organizes the terms in groups with the same order of magnitude μ^s and total degree $l/2$ (possibly semi-odd) in the variables x and $\mathscr{Y} = \frac{\xi^2+\eta^2}{2}$.

Let R_1 and R_2 be the maximum orders considered for the normalization scheme, thus the algorithm requires $R_1 \cdot R_2$ normalization steps, constructing the finite sequence of Hamiltonians $H^{(1,0)} = H_b, H^{(1,1)}, \ldots, H^{(R_1,R_2)}$. We remark here that $H^{(r_1+1,0)} = H^{(r_1,R_2)}\ \forall\ 1 \leq r_1 \leq R_1$. Hence, the final Hamiltonian, reads

$$H^{(R_1,R_2)}(x, \xi, \tau, \eta, Y_p) = Z^{(R_1,R_2)}\left(x, \frac{(\xi^2+\eta^2)}{2}, \tau, Y_p\right) + R^{(R_1,R_2)}(x, \xi, \tau, \eta)\ , \tag{44}$$

where we distinguish the normal form $Z^{(R_1,R_2)}$ from the remainder $R^{(R_1,R_2)}$. While the dependence of $Z^{(R_1,R_2)}$ on x and τ remains generic, it depends on ξ and η *only* though the form $\frac{\xi^2+\eta^2}{2}$. Thus, we have

$$H^{(R_1,R_2)}(x,\tau,\mathscr{Y},\phi_f,Y_p) = Z^{(R_1,R_2)}\left(x,\tau,\mathscr{Y},Y_p\right) + R^{(R_1,R_2)}(x,\tau,\mathscr{Y},\phi_f)\ . \tag{45}$$

The key remark is that ϕ_f becomes ignorable in the normal form and, therefore, $\mathscr{Y}$ becomes an integral of motion of $Z^{(R_1,R_2)}$. Then, the normal form can be viewed as a Hamiltonian of one d.o.f. depending on two constant actions $\mathscr{Y}$ and Y_p, i.e. $Z^{(R_1,R_2)}$ represents now a formally integrable dynamical system. In all subsequent computations, we fixed the values $R_1 = 2$ and $R_2 = 4$, corresponding to a second order expansion and truncation on the mass parameter μ and fourth order for the polynomial degree of ξ and η (second order expansion in the eccentricity e; note also that the expansion is of second order as well in the primary's eccentricity e'). In the following, these normalization orders are shown to be sufficient for the normal form to provide a good representation of the original Hamiltonian in the domain of regular motions.

4 Application: Location of the Resonances by Means of the $\langle H_b \rangle$

The obtention of a normal form by averaging the basic Hamiltonian allows to extract information of the resonant structure by pure analytical means. In this section, we focus on the use of the normal form approximation $Z^{(R_1,R_2)}$ in (45) for the computation of the values of the three main frequencies of motion. With these values, it is possible to locate the position of the most important resonances for a certain combination of physical parameters.

Consider an orbit with initial conditions as specified in terms of the two parameters Δu and $e_{p,0}$ as in the stability map of Fig. 2. The computation proceeds by the following steps.

1) We first evaluate the synodic frequency ω_s, i.e., the frequency of libration of the synodic variables τ and x. The normal form $Z^{(R_1,R_2)}$ leads to Hamilton's equations:

$$\frac{\mathrm{d}x}{\mathrm{d}t} = f(x,\tau;\mathscr{Y}) = -\frac{\partial Z^{(R_1,R_2)}}{\partial \tau} \tag{46}$$

and

$$\frac{\mathrm{d}\tau}{\mathrm{d}t} = g(x,\tau;\mathscr{Y}) = \frac{\partial Z^{(R_1,R_2)}}{\partial x}\ . \tag{47}$$

For every orbit we can define the constant energy

$$Z^{(R_1,R_2)}(x,\tau;\mathscr{Y},Y_p) - Y_p \equiv \zeta^{(R_1,R_2)}(x,\tau;\mathscr{Y}) = \mathscr{E} \ . \tag{48}$$

Note that since Y_p appears only as an additive constant in $Z^{(R_1,R_2)}$, the function $\zeta^{(R_1,R_2)}$ does not depend on Y_p. Also, according to (19) and (36), we have $\mathscr{Y} = \frac{e_{p,0}^2}{2}$. Then, for a fixed value of $\mathscr{E}$, if $\frac{\partial \zeta^{(R_1,R_2)}}{\partial \tau} \neq 0$, we can express τ as an explicit function of x,

$$\zeta^{(R_1,R_2)}(x,\tau;\mathscr{Y}) = \mathscr{E} \quad \Longrightarrow \quad \tau = \tau(\mathscr{E},x;\mathscr{Y}) \ . \tag{49}$$

Thus, replacing in (46),

$$\frac{\mathrm{d}x}{\mathrm{d}t} = f(x,\tau(\mathscr{E},x;\mathscr{Y});\mathscr{Y}) \quad \Longrightarrow \quad \mathrm{d}t = \frac{\mathrm{d}x}{f(x,\tau(\mathscr{E},x;\mathscr{Y});\mathscr{Y})} \ , \tag{50}$$

by which we can derive an expression for the synodic period T_{syn}

$$T_{syn} = \oint \frac{\mathrm{d}x}{f(x,\tau(\mathscr{E},x;\mathscr{Y});\mathscr{Y})} \ , \tag{51}$$

and thus the synodic frequency is $\omega_s = \frac{2\pi}{T_{syn}}$. In practice, (49) is hard to invert analytically, and likewise, the integral (51) cannot be explicitly computed. We thus compute both expressions numerically on grids of points of the associated invariant curves on the plane (τ, x), or by integrating numerically (50) as a first order differential equation (we found that the latter method is more precise than the former).

2) We now compute the fast and secular frequencies ω_f, g. From Eq. (48), we find $\dot{\theta} = \frac{\partial \mathscr{Z}^{(R_1,R_2)}}{\partial Y_p} = 1$, implying $g = 1 - \omega_f$. To compute ω_f, we use the equation

$$\omega_f = \frac{1}{T_{syn}} \int_0^{T_{syn}} \frac{\mathrm{d}\phi_f}{\mathrm{d}t}\,\mathrm{d}t = \frac{1}{T_{syn}} \int_0^{T_{syn}} \frac{\partial Z^{(R_1,R_2)}(x,\tau;\mathscr{Y})}{\partial \mathscr{Y}}\,\mathrm{d}t \ . \tag{52}$$

Replacing (50) in (52), we generate an explicit formula for the fast frequency

$$\omega_f = \frac{1}{T_{syn}} \oint \frac{1}{f(x,\tau(\mathscr{E},x;\mathscr{Y});\mathscr{Y})} \frac{\partial Z^{(R_1,R_2)}(x,\tau(\mathscr{E},x;\mathscr{Y});\mathscr{Y})}{\partial \mathscr{Y}}\,\mathrm{d}x \ . \tag{53}$$

Both frequencies ω_f and ω_s are functions of the labels $\mathscr{E}$ and $\mathscr{Y}$, which, in the integrable normal form approximation, label the proper libration and the proper eccentricity of the orbits. In the normal form approach one has $e_{p,0} = e_p = \mathrm{const}$, implying $\mathscr{Y} = e_p^2/2$. If, as for the FLI maps in Fig. 2 (see Páez and Efthymiopoulos 2015), we fix a scanning line of initial conditions of the form $x_{in} = B\,u_{in} = B\,(\tau_{in} - \tau_0)$, with B a constant, the energy $\mathscr{E}$, for fixed e_p, becomes a function of the initial condition u_{in} only. Thus, u_{in} represents an alternative label of the proper libration (Érdi 1978). With these conventions, all three frequencies become functions of the labels (u_{in}, e_p). A generic resonance condition then reads

$$\Phi_{m_f,m_s,m} = m_f\omega_f(e_p, u_{in}) + m_s\omega_s(e_p, u_{in}) + mg(e_p, u_{in}) = 0 \ . \tag{54}$$

For fixed resonance vector (m_f, m_s, m), (54) can be solved by root-finding, thus specifying the position of the resonance in the plane of the proper elements (u_{in}, e_p).

In order to test the accuracy of the above method, we compare the results of the analytical estimation with the position of the resonances extracted from the FLI map. Under the assumption that the local minimum of the FLI in the vicinity of a resonance gives a good approximation of the resonance center, we study the curves of the FLI Ψ as a function of the libration amplitude Δu, for a fixed value of $e_{p,0}$. The confirmation of the resonant nature of the candidate initial conditions is done by means of Frequency Analysis (Laskar 2004). By changing the value of $e_{p,0}$ along the interval $[0, 0.1]$, we can depict the centers of the resonances on top of the FLI map.

Figure 3 shows an example of these computations, for the parameters $\mu = 0.0024$ and $e' = 0.06$. The normal form predictions are superposed as yellow lines upon the underlying the FLI stability map (panel (B), Fig. 2) and the resonant candidates extracted from the FLI maps denote the green curves. Due to the noise in the FLI curves, it is not possible to clearly extract the position of the resonance

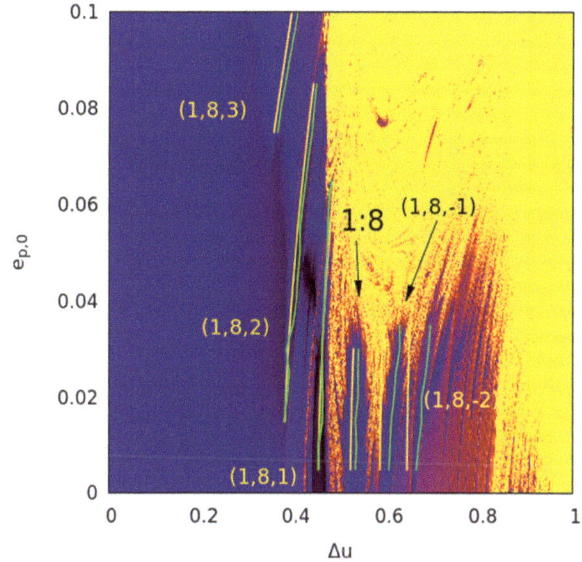

Fig. 3 Location of the center of different resonances by means of the normal form $\langle H_b \rangle$ (*yellow*) and the minima of the FLI indicator (*green*), on top of the corresponding FLI stability map for $\mu = 0.0024$ and $e' = 0.06$

Table 1 Averaged values of $u_{\mathscr{Z}}$, u_{Ψ} and δu_{in} for the resonances in Fig. 3 for $\mu = 0.0024$ and $e' = 0.06$

Resonance	$\overline{u_{\mathscr{Z}}}$	$\overline{u_{\Psi}}$	$\overline{\delta u_{in}}$
1:8	0.524485	0.535153	$1.993063{\times}10^{[-2]}$
(1, 8, 1)	70.465475	0.464924	$6.377401{\times}10^{[-3]}$
(1, 8, 2)	0.406439	0.412246	$1.605145{\times}10^{[-2]}$
(1, 8, 3)	0.374879	0.385020	$2.617987{\times}10^{[-2]}$
(1, 8, −1)	0.587834	0.616093	$4.572688{\times}10^{[-2]}$
(1, 8, −2)	0.646464	0.679154	$4.796435{\times}10^{[-2]}$

centers for all values of $e_{p,0}$, while an analytic estimation (with varying levels of accuracy) is always possible. At any rate, in Fig. 3 we plot the values of the centers only in the cases when both methods provide clear results.

Table 1 summarizes the results for the location of the centers ($u_{\mathscr{Z}}$, u_{Ψ}) and the relative errors ($\delta u_{in} = \frac{|u_{\mathscr{Z}} - u_{\Psi}|}{u_{\Psi}}$), on average, for the resonances shown in the figure. We can note that the level of approximation is very good for relatively low values of u_{in}, while the error in the predicted position of the resonance increases to a few percent for greater values. Nevertheless, we demonstrate the overall efficiency of the normal form approach in order to analytically determine the locus of resonances in the space of proper elements. More detailed presentations of the above methods will be given in forthcoming publication.

Acknowledgements During this work, RIP was fully supported by the Astronet-II Marie Curie Initial Training Network (PITN-GA-2011-289240).

References

Brown, E.W., Shook, C.A.: Planetary Theory. Cambridge University Press, New York (1933)
Deprit, A., Delie, A.: Trojan orbits I. d'Alembert series at L_4. Icarus **4**, 242–266 (1965)
Érdi, B.: An asymptotic solution for the Trojan case of the plane elliptic restricted problem of three bodies. Celest. Mech. Dyn. Astron. **15**, 367–383 (1977)
Érdi, B.: The three-dimensional motion of Trojan asteroids. Celest. Mech. Dyn. Astron. **18**, 141–161 (1978)
Érdi, B.: The Trojan problem. Celest. Mech. Dyn. Astron. **65**, 149–164 (1996)
Froeschlé, C., Guzzo, M., Lega, E.: Graphical evolution of the Arnold web: from order to chaos. Science **289**(5487), 2108–2110 (2000)
Garfinkel, B.: Theory of the Trojan asteroids, Part I. Astron. J. **82**(5), 368–379 (1977)
Gascheau, G.: Examen d'une classe d'équations difféntielles et applicaction à un cas particulier du problème des trois corps. Compt. Rendus **16**(7), 393–394 (1843)
Gopalswamya, N. et al.: Earth-affecting solar causes observatory (EASCO): A potential international living with a star mission from Sun-Earth L5. J. Atmos. Sol. Terr. Phys. **73**(5–6), 658–663 (2011)
Laskar, J.: Frequency map analysis and quasiperiodic decompositions. In: Benest, D., Froeschlé, C., Lega, E. (eds.) Hamiltonian Systems and Fourier Analysis, pp. 99–134. Cambridge Scientific, Cambridge (2004)
Morais, M.H.M.: Hamiltonian formulation on the secular theory for a Trojan-type motion. Astron. Astrophys. **369**, 677–689 (2001)

Murray, C.D., Dermott, S.F.: Solar Systems Dynamics. Cambridge Universiy Press, Cambridge (1999)
Namouni, F.: Secular interactions of coorbiting objects. Icarus **137**(2), 293–314 (1999)
Namouni, F., Murray, C.D.: The effect of eccentricity and inclination on the motion near the Lagrangian points L_4 and L_5. Celest. Mech. Dyn. Astron. **76**(2), 131–138 (2000)
Páez, R.I., Efthymiopoulos, C.: Trojan resonant dynamics, stability and chaotic diffusion, for parameters relevant to exoplanetary systems. Celest. Mech. Dyn. Astron. **121**(2), 139–170 (2015)
Páez, R.I., Locatelli, U.: Trojan dynamics well approximated by a new Hamiltonian normal form. Mon. Not. R. Astron. Soc. **453**(2), 2177–2188 (2015)

On Distributed Control Strategies for Spacecraft Formation Flying

Fabrizio Paita, Gerard Gómez, and Josep J. Masdemont

Abstract In this paper we consider the problem of designing control strategies for formation flying. Although stemming from particular goals and controls, the analysis performed here outlines several aspects independent from the dynamics or the control objective. First, we describe some of the difficulties that one faces when controlling formations under different vector fields. Secondly, we see how the dynamics can also be exploited to design controls in an advantageous way. Finally, we introduce a statistical approach useful to derive information on the efficiency of these kind of controls.

1 Introduction

Spacecraft formation flying is a well established concept, with several missions having been planned (examples include Darwin and Proba 3 from ESA, TPF, Grace and Grail from NASA, and the Swedish Prisma project). The key advantage lies in the possibility of dividing the payload among the spacecraft, which in turn allows to employ simple designs for the spacecraft and to achieve a great mission stability.

Of course, the larger the formation is the more activities it can perform. However, this also leads to a greater involvement of ground control in controlling the spacecraft. From this arise the great effort of designing control laws which allow the spacecraft to operate in an autonomous manner.

F. Paita (✉)
Universitat Politècnica de Catalunya, Avinguda Diagonal 647, Barcelona, Spain
e-mail: fabrizio@maia.ub.es

J.J. Masdemont
Institut d'Estudis Espacials de Catalunya and Universitat Politécnica de Catalunya, Diagonal 647, 08028 Barcelona, Spain
e-mail: josep.masdemont@upc.edu

G. Gómez
Institut d'Estudis Espacials de Catalunya and Universitat de Barcelona, Gran Via 585, Barcelona, Spain
e-mail: gerard@maia.ub.es

G. Gómez, J.J. Masdemont (eds.), *Astrodynamics Network AstroNet-II*,
Astrophysics and Space Science Proceedings 44,
DOI 10.1007/978-3-319-23986-6_15

To achieve this, a natural solution is to design algorithms relying on the relative data of the spacecraft to determine the control maneuvers. This approach, sometimes referred to as behavioural (Lawton and Beard 2002; Ren and Beard 2004), can be actually seen as part of the extensive research on the consensus dynamics of multiple agents (Vicsek and Zafeiris 2012).

That said, many control algorithms based on consensus dynamics usually focus on the asymptotic convergence of the formation to a common status. With the goal of giving a broader view on the various aspects of these algorithms, in this paper we present in an unified fashion results coming from different works (Paita et al. 2013, 2014, 2015).

In particular, in Sect. 2 of this paper we first introduce the language of graphs (necessary to modellize the informations exchange among the spacecraft) and then a particular translational consensus model, suggested by Cucker and Smale (2007), which will serve as test bench for our study.

Then, in Sect. 3 we transform this model in a control for the translational dynamics of a formation. By employing it in a keplerian context, we are able to outline the limitations imposed on formation flying controls by the dynamics and the propulsion used for the spacecraft.

Instead, in Sect. 4 we consider distributed control strategies in the context of attitude synchronization, where we introduce a statistical approach aimed to unravel the various dependencies of the controlled dynamics, and we see how to exploit the dynamics in order to surpass some of the given limitations.

Finally, in Sect. 5 we summarize the results presented in this paper. We remark that, although a lot things are taken for granted, references are given where these aspect are studied more in-depth.

2 Analytical Background

2.1 Graph Theory

In the present subsection we recall the meaning of graph and of some related concepts. These offer a natural way to model interactions among multiple agents (regardless of the dynamics) and are therefore used throughout the paper, particularly in Sect. 4. A graphical example to couple with these notions is given in Fig. 1.

First, we define a *graph* as a pair (X, Ω), where X is a finite non-empty set of indexed nodes (the agents) and Ω is a set of pairs of nodes, which we call *edges* (the links). These edges can have some weights attached to them, which indicate the strength of the link. Graphs with weighted edges are referred to as *weighted graphs*.

In an *undirected graph* the nodes connected by an edge (x_i, x_j) both pass informations to the other. In contrast, in a *digraph* (directed graph) every edge has a direction determined by the order of the pair (e.g., (x_i, x_j) means that x_i is passing information to x_j but not vice-versa).

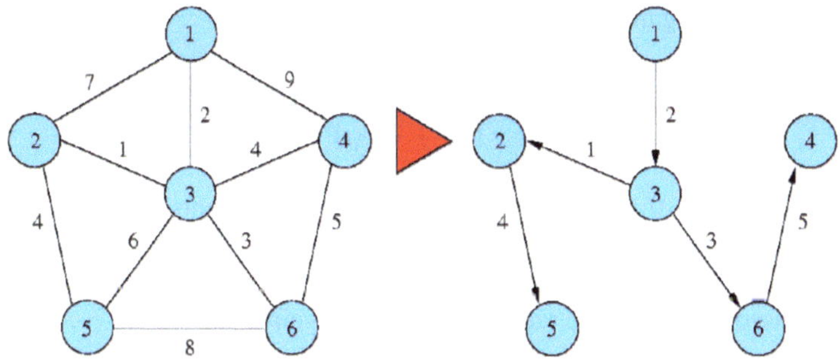

Fig. 1 Examples of weighted graphs, with an undirected connected one on the *left* and a directed one, possessing a directed spanning tree and derived from the first by assigning directions to the edges, on the *right*

We say that a graph possesses a *path* if there exists a subset of Ω comprised of consecutive edges $(x_{i_1}, x_{i_2}), (x_{i_2}, x_{i_3}), \ldots, (x_{i_{N-1}}, x_{i_N})$, with $\{i_1, \ldots, i_N\}$ denoting the indexes of the nodes. Additionally, we define as *root* the first node of the path.

With the concept of path we can describe some ways in which informations spread across the graph. In particular, we say that an undirected graph is *connected* if there is an undirected path between any distinct pair of nodes. Moreover, we call *directed tree* a digraph where every node has exactly one parent, except for the root which we assume to have none. Finally, we say that a *directed spanning tree* of a digraph is a directed tree formed by edges that connect all the nodes in the graph.

2.2 A Motivating Example: The Cucker-Smale Model

In the present subsection we introduce a translational model to be used both as a test bench for the effect of the dynamics on the strategies under study and as an inspiration for improving their design.

First introduced in 2007 (Cucker and Smale 2007), this model is inspired by the behaviour of flocks of animals and has seen a lot of additional work, ranging from modifications of the control itself like Park et al. (2010) and Motsch and Tadmor (2011) to direct applications in astrodynamics like Paita et al. (2014) and Perea et al. (2009). Here we are interested in its original formulation where, if we denote with x_i and v_i the three-dimensional position and velocity of the ith element of a flock of $N+1$ agents, the equations of motion can be written as

$$\dot{x}_i = v_i, \quad \dot{v}_i = -\sum_{j \in \mathscr{L}(i)} a_{ij} \left(v_i - v_j \right), \tag{1}$$

where the leadership sets $\mathscr{L}(i)$ and the weights a_{ij} determine respectively the communication graph and the weights associated to the edges.

In particular, the gains a_{ij} are defined as

$$a_{ij} = a_{ij}(t) = \frac{K}{\left(\sigma^2 + \|x_i(t) - x_j(t)\|^2\right)^{\beta}}, \tag{2}$$

where $K > 0$, $\sigma^2 > 0$ and $\beta \geq 0$ are a given set of constants. Scaling the weights with respect to the relative distances ensures that every agent pays more attention to the companions which are closest to it. The term σ^2 helps avoiding singularities, while β is used to control the rate of decay of the gains.

In the related literature, we can find either an all-to-all communication structure (Cucker and Smale 2007) (where $\mathscr{L}(i) = N$ for every i) or a hierarchical one (Shen 2008) defined by

$$a_{ij} \neq 0 \Rightarrow j < i, \quad \forall i > 0, \ \mathscr{L}(i) \neq \emptyset. \tag{3}$$

The second condition says that every element has to communicate with someone, while the first, given the natural order induced by the indexation of the flock as $[0, 1, \ldots, N]$, asks for every element i to communicate only with the elements j such that $j < i$ (only exception is agent 0). The structures correspond respectively to an undirected connected graph and a digraph with a spanning tree.

Under this model, the agents velocities converge to a common value exponentially fast (which corresponds to the root one in case of hierarchy). This implies that the formation will move as a single rigid body, potentially led about a predefined trajectory. A similar result has been proved for an accelerating leader (Shen 2008) (with the acceleration decaying sufficiently fast), thus pointing to the possibility of steering the formation from the original trajectory without spreading.

More importantly, the works mentioned here contain results outlining the dependence of the dynamics from elements inherent in the formation (dimension, relative initial conditions and graph structure). It is in light of these results that the analysis performed in Sect. 4 is actually conducted.

3 Translational Dynamics

3.1 Implementation

The simplest way to implement the Cucker-Smale model as a control for spacecraft formation flying has been suggested in Perea et al. (2009). If $x_i = x_i(t)$ and $v_i = v_i(t)$ denote, respectively, the three-dimensional *relative* position and velocity of the ith satellite with respect to a reference inertial orbit (x_0, v_0), and we define

$$x = (x_1, \ldots, x_N)^T, \quad v = (v_1, \ldots, v_N)^T, \tag{4}$$

with N being the dimension, then the dynamics reads as

$$\dot{x} = v, \quad \dot{v} = f(x, v) + u(x, v), \tag{5}$$

where u denotes the adaptation of the Cucker-Smale model as

$$[u(x,v)]_i = -\sum_{j\in\mathscr{L}(i)} a_{ij}\left(v_i - v_j\right) - [f(x,v)]_i, \tag{6}$$

and f the relative vector field.

As already discussed in Perea et al. (2009), a continuous application of a control depends on continuous thrusters and continuous availability of relative data. Since these may not be true for a real space mission, it is necessary to discretize the control. To do so, let t_{i-1} and t_i be the instants of two consecutive maneuvers and

$$\pi_n = \{t_{i-1} = t_{i,0}, t_{i,1}, \ldots, t_{i,n} = t_i\}, \tag{7}$$

a partition of the interval defined by them, where, for each epoch $t_{i,j}$, each satellite can measure relative positions, velocities and accelerations. Then, denoting by $\Delta t_{i-1,j} = t_{i,j} - t_{i,j-1}$ the time span between two consecutive estimates of the relative data, we can consider an approximated impulsive maneuver Δu_i at time t_i as

$$\Delta u_i = \int_{t_{i-1}}^{t_i} u(t)\,dt \approx \sum_{j=1}^{n} u\left(t_{i,j}\right)\Delta t_{i,j}. \tag{8}$$

While the previous note on propulsion limits remains generally valid, the performance may vary depending on the vector field acting on the spacecraft. In the present paper we operate in a keplerian context which, in relative coordinates, can be described by the Hill's equations

$$\ddot{x} - 2n\dot{y} - n^2(x + r_l) = -\frac{\mu}{r_f^3}(x + r_l),\quad \ddot{y} + 2n\dot{x} - n^2 y = -\frac{\mu}{r_f^3}y,\quad \ddot{z} = -\frac{\mu}{r_f^3}z. \tag{9}$$

Here r_l and r_f denote respectively the inertial radius of the reference (moving in a circular orbit) and of the agent, n the mean motion of the former and μ the gravitational parameter. For the interested reader, a more complete analysis of the vector field effect can be derived from Perea et al. (2009) (where the control is employed to drive a formation along an Earth-Moon transfer orbit) and Paita et al. (2014).

3.2 Numerical Discussion

In this subsection, we employ the Cucker-Smale control with Eqs. (9) in order to understand what kind of difficulties can be encountered in this context. Many details are skipped, but they can be found in Paita et al. (2014).

As a first step, in Fig. 1, we show the 30 days evolution of a six spacecraft formation around a reference orbit (where a leader is supposed to be moving). As seen in the frame relative to this orbit (left), the spacecraft appear somewhat fixed, since they all end up moving at the same velocity. In a frame centered in the central body (Earth) center of mass (middle) this translates to the fact that, similarly to the reference, they move in circular orbits either displaced or stretched (depending on their initial positions). Of course, as shown on the right, this is an asymptotical process, which requires the velocities to align and that is maintained by the fact that the spacecraft are periodically cancelling their relative acceleration.

However, when the control is discretized (here we have an impulsive maneuver every 10^{-1} s), this process holds true only for short periods of time. In fact, as shown in Fig. 3 (where three spacecraft initially placed on the relative semiaxes at the same distance from the reference are acting for 30 days under the controlled dynamics), this dependence gets worse when the relative acceleration with respect to the reference increases (as given by the reference semi-major axis and the initial relative distance), with the spacecraft gradually losing their relative position with respect to the reference.

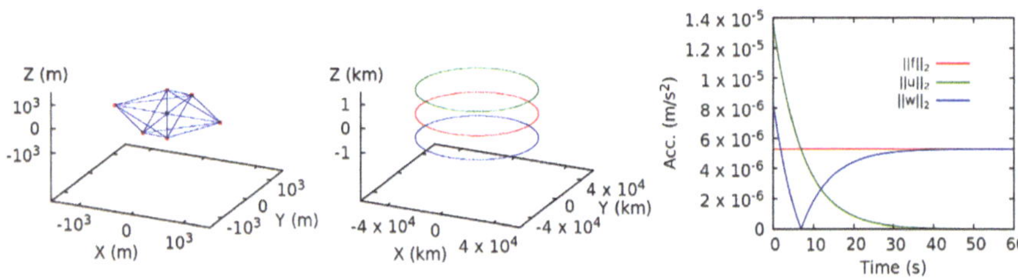

Fig. 2 Hill's frame formation evolution under 30 days of integration (*left*), inertial frame evolution for the leader and the "polar" spacecraft (*center*) and time profile of the thrust components for the $Z > 0$ spacecraft

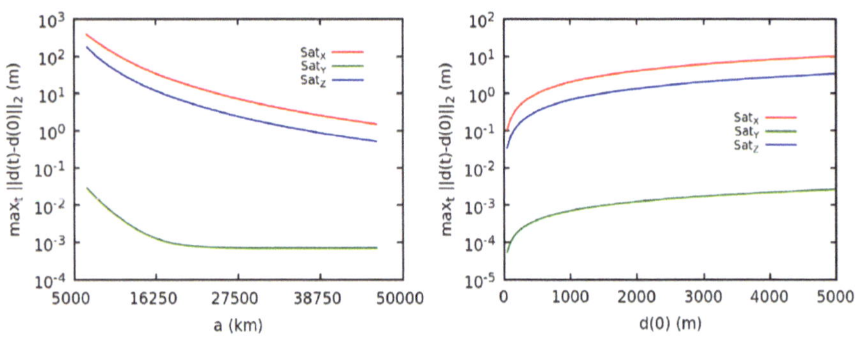

Fig. 3 *Left:* Maximal variation from initial leader-follower distance w.r.t. semi-major axis of the leader orbit. *Right:* Maximal variation from initial leader-follower distance w.r.t. the latter. Integration time is 30 days

This, as better explained in Paita et al. (2014), is actually caused by the fact that, by changing the velocities of the spacecraft, one is changing their energies. Thus, unless the control is applied in a continuous manner, all sorts of bad things happen (like satellites falling on the reference orbit). Therefore, employing such a control in a keplerian context requires either a very weak gravitational field (like an asteroid) or a continuous propulsion (like solar sails).

4 Attitude Synchronization

4.1 Formalism

In this subsection we introduce some notations regarding attitude dynamics to be used in the rest of the section. To describe the attitude of a spacecraft, we parametrize the rotation group through unitary quaternions $q = (\bar{q}, \hat{q})$ (where $\bar{q} = \cos\left(\frac{\theta}{2}\right)$ represents the scalar part of the quaternion and $\hat{q} = \hat{n}\sin\left(\frac{\theta}{2}\right)$ the vectorial part), with its conjugate denoted as $q^\star = (\bar{q}, -\hat{q})$.

Additionally we describe the attitude dynamics of a spacecraft in its own principal frame via the Euler's equations (Hughes 1986)

$$\dot{\bar{q}} = -\frac{1}{2}\omega \cdot \hat{q}, \quad \dot{\hat{q}} = -\frac{1}{2}\omega \times \hat{q} + \frac{1}{2}\bar{q}\omega, \quad I\dot{\omega} + \omega \times I\omega = \tau, \tag{10}$$

where I denotes the tensor of inertia of the spacecraft, ω its angular velocity and τ the torque due to external and control forces. In what follows, we assume that, where appropriate, all the vectors have been transformed and represented in the same coordinate frame.

4.2 Control Laws Examples

In this section we propose consensus control strategies to tackle the problem of attitude synchronization. In particular we see that, albeit we do not have to deal with gravity in this context (at least for very simple models), we still have to take into account propulsion limitations. It is in this and in other aspects that some aspects of the Cucker-Smale model can be recycled in order to improve the consensus process.

We start with the PD-like control law (Ren 2007) defined via the ith control torque

$$\tau_i = \omega_i \times I_i\omega_i - I_i \sum_{j=0}^{N} g_{ij} \left[a_{ij}\widehat{q_j^\star q_i} + b_{ij}\left(\omega_i - \omega_j\right)\right], \tag{11}$$

where the weights a_{ij} and b_{ij} are real positive constants, while $g_{ij} = 1$ if agents i and j are connected by an edge and $g_{ij} = 0$ otherwise. As you can see, this control follows the same implementation $u = w - f$ of the translational case. Differently from that case however, the term $f = \omega_i \times I_i \omega_i$ is not that significant (as long as the spacecraft is not spinning too fast). Furthermore, we can assume a continuous implementation of the control torque, since the simplest spacecraft (useful for a formation, since they can be mass-produced) are usually equipped with actuators like reaction wheels. This avoids geometrical problems related with the vector field that we are erasing, and moves the focus on dealing with limited propulsion capabilities.

It is by considering the latter problem that we can reuse the idea of Cucker and Smale of scaling the gains with respect to some metric. In particular, we impose

$$b_{ij} = b_{ij}(t) = K_\omega \left(\sigma^2 + \phi^2 \left(q_i, q_j\right)\right)^\beta , \tag{12}$$

where $K_\omega > 0$, $\sigma^2 > 0$ and $\beta \in \mathbb{R}$ are constants, while $\phi\left(q_i, q_j\right) = 2 \arccos\left(|q_i \cdot q_j|\right)$ denotes the geodesic distance between spacecraft i and j. Here it is supposed that scaling with respect to the geodesic distance the b_{ij}'s helps to balance the effect of the attractive terms $a_{ij}\widehat{q_j^\star q_i}$, thus limiting torque accumulation problems due to saturation (which represent the main problem with our assumptions).

Of course, under these controls we achieve attitude synchronization, with the conditions required listed in the following result.

Theorem 1 *Consider the control torque defined by Eq.* (11)*, with a_{ij} positive constants and $b_{ij} = b_{ij}(t) > 0$. Furthermore, assume that one of the following hypotheses on the underlying communication graph hold true*

1. *The graph is undirected and connected*
2. *The graph is directed and it possesses a spanning tree (only numerical evidence)*

Then, there exist $\tilde{q}$ and $\tilde{\omega}$ such that $q_i \to \tilde{q}$ and $\omega_i \to \tilde{\omega}$, $\forall i \in \{0, \ldots, N\}$.

Obviously, in the second case, the final states correspond to that of the root.

4.3 Numerical Discussion

In this subsection, we aim to cover three different aspects of our analysis. First, we propose a statistical approach which can be used in general to evaluate the efficiency of formation flying control strategies. From this, we also outline, similarly to Cucker and Smale (2007), the most important elements affecting the synchronization process and, more generally, any consensus dynamics. Finally, we show the goodness of the adaptive design (12) (which in turn validates the idea of exploiting the dynamics in the gains design).

To achieve these goals, in Fig. 4 we plot, for the same hierarchical graph structures considered in Paita et al. (2014) and a certain baseline configuration, the

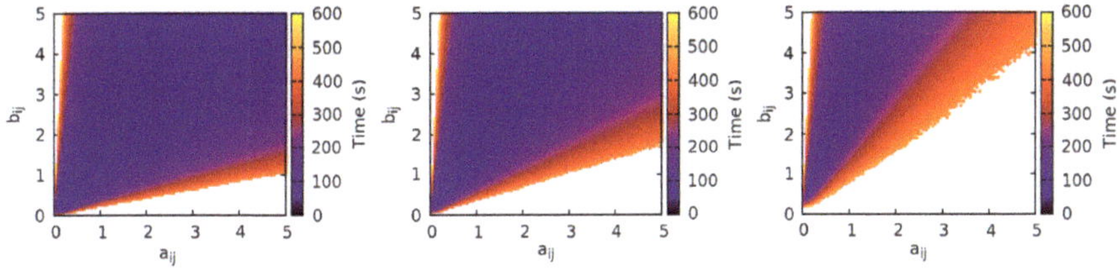

Fig. 4 Average synchronization time for a 20 spacecraft formation as a function of the gains a_{ij} and b_{ij} (equal for all the agents) via control (11), with a 100×100 uniformly spaced grid (from 10^{-1} to 5 in both directions). From *left* to *right*, spacecraft receive info from 7, 3 or 1 preceding agents or the maximal number available

average synchronization time for a 20 spacecraft formation as a function of the gains a_{10} and b_{10}, with the average done on 500 different sets of initial conditions and a maximum of 50 sets allowed to surpass a time threshold of 600 s. However, if more than 50 sets fail to synchronize within this threshold, the average is not computed. Additional details about these plots (particularly on the reasons to work with hierarchical graphs) are given in Paita et al. (2015).

Besides offering an overview of the parametrical regimes that can be taken during a synchronization process (with the optimal one lying somewhat in the middle), these plots show us an apparent dependence of the synchronization time from the denseness of the graph, with the former getting higher the less dense the graph becomes (although the general behaviour remains the same).

A similar analysis (not shown here, but can be found in Paita et al. 2015) can be conducted for the adaptive control (12), where one looks for couples (β, σ^2) which improve the results obtained in Fig. 2 for an optimal choice (in terms of synchronization time) of K_q and K_ω (with β and σ^2 assumed equal across the agents). Besides showing the same behaviour with respect to the denseness of the graphs, this analysis allows us to determine an optimal regime also for the adaptive control proposed before.

Therefore, we can compare our controls when at their best and this is done in Fig. 5, where for each graph, we plot the synchronization time against the dimension of the formation. Additionally, for every dimension, we consider 400 (or the maximal number allowed by the dimension) shuffles of the baseline formation (indicated in black), but we keep the leader fixed. Then, for every permutation, we consider again the average following the same rules of the previous plots.

Now, keeping in mind our assumptions, not only these plots suggest an advantage in using our adaptive design (which however shrinks to nothing seemingly the sparser the graph is), but they also complete our vision of the parameters influencing the dynamics by giving us information on the growth of the synchronization time with respect to the dimension of the formation and the relative initial conditions of the spacecraft (as indicated by the various permutations). Again, for a more in depth discussion of these results, please refer to Paita et al. (2015).

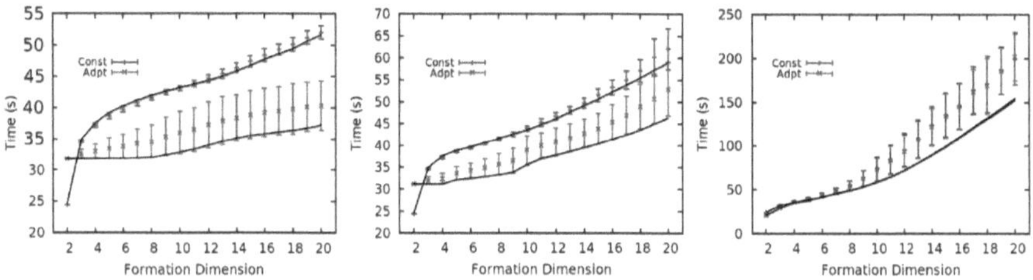

Fig. 5 Synchronization time against formation dimension for 400 permutations of the baseline configurations considered in the plots of the previous subsection. For every picture, we plot in *red* the results obtained with control (11), in *green* those given by the adaptive design (12) while in *black* we indicate the corresponding results obtained with the baseline configuration. As for the gains parameters, we have $K_q = 0.5$ and $K_\omega = 1$. Instead, β and σ^2 are: $\beta = 0.17$ and $\sigma^2 = 0.032$ (*left*), $\beta = 0.13$ and $\sigma^2 = 0.02$ (*center*), $\beta = 0.2$ and $\sigma^2 = 0.094$ (*right*)

5 Conclusion

In this paper we have offered a partial overview of the challenges and techniques that one may have to deal with when designing control strategies for formation flying. Starting from a well studied translational dynamical model, we have adapted it to a formation control and, by employing it in a keplerian context, we have discussed what kind of limitations dynamics and propulsion impose on such a control. We have then moved to the problem of attitude synchronization, where we have discussed the differences with the translational context and introduced appropriate controls. Furthermore, with the help of a statistical approach, we have described the effect on the controlled dynamics of the parameters inherent in a formation. Finally, by moving again from the model introduced at the beginning, we have seen how the dynamics can be exploited to design advantageous control strategies.

References

Cucker, F., Smale, S.: Emergent behavior in flocks. IEEE Trans. Autom. Control **52**(5), 852–862 (2007)

Hughes, P.C.: Spacecraft Attitude Dynamics. Wiley, New York (1986)

Lawton, J.R., Beard, R.W.: Synchronized multiple spacecraft rotations. Automatica **38**(8), 1359–1364 (2002)

Motsch, S., Tadmor, E.: A new model for self-organized dynamics and its flocking behavior. J. Stat. Phys. **144**(5), 923–947 (2011)

Paita, F., Gómez, G., Masdemont, J.J.: On the Cucker–Smale model applied to a formation moving in a central force field. In: Paper Presented at the 64th IAC, Beijing, 23–27 September 2013

Paita, F., Gómez, G., Masdemont, J.J.: A distributed attitude control law for formation flying based on the Cucker-Smale model. In: Paper Presented at the 65th IAC, Toronto, 29 September–3 October 2014

Paita, F., Masdemont, J.J., Gómez, G.: On the rotational cucker-smale model: optimal formation configuration and adaptive gains design. In: Paper Presented at the 66th IAC, Jerusalem, 12–16 October 2015

Park, J., Kim, H.J., Ha, S.Y.: Cucker-Smale flocking with inter-particle bonding forces. IEEE Trans. Autom. Control **55**(11), 2617–2623 (2010)

Perea, L., Gómez, G., Elosegui, P.: Extension of the Cucker-Smale control law to spaceflight formations. J. Guid. Control Dyn. **32**(2), 527–537 (2009)

Ren, W.: Distributed attitude alignment in spacecraft formation flying. Int. J. Adapt. Control **21**(2–3), 95–113 (2007)

Ren, W., Beard, R.W.: Decentralized scheme for spacecraft formation flying via the virtual structure approach. J. Guid. Control Dyn. **27**(1), 73–82 (2004)

Shen, J.J.: Cucker-Smale flocking under hierarchical leadership. SIAM J. Appl. Math **68**(3), 694–719 (2008)

Vicsek, T., Zafeiris, A.: Collective motion. Phys. Rep. **517**(3–4), 71–140 (2012)

Formation Flying Guidance for Space Debris Observation, Manipulation and Capture

Thomas V. Peters

Abstract This article provides a brief overview of the space debris population, debris attitude dynamics, technologies for debris removal, followed by a more in-depth discussion of robotic arm based capture of debris. Guidance aspects of active debris removal missions are discussed. Mission phases for active debris removal missions are rendezvous, inspection, attitude synchronization and capture and de-tumbling. The need for attitude synchronization is driven by recent observations of Envisat which exhibits a fairly high rotation rate.

1 Introduction

This article combines the results of several different projects currently ongoing or conducted at GMV within the past few years in the area of active debris removal (ADR). The objective of the article is to provide an outline of a debris removal mission and address engineering issues related to the GNC system, in particular the guidance function. Material is discussed from the following ESA funded projects:

- Detumbling: detumbling space debris after capture
- Patender: net capture tests
- COBRa: influencing debris (orbit and) attitude by plume impingement
- Android: demonstrate robotic and net capture of space debris
- eDeorbit: de-orbit Envisat

The article is structured as follows. The rest of the introduction will address three issues; an analysis of the debris distribution, dynamics of space debris objects and an overview of ADR technologies that are currently being studied or developed. The debris distribution provides means for selecting the most promising candidates and it will also provide important boundary conditions for the selection of the approach strategy. The ADR technologies are mapped to the selected candidates and the boundary conditions to see which technologies provide the best match. The

T.V. Peters (✉)
GMV Spain, Calle de Isaac Newton 11, Tres Cantos, 28760 Madrid, Spain
e-mail: tpeters@gmv.com

G. Gómez, J.J. Masdemont (eds.), *Astrodynamics Network AstroNet-II*, Astrophysics and Space Science Proceedings 44,
DOI 10.1007/978-3-319-23986-6_16

technology readiness level (TRL) is a major concern for selecting technologies. The robotic arm is selected, because this technology has the highest TRL.

The next portion of the article focuses on ADR mission phases. Many of the phases are common to all technologies for ADR, up to the proximity operations. For example, net capture systems tend to be deployed at a distance of at least 20 m, while a robotic arm requires operations at a distance of a few meters or less.

1.1 Space Debris Distribution and Target Selection

Due to the intensive activities in the space during the last half century, the population of man-made space objects is playing an increasingly important role in the space environment. Today more than 6000 satellites are orbiting around the Earth but only 900 are operational and the problem is going to grow in the future: almost 1200 new satellites are expected to be launched in the next 8 years based on a forecast by Euroconsult. The population of man-made space objects consists of approximately 6 % operational spacecraft, 22 % non-functional spacecraft, 17 % rocket upper stages, 13 % mission-related debris and 42 % fragments from explosions or collisions (Committee on Space Debris 1995). Table 1 provides a classification of the major types of debris and their characteristics. Currently, the removal of small debris objects is not practical (Kaplan et al. 2010). A commonly proposed strategy consists of mitigation on one hand and removal of the largest objects on the other, which would remove the largest sources of potential new small debris.

The total mass of the population is estimated at 6300 tons. Figure 1 shows the distribution of debris in LEO. The highest concentration can be found at an inclination of 82–83° and around the sun-synchronous inclination. A large portion of the population at inclination 82–83° consists of objects launched from Plesetsk using the Cosmos-3M launch vehicle. The sun-synchronous orbit is of particularly

Table 1 Debris classification (from Committee on Space Debris 1995, Levin et al. 2012)

Type	Characteristics	Hazard
Tiny	Not tracked, < 1 cm	Shielding exists, damage to satellites may occur
Small	Not tracked, diameter 1–10 cm, 98 % of lethal objects, ~400.000 objects in LEO	Too small to track and avoid, too heavy to shield against
Medium	Tracked, diameter > 10 cm, < 2 kg, 2 % of lethal objects, ~24.000 objects in LEO, > 99 % of mass (incl. large objects)	Avoidance manoeuvres performed most often for this category
Large	Tracked, > 2 kg, < 1 % of lethal objects, > 99 % of mass (incl. medium objects)	Primary source of new small debris, 99 % of collision area and mass

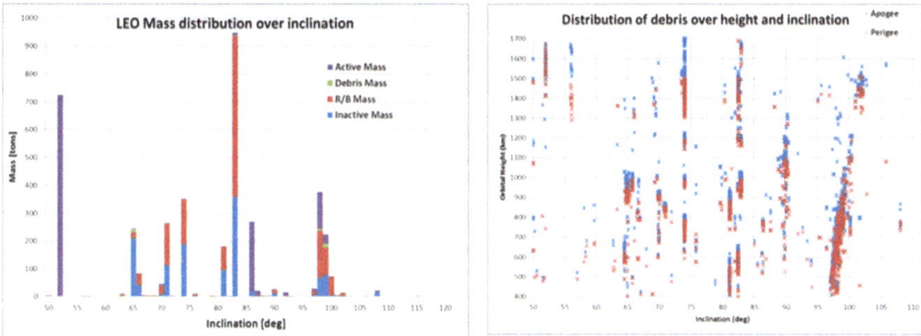

Fig. 1 Debris distribution in LEO orbits

high importance because of its usefulness for remote sensing and Earth observation purposes.

Currently it is not practical to remove anything but the largest debris objects. Such an approach would nevertheless make sense, because the large objects tend to be the primary source of new small debris and because 99 % of the total mass of the debris is concentrated in the large objects. For the design of a European ADR mission a target object needs to be selected. To do this the debris database was filtered using two sets of criteria:

- European build, high mass, SSO, lifetime greater than 25 years.
- Many high mass, similar objects in similar orbits, lifetime greater than 25 years

The following objects are identified:

- Envisat, ERS family, Spot family, MetOp-A (still operational, but at end of life and heavy)
- Rocket boosters:
  - (Ariane 4) H-10, 1780 kg, nine objects in SSO, two of which have a predicted lifetime < 25 years
  - Cosmos-3M, 1420 kg, approximately 236 objects, 210 of which with predicted lifetime greater than 25 years

The rationale for selecting defunct European satellites as targets for removal missions is that design information (such as mass, inertia and shape) is more easily available. It is also expected that legal issues can be addressed more easily for European objects. Another selection criterion is to opt for objects of civilian build. It will likely be easier to obtain design information on the object and more importantly it will be easier to obtain permission to approach to and remove it. Rocket boosters are particularly interesting targets for debris removal missions because of their high similarity as a group (Isakowitz et al. 1999). This means that a (possibly multi-object) debris removal mission can use a similar interface to handle the debris objects. It should also be noted that the 82–83° (where many of the Cosmos-3M upper stages can be found) orbit is inclination-paired with SSO (Levin et al. 2012), that is, the orbits precess in opposite direction and when the orbital planes meet, the debris meets head-on which increases the probability of collision. Removing rocket boosters and other large objects from the 82–83° inclination orbit helps to protect Sun-synchronous orbit. Other studies (Liou 2010; Peterson 2012) apply different selection criteria but arrive at broadly similar conclusions, namely, to remove rocket boosters and objects in SSO.

1.2 Space Debris Dynamics

Space can be divided into three regions, LEO, MEO and GEO. Remote sensing and Earth observation satellites are often placed in LEO, more in particular, in sun-synchronous near polar orbits. MEO is used for navigation and communication satellites, such as GPS, Glonass and Galileo, most commonly using orbits with a period of around 12 h. GEO is used for communications and weather satellites. Spacecraft in each of these types of orbits are faced with different perturbation environments, and different strategies are used to counter perturbations and stabilize the attitude.

Following Carroll (2002), torques can be grouped into three categories; random torques, dissipative torques and orienting torques. Examples of random torques are torques due to leakage, hypervelocity impacts, momentum transfer, spurious activation of thrusters or reaction wheels (if S/C is non-passivated) and wind milling torques under the action of SRP. These torques may either spin-up or spin-down an object. Dissipative torques tend to dissipate energy and spin down objects. The most important dissipative torques are due to magnetic eddy currents in non-ferromagnetic metals such as aluminium and copper (Williams and Meadows 1978). Orienting torques tend to orient an object towards a preferred attitude, for example, aligned with the gravity gradient or the Earth magnetic field. Examples of such torques are gravity gradient torque, magnetic torques due to the presence of permanent magnets, torques due to aerodynamic or solar radiation pressure forces acting at a centre of pressure offset from the centre of mass. These torques are different for the different orbital regions. The perturbation environment in LEO differs qualitatively from the perturbation environment in MEO and GEO. Magnetic and gravity gradient torques decrease to third power of the orbit radius, which means that compared to LEO these torques are a factor of 50 to 200 lower in MEO and GEO, respectively. Many sources (Carroll 2002; Praly et al. 2011; Smith 1962; Ojakangas and Cowardin 2012) point out that in the long term the attitude dynamics of LEO space debris is dominated by dissipative terms. The expectation is that debris objects will eventually settle down in an attitude motion that is either coupled to the gravity gradient or to the magnetic field.

A theoretical study of the spin dynamics of Ariane 4 upper stages in 600–800 km altitude orbits (Praly et al. 2011) found an exponential decay of the spin rate due to eddy current damping, with a decay time of less than 250 days. Shorter decay times are reported for upper stages in an axial spin than in a flat spin. The upper stages start out with a spin rate of between 45 and 60°/s. The study points out that amateur observations show a steady spin rate of the upper stages under consideration during, or towards the end of, the observation periods. The expected spin rates are below 1°/s. Observations made from 1974 to 1976 of a Kosmos upper stage indicated a spin-up from 72°/s to 300°/s over a period of approximately two weeks, starting about 25 days after launch (Boehnhardt et al. 1989) . After the initial spin-up, the spin rate was observed to decay exponentially to 6°/s 770 days after launch. The characteristic decay time is calculated as 183 days, which compares quite well with the theoretical data for the Ariane 4 upper stage.

The defunct Envisat satellite has recently been the subject of studies (Bastida Virgili et al. 2014; Kucharski et al. 2014) to determine its rotational state. Observations made in early 2012 indicated a rotation rate of about 0.4°/s. However, observations made from March 2013 onward, using ISAR and SLR measurements, indicate a rotation rate of 2.6–3.5°/s. The SLR measurements indicate that the rotation period is increasing by 36.7 ms/day, indicating that some form of damping torque is acting on Envisat. An extrapolation of the data using both the quadratic curve fit used by the authors of the study and an exponential decay curve fit indicates that Envisat may recover the original rotation rate of 0.4°/s by 2019 (quadratic fit) or 2026 (exponential fit). Based on the available data the characteristic decay time for Envisat can be estimated at 3–6 years.

In MEO and GEO the magnetic and gravity gradient torques are far less effective in spinning down debris objects. In this environment, mechanical damping and orienting torques act together to put objects that originally were spinning axially into a flat spin about the major axis (Carroll 2002). Some upper stages are put in an axial spin prior to or after payload release. In addition, it is estimated that over a hundred defunct geostationary satellites were originally spin-stabilized with rates of up to 360°/s. Light curve observations seem to confirm a fast, flat spin for two objects investigated, with rotation rates of 38°/s and 212°/s. Both objects were observed over a year after launch (Santoni et al. 2013).

In summary, it is expected that most debris in LEO (with some exceptions) is in a slow rotation state, while in MEO and GEO a large amount of objects is in a fast rotation state.

1.3 Debris Capture Options

Many options exist for capturing space debris objects. Table 2 provides an overview of methods identified at GMV. The robotic arm, nozzle docking and the tentacle option are considered rigid-link methods. The net, harpoon, pushing sock options are flexible link methods and the rest are contactless methods.

The robotic arm is a method that can partially be tested and that has the highest TRL of all capture and de-tumbling techniques. This means that the least amount of development would be required before an ADR mission featuring a robotic arm could be launched.

Table 2 Capture methods

Capture method	Sensitivity to rotation rate	Rotation related issues	Structural issues
Robotic arm	High	Synchronization required	Requires structural hard point
Docking with nozzle	High	Synchronization required	Requires non-steerable nozzle
Tentacles	High	Synchronization required	May require structure not covered by MLI for firm grip
Net	Low	Fast de-spin required to avoid tether wind-up around target	May require measures to avoid breaking off pieces of target
Harpoon (Rigid)	High	Synchronization required	Requires strong structure for contact (e.g., honeycomb panels) and avoidance of propellant tanks
Harpoon (Non-rigid)	Low	Fast de-spin required to avoid tether wind-up around target	Requires strong structure for contact (e.g., honeycomb panels) and avoidance of propellant tanks
Pushing sock air-bag	High	Requires pre-capture de-spin	May require measures to avoid breaking off pieces of target
Foam projection	High	Centrifugal forces may disrupt foam; requires pre-capture de-spin	May require structure not covered by MLI for firm grip (i.e., MLI may tear off)
Ion-beam Shepherd	Low	Low sensitivity to spin rate; method may be used to control rotation	None
Electrostatic tractor (only for GEOs)	Low	Low sensitivity to spin rate; method may be used to control rotation	None
Magnetic tractor	Low	Low sensitivity to spin rate; method may be used to control rotation	None

Net and harpoon based options are considered relatively risky. The control of the chaser—target combination is difficult, which may lead to serious failures during capture, de-tumbling or de-orbiting. Also, the initial interaction of the capture device may lead to additional debris generation.

The contactless methods tend to have a fairly low risk because no physical contact is made and no attitude synchronization is required. On the other hand, all contactless methods have a low TRL. Contactless methods also tend to have important restrictions such as only being useful in GEO in case of electrostatic tractor or requiring very large magnets in case of the magnetic tractor.

For these reasons the robotic arm is selected as the most promising option for a near-term debris removal mission.

2 ADR Mission Phases

2.1 Mid-Range Rendezvous

The definition of the mid-range rendezvous strategy to be implemented in the guidance depends on the type of sensor used and the requirements imposed by the navigation. It is assumed that just before the mid-range rendezvous the navigation is based on GPS measurements of the chaser and TLE's of the target. Furthermore, the mid-range rendezvous is performed using a wide-angle camera with a 28° field of view and a 2048×2048 pixels CCD element. The mid-range rendezvous starts with a search phase that aims to point the camera at the target object. Table 3 shows the detection range, the range at which 90 % of the field of view of the camera is filled

Table 3 Wide angle camera ranges

Range (m)	# pixels/2 m diameter	# pixels/4 m diameter	# pixels/9 m diameter	Comment
151,000	0.06	0.11	0.25	Max range WAC for 9 m diam. object
67,000	0.13	0.25	0.56	Max range WAC for 4 m diam. object
33,500	0.25	0.50	1.13	Max range WAC for 2 m diam. object
2000	4	8	19	Handover distance
20	419	835	1855	90 % of WAC FOV filled (9 m diam.)
9	927	1833	3886	90 % of WAC FOV filled (4 m diam.)
4.5	1833	3505	6583	90 % of WAC FOV filled (2 m diam.)
2	3886	6583	9660	Working distance

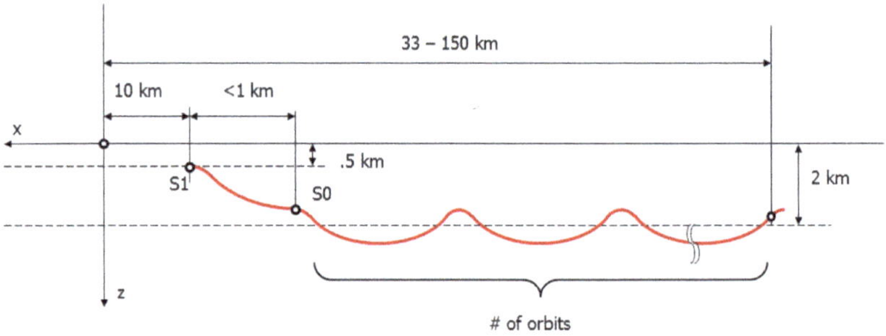

Fig. 2 Mid-range rendezvous strategy, first part

and the operating distance for a robotic arm. A projected size of the target of 0.25 pixels is sufficient to achieve detection. The table shows that depending on the size of the target, detection can occur at ranges from 33.5 to 151 km.

Figure 2 shows the first part of the rendezvous strategy. The first part of the rendezvous is based on drifting orbits with a slight difference in eccentricity with respect to the target orbit. The chaser is inserted into an orbit about 150 km behind and 2 km below the target orbit either by means of ground commands or by means of autonomous manoeuvre planning based on GPS measurements and the known two-line elements of the target. During the first phase, before point S0, the chaser points its camera in the direction of the most likely location of the target. After detection the chaser remains in a drift orbit 2 km below the target while the navigation function estimates the orbit of the target. The slight difference in eccentricity aids the navigation to reach convergence more quickly.

Figure 3 shows the second part of the rendezvous strategy. The rendezvous strategy is based on drifting trajectories and a safe orbit based on eccentricity / inclination separation (D'Amico and Montenbruck 2006). The second part of the rendezvous starts at a distance of about 10 km behind the target and 500 m below. This location is labelled as S1. At this point the chaser modulates its drift rate while maintaining the relative apogee at 50 m below the target orbit. This sequence occurs between points S2 and S4. Out-of-plane control is performed early, at point S3, such that no major manoeuvres occur in the drift phase between points S3 and S4. This allows the navigation to reach the best possible estimate of the target orbit. At point S4 the chaser enters into a co-elliptic drift orbit 50 m below the target, and at point S5 the chaser enters into a safe orbit with a closest along-track approach distance of about 100 m. (This means that the centre of the safe orbit ellipse is located at an along-track distance of about 200 m).

The second part of the mid-range rendezvous strategy and the inspection from a spiral orbit were simulated in a simplified simulator, which included J2 as an orbital perturbation, but which featured ideal navigation and ideal actuation. This simplified simulator has been used to develop and test similar approach strategies that have been validated in higher-fidelity simulators. Figure 4 shows the x-component of the guidance error of the LVLH state vector. Errors in the

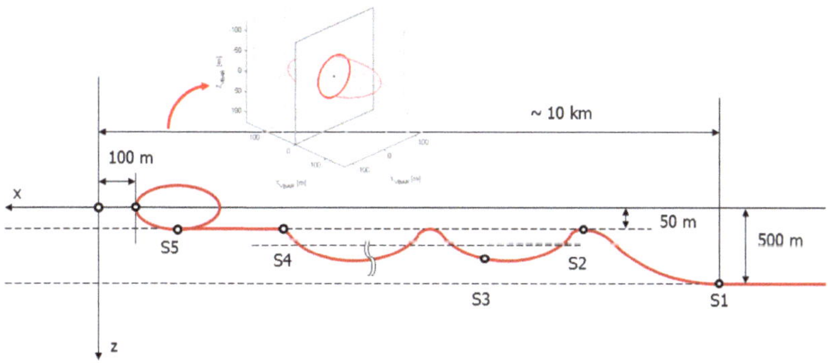

Fig. 3 Mid-range rendezvous strategy, second part

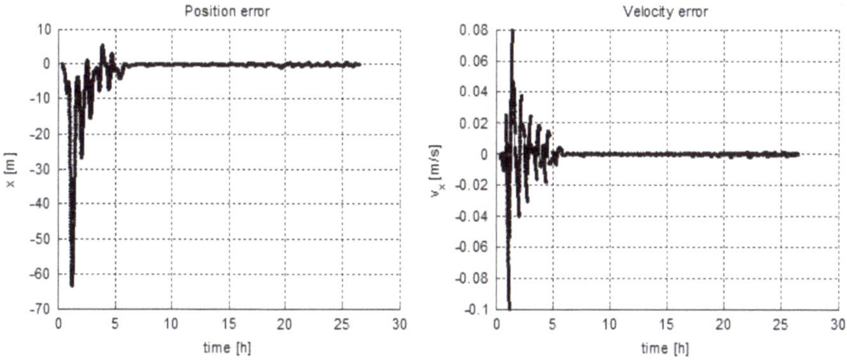

Fig. 4 State vector errors

guidance reference are due to not including J2 and nonlinearity effects into the guidance OBSW models. The initial errors are quite large, up to 62 m in, and velocity errors of up to 10 cm per second. After the first 5 h of rendezvous (or about three orbits), the errors sharply decrease. This decrease in the errors occurs when the chaser enters the drift orbit linking to the safe orbit at the end of rendezvous. That is to say, the large errors in position and velocity all occur at large distances. It should be noted that the errors in position and velocity show a slight increase over time, up to at most 80 cm in position at the end of the simulation. This effect is due to the fact that an unperturbed Keplerian propagator is used as a reference orbit to propagate the relative trajectories. The Keplerian orbit is initialized at the start of the simulator and it starts diverging from the true orbit over time. In addition, the large errors at the start are caused in part by this mismodelling, and in part by the fact that the relative equations are linearized. These two causes of errors are known, and could in principle be improved. Firstly, a J2-based relative propagator could be used to improve the reference trajectory. Secondly, the guidance could be made to operate on linearized

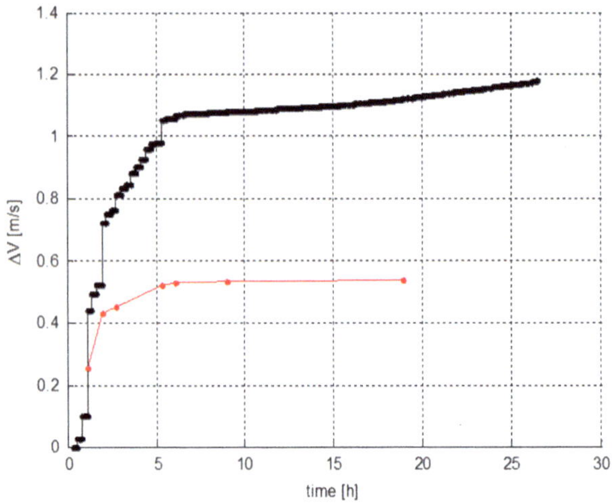

Fig. 5 ΔV for the rendezvous and inspection

differential orbital elements, which suffer less from linearization errors. Thirdly, even if these improvements are not implemented, the guidance could periodically update its plan, re-initializing the Keplerian orbit that is used to generate the reference trajectory. In this case, the reference orbit will be closer to the true orbit and the reference trajectory generated in this way will be closer to the truth.

Finally, Fig. 5 shows the ΔV required to follow the reference trajectory. The red line indicates the ideal guidance ΔV while the black line indicates the true ΔV. The true ΔV is over twice the guidance ΔV although the overall total ΔV is quite low to other close-range rendezvous strategies. For example, a rendezvous based on radial hops would require $\frac{1}{2} \cdot n \cdot D = \frac{1}{2} \cdot 0.0011 \cdot 10000 = 5.5$ m/s, not taking into account correction manoeuvres. That is, the current rendezvous strategy uses only 1/10th of the *DeltaV* required for a rendezvous strategy based on radial hops. Note in this figure that the *DeltaV* required to control the inspection flight from about 10 h onwards steadily increases over time. This is due to initialization of the Keplerian orbit. This ΔV could be brought down by using a J2 propagator in the guidance model.

2.2 Inspection from Spiral Orbit

The target debris object is inspected from a spiral orbit to determine mass properties, configuration and possible damage to the structure. It is expected that information on the mass properties (mass, location of centre of mass and moment of inertia matrix) and the configuration are available from design documents a priori. The inspection phase serves to update this information for use in the subsequent phases. It is assumed that inspection is performed mainly by means of on board cameras. The relative state information will be extracted via image processing, filtering, and combined with the spacecraft telemetry. In addition, images will be sent to ground for extracting the characteristics of the target. The camera constraints, illumination conditions and appearance of the Earth in the collected images are assessed. The effects that have been considered are the following:

- Earth/Space in the FoV of the camera (blue/cyan respectively);
- Eclipse (black);
- Sun inside the exclusion angle of the camera (yellow).

Figure 6 shows a linear relative spiral orbit trajectory in the LVLH frame together with the visualization of the effects described above for a target object in a sun-synchronous reference orbit. The figure suggests the following considerations:

- No eclipse is present.
- The Earth will be in the background during 16.5 % of the orbits, and will be inside the camera field of view for 29.5 % of the orbit period.
- The Sun will be inside the exclusion angle of the camera for a very small amount of time during only five orbits.

Figure 7 shows the true reference trajectory that was planned for the COBRA-IRIDES experiment (Peters et al. 2014). In this experiment the MANGO satellite of the PRISMA configuration would perform rendezvous with the Picard spacecraft in order to study the effect of plume impingement on the attitude dynamics of Picard. The reference trajectory of the COBRA IRIDES experiment is based on the trajectory of the IRIDES experiment with a characteristic dimension of 10 m and a drift rate of 5 m per orbit. The date of the experiment has been selected as the 15th of October 2014.

Analysis has been carried out including SRP and drag perturbations (considering data on Mango and Picard) over the reference orbit for the selected date. The impact

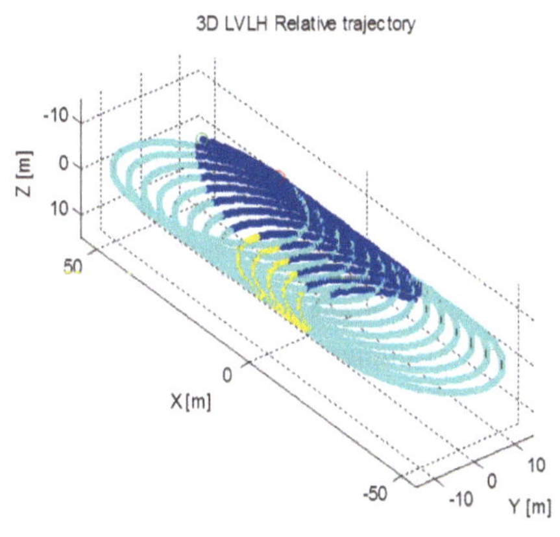

Fig. 6 LVLH inspection trajectory

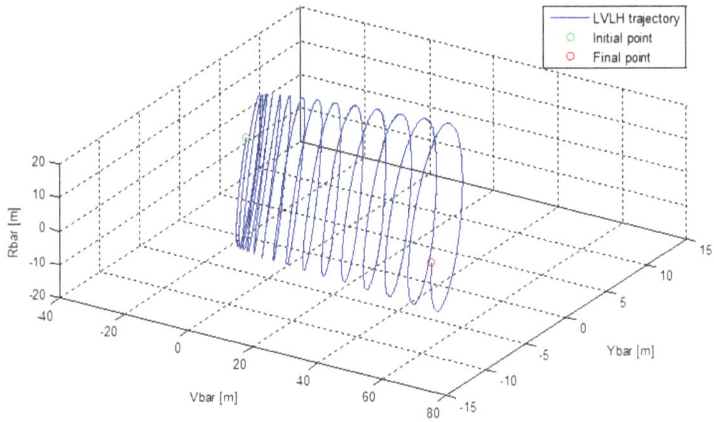

Fig. 7 True reference in LVLH

of these perturbations is not negligible, as can be seen in the figure above. The main effect is on the drift rate, which increases over time. This orbit has been obtained in order to have the desired relative position conditions at the nominal time for the experiment execution.

2.3 *Attitude Synchronization*

Envisat exhibits a relatively high rotation rate of 2.5-3°/s, such that an attitude synchronization manoeuvre is required. The attitude synchronization manoeuvre takes the chaser from an initial position in the LVLH frame with a target-pointing attitude to a position in the target body frame, with an attitude defined with respect to the target body frame. The synchronization sequence consists of the following steps:

1. transfer to the angular momentum vector
2. Approach to 10 m and attitude synchronization with target
3. Transfer into body frame from the instantaneous angular momentum vector, maintaining attitude synchronization
4. Transfer to approach axis in target body fixed frame
5. Approach from approach axis to terminal approach point

Figure 8 shows an example of an attitude synchronisation trajectory with Envisat. In this example, the chaser makes its initial approach over the angular momentum vector of the target. The projection of the angular velocity vector is shown in green, the projection of the angular momentum vector in magenta, the guidance trajectory in blue, and the real world trajectory in red. The figure shows that the motion of the projection of the angular momentum vector is a circular arc in the LVLH

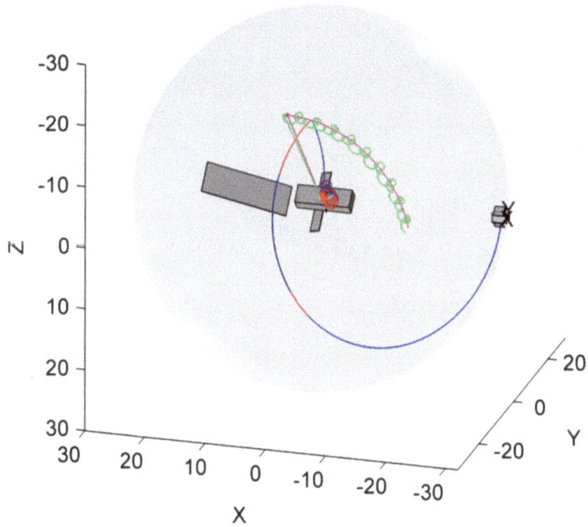

Fig. 8 Attitude synchronisation

frame, while the motion of the projection of the angular velocity vector also exhibits precession around this arc. The angular momentum vector is nearly constant in the inertial frame, but the angular velocity vector is not. The tip velocity of the projected angular velocity vector is higher than the tip velocity of the angular momentum vector. This means that approaching over the angular momentum vector generally requires a lower ΔV than approaching over the angular velocity vector. For this reason the attitude synchronisation with Envisat will be performed over the angular momentum vector.

The ΔV required for synchronization is fairly high compared to other phases of the rendezvous, especially for large debris objects. Especially steps 3 and 4 are fairly costly, because the chaser is moving in the target body frame at a comparatively large distance from the centre of mass. The terminal point is selected such that the chaser can access the capture point with the robotic arm, but also such that the centrifugal accelerations are reduced as much as possible.

2.4 *Capture and De-tumbling*

The capture and de-tumbling phase are currently being studied in the de-tumbling project. This phase is again driven by the high rotation rate of Envisat; before the de-orbiting phase the combination of target and chaser needs to be de-tumbled and stabilised in an inertial pointing attitude. The model presented in Dimitrov and Kazuya (2004), shown in Fig. 9 will be used to model the contact between the chaser and the target. The segments and the joints of the robotic arm are considered to be flexible.

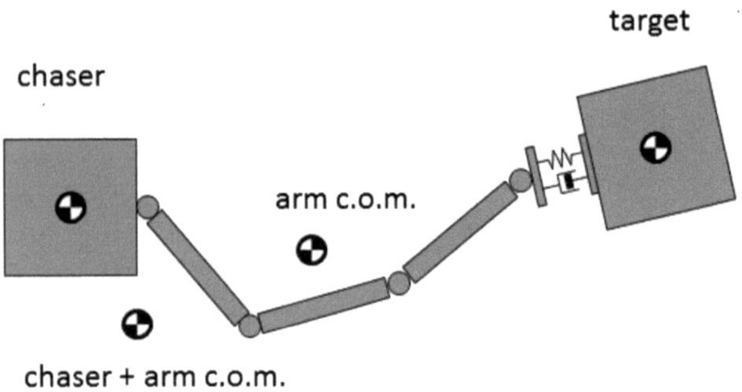

Fig. 9 Capture model

It is assumed that the relative orbit and attitude control system will remain active during the de-tumbling. The rationale for this approach is that it enables the forces and torques on the capture point to be controlled more precisely. Just before the chaser grabs the target, the forces and torques on the capture point are zero (that is, the capture has not taken place yet). Just after capture, the forces and torques on the interface point should remain small as the chaser continues to control its relative position and attitude with respect to the target.

After capture has been verified, the chaser starts computing the feed-forward torque required to de-tumble the target while respecting a maximum force and torque limit on the capture point. The feed-forward torque is computed as if the chaser were not attached to the target. The chaser GNC provides the thruster acceleration required to remain stationary in the target body frame plus the thruster acceleration required to provide the desired force and torque at the capture point.

3 Conclusion

This article has provided an overview of the GNC issues associated with ADR missions, with an emphasis on guidance strategies for major mission phases. Feasible guidance strategies are available that can form a part of a GNC system to perform active debris removal.

Current thinking is driven by the fact that Envisat is the most likely target for removal, and Envisat is unusual in the sense that it has experienced a spin-up event a number of years ago and is consequently spinning at a rate of about 2.5-3°/s. This means that special measures have to be taken to perform attitude synchronization manoeuvre before capture can take place.

The complexity of ADR missions in general and of a mission to deorbit Envisat in particular makes it highly desirable to perform an ADR mission on a small target first in order to test the technologies and GNC approaches to debris removal on a smaller scale.

References

Boehnhardt, H., Koehnhke, H., Seidel, A.: The acceleration and the deceleration of the tumbling period of Rocket Intercosmos 11 during the first two years after launch. Astrophys. Space Sci. **162**(2), 297–313 (1989)

Carroll, J.A.: Space Transport Development Using Orbital Debris. Final Report on NIAC Phase I, Research Grant (07600-087) (2002)

Committee on Space Debris: Orbital Debris: A Technical Assessment. National Academies Press, Washington (1995)

D'Amico, S., Montenbruck, O.: Proximity operations of formation-flying spacecraft using an eccentricity/inclination vector separation. J. Guid. Control Dyn. **29**(3), pp. 554–563 (2006)

Dimitrov, D.N., Kazuya, Y.: Momentum distribution in a space manipulator for facilitating the post-impact control. In: Proceedings of 2004 IEEE/RSJ International Conference on Intelligent Robots and Systems, 2004 (IROS 2004), vol. 4, pp. 3345–3350. IEEE, Sendai (2004)

Isakowitz, S.J., Hopkins, J.B., Hopkins Jr., J.P.: International Reference Guide to Space Launch Systems, 3rd edn. AIAA, Washington (1999)

Kaplan, M.H., Bradley Boone, B., Brown, R., Criss, T.B., Tunstel, E.W.: Engineering issues for all major modes of in situ space debris capture. In: AIAA SPACE 2010 Conference & Exposition, Anaheim, 30 August–2 September 2010

Kucharski, D., Kirchner, G., Koidl, F., Fan, C., Carman, R., Moore, C., Feng, Q.: Attitude and spin period of space debris envisat measured by satellite laser ranging. IEEE Trans. Geosci. Remote Sens. **52**(12), 7651–7657 (2014). doi:10.1109/TGRS.2014.2316138.

Levin, E., Pearson, J., Carroll, J.: Wholesale debris removal from LEO. Acta Astronaut. **73**, 100–108 (2012)

Liou , J.C.: A parametric study on using active debris removal for LEO environment remediation. In: 61st International Astronautical Congress, Prague, Paper IAC-10.A6.2.5 (2010)

Ojakangas, G., Anz-Meador, P., Cowardin, H.: Probable rotation states of rocket bodies in low earth orbit. In: Proceedings of the 14th Advanced Maui Optical and Space Surveillance Technologies Conference, held in Wailea, Maui, 11–14 September 2012

Peters, T.V., Olmos, D.D.E., Lavagna, M., Benvenuto, R., Attina, P., Parissenti, G., Cropp, A.: The COBRA IRIDES experiment. IAC-14-A6.6.9. In: Proceedings of the 65th International Astronautical Congress, Toronto (2014)

Peterson, G.E.: Target identification and delta-V sizing for active debris removal and improved tracking campaigns. In: Proceedings of the 23rd International Symposium on Spaceflight Dynamics, Pasadena, Paper No. ISSFD23-CRSD2-5, 29 October–2 November 2012

Praly, N., Petit, N., Bonnal, C., Laurent-Varin, J.: Study on the eddy current damping of the spin dynamics of spatial debris from the ariane launcher. In: 4th European Conference For Aerospace Sciences (2011)

Santoni, F., Cordelli, E., Piergentili, F.: Determination of disposed-upper-stage attitude motion by ground-based optical observations. J. Spacecr. Rocket. **50**(3), 701–708 (2013). doi:10.2514/1.A32372.

Smith, G.L.: A theoretical study of the torques induced by a magnetic field on rotating cylinders and spinning thin wall cones, cone frustrums, and general bodies of revolution, NASA. TR-R-129 (1962)

Virgili, B., Lemmens, S., Krag, H.: Investigation on Envisat attitude motion. In: e.Deorbit Workshop (2014)

Williams, V., Meadows, A.J.: Eddy current torques, air torques and the spin decay of cylindrical rocket bodies in orbit. Planet. Space Sci. **26**, 721–726 (1978)

Low Thrust Relative Motion Control of Satellite Formations in Deep Space

Claudiu-Lucian Prioroc and Seppo Mikkola

Abstract The problem of placing and controlling a formation of satellites on a Halo orbit is studied. The Earth-Sun circular restricted three body problem is considered. A family of artificial Halo orbits with the same periods, around the L_1 and L_2 Lagrange points in the Earth-Sun system is found using the pseudo-arc-length continuation method. The orbits are used are reference trajectories for satellites to track. The problem of orbit stability, bounding and controlling the relative motion by means of nonlinear control is addressed.

1 Introduction

In this paper a spacecraft formation using electric propulsion and flying in artificial halo orbits around the L_1 and L_2 libration points is proposed. This type of satellite formations can be used as a platform for continuous remote sensing over large areas of the Earth (King et al. 1999). This type of missions are not possible from LEO and GEO satellites. The Triana mission concept (Watzin 2000) proved that Earth observations missions like Earth imaging in a broad spectral range, Earth radiance measurements at different wavelengths, Solar wind monitoring and Solar observations (Martin et al. 1992) and magnetic field monitoring from Lagrange points are of scientific use and feasible from an operational point of view. According to Chepfer et al. (2002) and Biggs and McInnes (2009), another mission that can only be done by a formation of satellites is the continuous measurement of atmospheric optical thickness and surface reflectance. This observations require at least two satellites. In Biggs and McInnes (2009) and Biggs et al. (2009) families of periodic orbits around the Lagrange points for solar sails were found. In Prioroc and Biggs (2015) it was showed that families of artificial orbits with the same period of low thrust propulsion solutions exist and can be used for formation flying in deep space. In the current study is shown how to use families of artificial periodic orbits for formation flying. The Floquet theory is used to study the linear stability

C.-L. Prioroc (✉) • S. Mikkola
Department of Physics and Astronomy, University of Turku, Tuorla Observatory, 21500 Piikkiö, Finland
e-mail: clpr@gmv.es; mikkola@utu.fi

G. Gómez, J.J. Masdemont (eds.), *Astrodynamics Network AstroNet-II*,
Astrophysics and Space Science Proceedings 44,
DOI 10.1007/978-3-319-23986-6_17

of periodic orbits. A tracking mechanism based on time-delay feedback control that was first addressed by Biggs et al. (2008) and Biggs and McInnes (2009) and later by Prioroc and Biggs (2015) is used for orbit tracking of the leader satellite. A nonlinear sliding mode controller for the relative motion between the satellites is used to control the formation.

2 The Continuous Thrust Circular Restricted Three Body Problem

A rotating frame of reference with fixed primary masses, where the XY plane rotates with the two primaries is considered. The two masses m_1 and m_2 are located at $P_1(\mu, 0, 0)$ and $P_2(\mu - 1, 0, 0)$ as is depicted in Fig. 1, where $\mu = m_2/(m_1 + m_2)$. The unit of time follows the agreement: the two masses complete on revolution around the common center of mass in one unit of time.

The CRTBP equations of motion with continuous thrust are Szebehely (1967):

$$\begin{aligned} -a_x + \ddot{x} + 2\dot{y} &= \frac{\partial \Omega}{\partial x} \\ -a_y + \ddot{y} - 2\dot{x} &= \frac{\partial \Omega}{\partial y} \\ -a_z + \ddot{z} &= \frac{\partial \Omega}{\partial z}. \end{aligned} \tag{1}$$

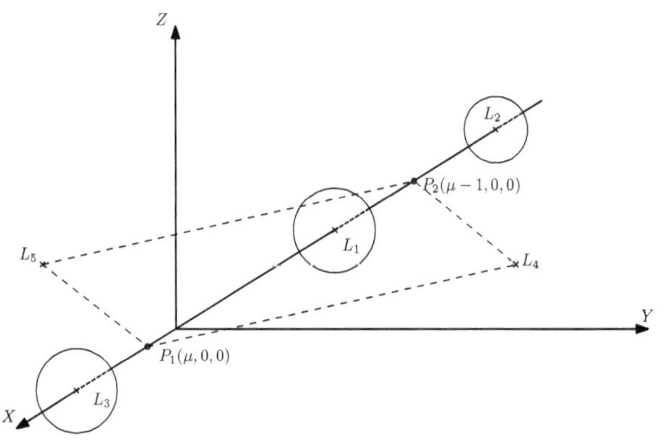

Fig. 1 Libration points in the CRTBP

where Ω is:

$$\Omega(x,y,z) = \frac{(x^2+y^2)}{2} + \frac{1-\mu}{\sqrt{(x-\mu)^2+y^2+z^2}} + \frac{\mu}{\sqrt{(x+1-\mu)^2+y^2+z^2}}. \tag{2}$$

The accelerations a_x, a_y and a_z from (1) are formulated as a function of thrust module T, the direction vector ϵ and satellite mass m_{sat}:

$$a_{x,y,z} = \frac{\epsilon \cdot T}{m_{sat}}, \tag{3}$$

where $\|\epsilon\| = 1$.

3 Fixed Period Artificial Halo Orbits

In this study families of artificial orbits perturbed in the X and Y axis are investigated. In order to obtain a family of fixed period orbits the MatCont (Dhooge et al. 2004) numerical continuation Matlab toolbox is used. The family of orbits is computed around the Lagrange points L_1 and L_2. This orbits have exactly the same orbit period. Artificial orbits around Lagrange points displaced on the Z axis where discussed by in Biggs and McInnes (2009), Biggs et al. (2009) and Biggs et al. (2008). This orbits, that are displace on the Z axis, are of interest for polar regions observation. To obtain periodic halo orbits, the Lindstedt-Poincaré method is used (Poincaré 1892; Gómez et al. 2001). The initial L_1 and L_1 halo orbits are given in the CRTBP:

$$L_1 = \begin{bmatrix} x(0) = -0.988764313048468 & \dot{x}(0) = 7.925532259891737 \cdot 10^{-5} \\ y(0) = 2.103434329704124 \cdot 10^{-4} & \dot{y}(0) = -0.008962876332213 \\ z(0) = 8.945994728522482 \cdot 10^{-4} & \dot{z}(0) = -3.721541455571416 \cdot 10^{-5} \\ a_x = 0, \quad a_y = 0, \quad a_z = 0 & \end{bmatrix} \tag{4}$$

$$L_2 = \begin{bmatrix} x(0) = -1.00843256653300 & \dot{x}(0) = -0.000874704552725653 \\ y(0) = -0.00341374582318268 & \dot{y}(0) = -0.00713173781085524 \\ z(0) = -0.000485470355353604 & \dot{z}(0) = 0.00126016114859147 \\ a_x = 0, \quad a_y = 0, \quad a_z = 0 & \end{bmatrix} \tag{5}$$

a_x and a_y accelerations are selected as the continuation parameters. The acceleration along the Z axis, a_z is fixed at 0. Where the direction vector is ϵ and T is the thrust magnitude.

In order to bound the maximum acceleration norm at 10^{-4} m/s^2, a value that corresponds to a thrust of 0.24 N on a 1000 Kg spacecraft was chosen, the continuation parameters a_x and a_y are bounded. The a_x and a_y accelerations are varied such that the orbit period is kept fixed. As an alternative for integrating a large number of initial conditions to solve the initial value problems (IVP) a continuation software is used (Doedel et al. 2006). In this study the MatCont Matlab Toolbox was used for the continuation of the initial solution. The initial orbit satisfies the nonlinear system $\dot{X}(t) = P \cdot f(Y(t), t)$, where the time is scaled $t \to \frac{t}{P}$, P is the orbit period thus $t \in [0, 1]$. The orbit length L is defined as the distance traveled in one complete orbit. The relation between length L and period P is satisfied:

$$P = \frac{L}{\int_0^1 \|f(Y(s))\| ds}.$$

The initial orbit $Y = (X(\cdot), L, a_x, a_y)$ is a solution of $F(Y) = 0$, for a given orbit period P and acceleration a_z,

$$F(Y) = \begin{cases} \dot{X} - P \cdot f(Y(t), t), \\ X(0) - P \cdot f(Y(1), 1), \\ P - \frac{L}{\int_0^1 \|f(Y(s))\| ds}. \end{cases}$$

The pseudo-arc-length continuation is used to find a family of solutions once the integration of the initial orbit is completed. The setup for the continuation is:

$$\begin{aligned} &F(Y_1) = 0, \\ &(Y_1 - Y_0)^T \dot{Y}_0 - \Delta s = 0, \\ &\|\dot{Y}_0\| = 1, \\ &Y = (X(\cdot), L, a_x, a_y). \end{aligned}$$

with the period P and the acceleration a_z fixed. By keeping the period P fixed, a time map is used in the continuation, instead of a Poincaré map when the orbit length L is fixed. For more details on pseudo arc-length continuation see Doedel et al. (2006) and Doedel et al. (2005).

In this study an orbit period of 181 days is used. The artificial L_1 periodic halo orbit family is shown in Fig. 2 and the artificial halo orbit family around L_2 in and Fig. 3.

The relation between the orbit amplitude, the required continuous acceleration and the thrust orientation in the XY plane can be observed in Figs. 4 and 5 for the L_1 halo family and Fig. 6 for the L_2. In order to maintain the artificial trajectory, the acceleration for very large orbits is by two orders of magnitude grater than the

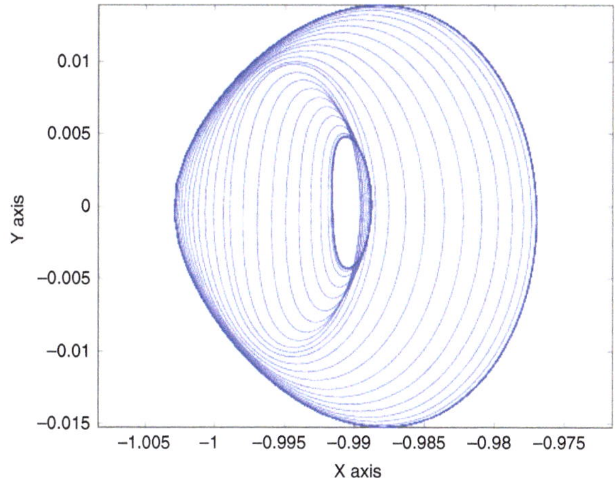

Fig. 2 *XY* view of the 181 days halo orbit family at libration point L_1

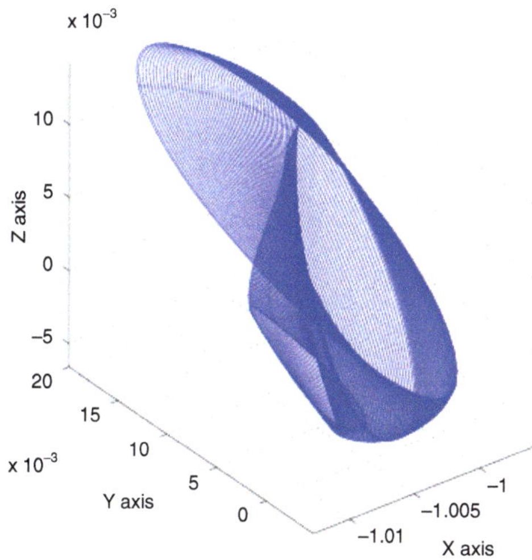

Fig. 3 3*D* view of the 181 days halo orbit family at libration point L_2

accelerations of the small orbit amplitude. In order to assess the formation flexibility by changing the amplitude and the orientation of the thrust vector, one can look at the thrust angle in the *XY* plane. The angle has a large variation for changes in orbit with small *X* amplitude, but it becomes very small for large orbits.

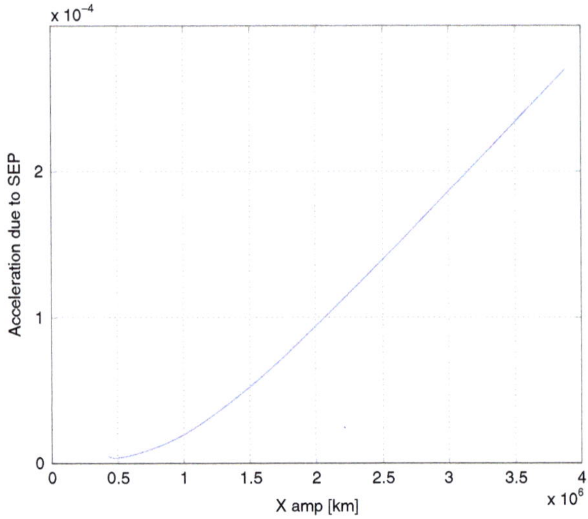

Fig. 4 The variation of thruster acceleration with respect to the X axis amplitude in the L_1 halo orbit family

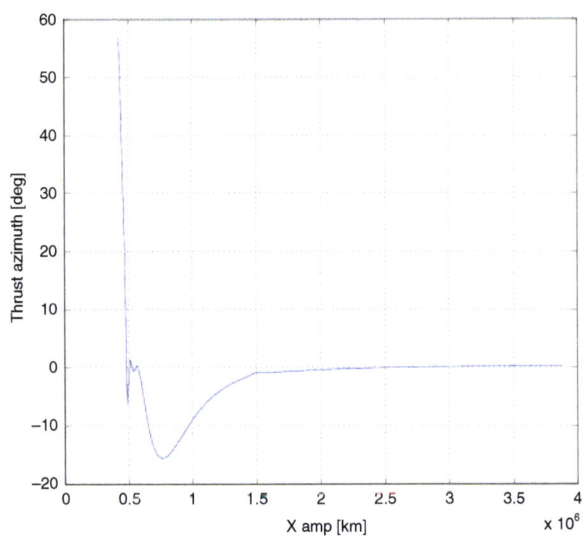

Fig. 5 The variation of the orientation of the acceleration vector in XY plane with respect to the X axis amplitude in the L_1 halo orbit family

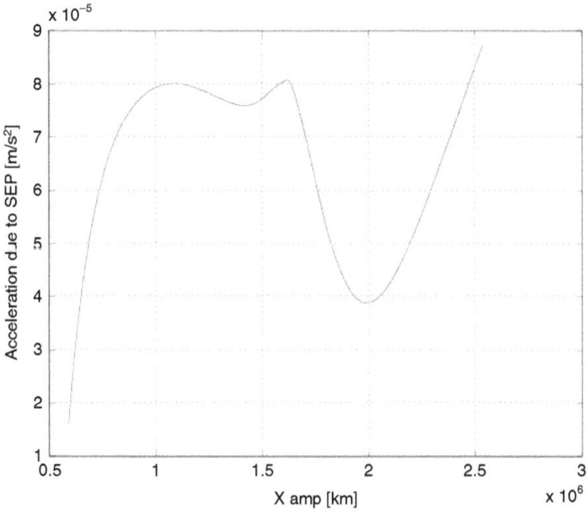

Fig. 6 The variation of thruster acceleration with respect to the X axis amplitude in the L_2 halo orbit family

4 Stability of Periodic Orbits in the Continuous Thrust CRTBP

Floquet theory is used to study the linear stability of the periodic orbit families in the nonlinear system The stability of the periodic orbits in the family depends on the eigenvalues, $\lambda_1, \lambda_2, \lambda_3, \lambda_4, \lambda_5, \lambda_6$, of the monodromy matrix M as follows:

For each eigenvalue λ_i there is a characteristic Floquet exponent α_i defined by:

$$\lambda_i = e^{\alpha_i T}.$$

Analyzing the eigenvalues of the monodromy matrices M for periodic orbits at the L_1 and L_2 libration points, one can see that they are: $\lambda_1 > 1$, $\lambda_2 < 1$, $\lambda_1 \cdot \lambda_2 = 1$, $\lambda_3 = \lambda_4 = 1$ and $\lambda_5 = \lambda_6$ are complex and of modulus 1. The theory says, that an orbit is linearly stable if and only if the real parts of all the characteristic Floquet exponents are less than or equal to zero. The real part of the characteristic exponents has the form: $\alpha_1 > 0, \alpha_2 < 0, \alpha_3 = \alpha_4 = \alpha_5 = \alpha_6 = 0$. The first two characteristic exponents λ_1, λ_2 are associated with the unstable character of the halo orbit. α_1 is the dominant exponent. To study the instability of the artificial halo orbits the stability indices K_1 and K_2 are defined as in Gómez et al. (2001):

$$K_i = \frac{1}{Re(\lambda_i)} + Re(\lambda_i),\ i = 1{,}2,$$

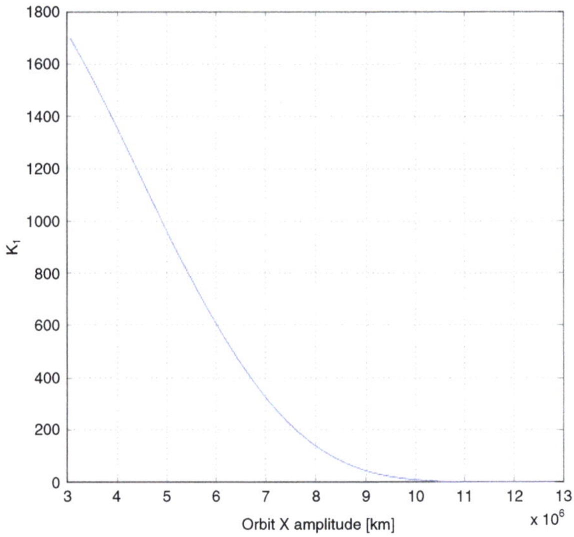

Fig. 7 Variation of the stability parameter K_1 with respect to the X axis amplitude in the L_1 halo orbit family

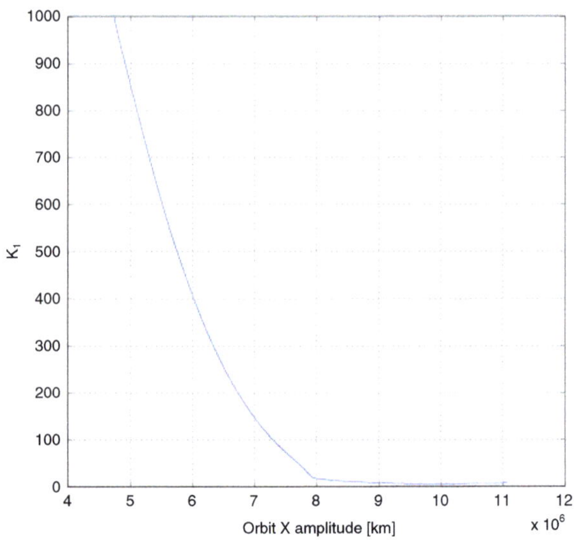

Fig. 8 Variation of the stability parameter K_1 with respect to the X axis amplitude in the L_2 halo orbit family

where $Re(\lambda_i)$ is the real part of λ_i. Figures 7 and 8 is depicted the K_1 stability parameter. This parameter corresponds to the dominant characteristic exponent and is plotted against the orbit amplitude on X axis. One can seen, that for small and medium X amplitudes the orbits are unstable. This means that the spacecrafts will not have a bounded relative motion without active control as can be seen from Fig. 9. One can observe that if the stability indices K_1 is decreasing the orbit amplitude is increasing and the orbit becomes stable. This is happening for large amplitudes and as stated before a larger thrust is required. High orbit amplitudes might not be

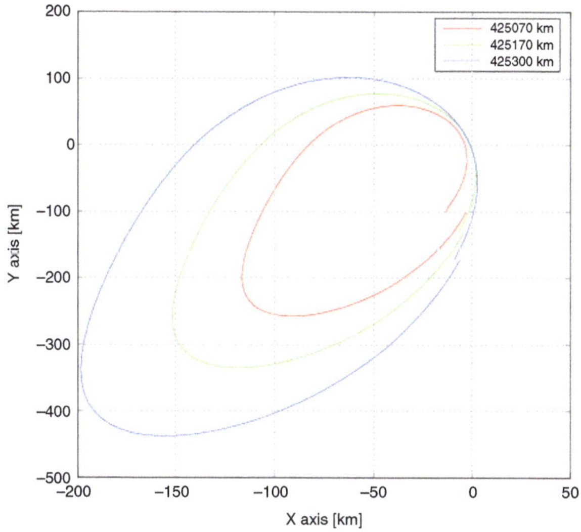

Fig. 9 Relative motion of spacecrafts for different orbit X axis amplitudes

feasible for libration point missions, as this type of orbits have so large orbits that they circle the Earth. But it is important to know that the degree of instability can be *reduced* by increasing the orbit amplitude.

5 Relative Motion Feedback Control

The families orbits determined with the continuation software are unstable due to numerical errors in the integration process. This means that the orbits are not closed. This is obvious when the trajectory is integrated for multiple revolutions. Tracking reference orbit that are not closed, will make the relative motion between satellites to diverge. To design reference orbits one can use a time delay feedback mechanism introduced in Pyragas (1992) and Chen and Yu (1999). Biggs et al. used the time delay feedback mechanism in Biggs et al. (2008) and obtained feasible reference orbits. Small feedback perturbation are used to stabilize chaotic systems. This method can obtain a closed orbit, of pre-specified period.

The feedback mechanism $\nu(t)$ is defined for the nonlinear system $\dot{X}(t) = f(X(t), t)$ as in Biggs et al. (2008):

$$\begin{aligned} \dot{X}(t) &= f(X(t), t) + \nu(t), \\ \nu(t) &= -C(X(t) - X(t - \tau)), \end{aligned} \tag{6}$$

where τ represents the delay time that is the orbit period $\tau = 3.119992528194654$ of 181 days. C is a 6×6 control gain matrix, to be determined. According to Biggs et al. (2008) a practical way to obtain the control gain C is to equal it to the identity matrix $I_{6\times6}$ times an experimentally determined scalar. One can observe that the Eq. (6) in the case of almost periodic numerically determined orbits i.e. $X(t) - X(t - \tau) = \varepsilon$, where ε is small, the feedback mechanism becomes $-C\varepsilon$ and is also small. In this way by using the time-delayed feedback mechanism, the trajectory is converging to a periodic orbit. To determine the control matrix C of the time-delay mechanism the *extremum seeking* method (Blackman 1962) is used. Extremum seeking, as the name implies, is a method to find a set of parameters that minimize a cost function $J(C)$. One candidate objective function is:

$$J(C) = \frac{1}{t_0 - t_f} \int_{t_0}^{t_f} \nu(C, \varepsilon)^2 dt, \tag{7}$$

Computing the gradient of the objective function $\nabla J(C)$, the control gains C can be improved with the update rule:

$$C(k + 1) = C(k) - \alpha \nabla J(C).$$

$C(k)$ are the updated control gains at iteration k. α is a parameter that is used to adjust the step. Thus for $k \to \infty$ the cost function $J(C)$ will approach a local minimum. A way to solve the unconstrained optimization problem, is to used the Matlab Optimization Toolbox. The *Quasi-Newton line search* algorithm of the *fminunc* routine with finite difference gradients was used.

By perturbing the initial condition of the artificial orbits determined earlier, an almost periodic orbit is obtained and by using the time-delay feedback mechanism the trajectory is stabilized about the halo orbit. After ten orbits the error between the position at orbit end and the position at orbit beginning is in the order of 10^{-10} as it can be seen in Fig. 10. Figure 11 shows the magnitude of the controls, that reduce with time.

To have a perfect reference orbit the numerical data that describe the new periodic orbit obtained with the time-delay feedback mechanism fitted to a Fourier function. A third order Fourier function showed a good precision.

$$X_{1j}(t) = a0 + a1 \cdot \cos t + b1 \cdot \sin t + a2 \cdot \cos 2t + b2 \cdot \sin 2t + a3 \cdot \cos 3t + b3 \cdot \sin 3t. \tag{8}$$

with $j = 1 : 6$.

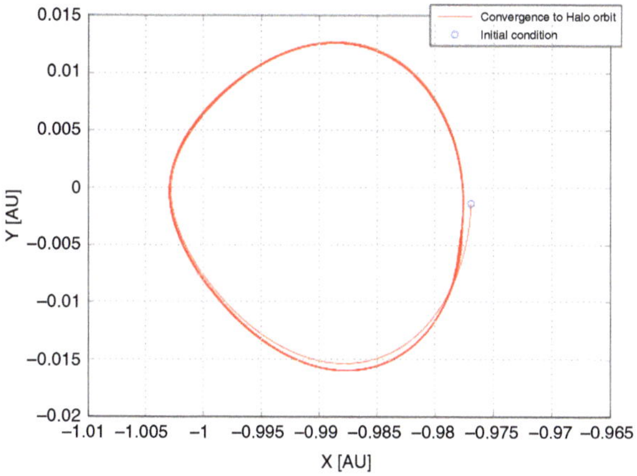

Fig. 10 Time delay feedback mechanism drives the perturbed trajectory to a periodic orbit in the vicinity of a Halo orbit

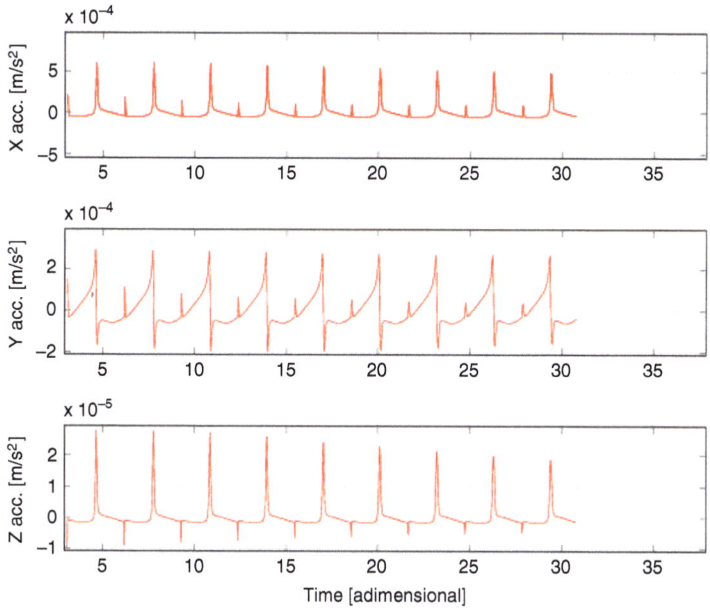

Fig. 11 The acceleration controls of the feedback mechanism needed to stabilize the perturbed orbit

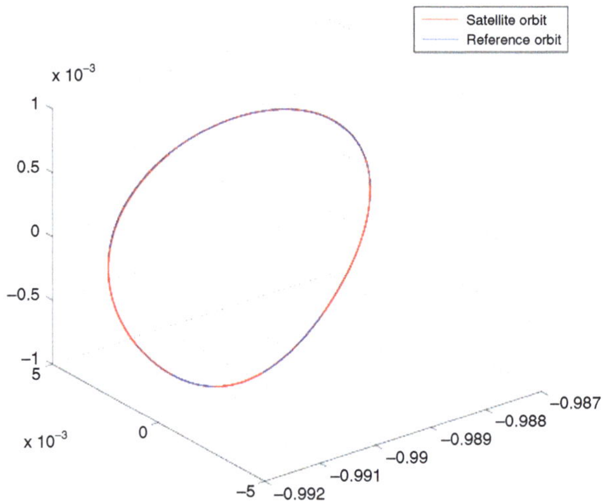

Fig. 12 PD controlled trajectory compared with the time-delay mechanism reference orbit, the *blue line* is the reference orbit while the *red line* is the controlled trajectory

6 Orbit Tracking

To track the reference orbit, a PD controller is used. The gains matrix of the PD controller is determined using the extremum seeking method. Prioroc and Biggs (2015) used the Fourier fit determined earlier as the reference orbit (8), because our aim is to track the reference orbit so that a desired relative motion between the satellites is preserved.

As it can be seen in Figs. 12 and 13, the PD controller is tracking the reference orbit and is able to stabilize the halo orbit. The control accelerations are small and any low thrust solution ca be used.

7 Sliding Mode Controller for Relative Motion

To control a formation of satellites at the Lagrange points on halo orbits active control is required. It is assumed that the satellites have information about the relative position between them, thus the system is observable. The variational equations are determined by numerical differentiation as in Prioroc and Mikkola (2015). A control force u is considered. The solutions of this equation can be used to approximate the relative motion of two satellites that are in the same orbit as showed by Prioroc and Mikkola (2015). The aim is to design a control law u that will drive the relative orbit of a follower satellite with respect to a leader satellite onto a desired relative orbit. A new variable in the state space of the system is

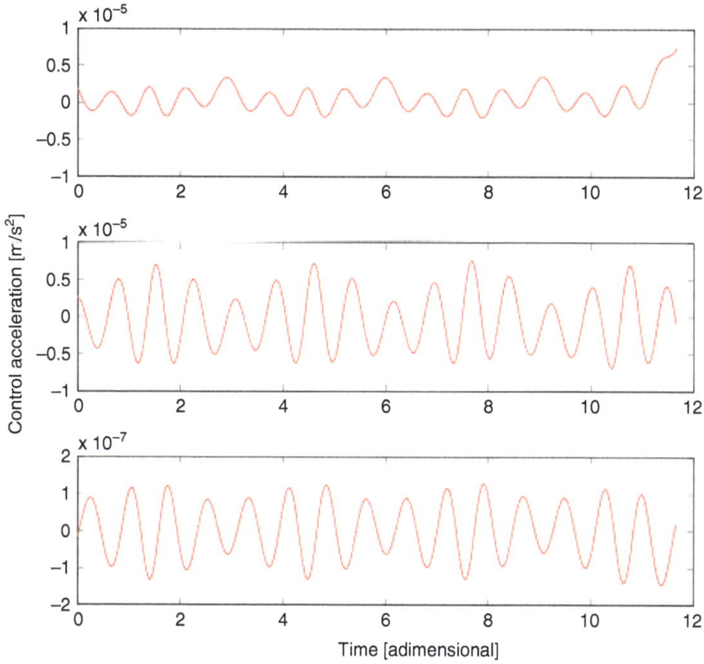

Fig. 13 PD control accelerations for each axis

introduced (Slotine and Sastry 1983):

$$\begin{aligned} \epsilon &= x - x_{ref}, \\ \epsilon &= w - w_{ref}, \\ \sigma(\epsilon, \dot{\epsilon}) &= \dot{\epsilon} - c\epsilon. \end{aligned} \tag{9}$$

In order to achieve asymptotically convergence of the state variables in the presence of bounded disturbances, the variable σ must be driven to zero by means of control u (Young et al. 1999). The σ dynamics are derived:

$$\dot{\sigma} = \dot{\epsilon}c + f(\epsilon, \dot{\epsilon}) + u. \tag{10}$$

The Lyapunov candidate function is chosen:

$$V = 0.5\sigma^2.$$

In order to provide asymptotic stability of the sigma dynamics, about the equilibrium point $\sigma = 0$ two conditions must be satisfied. The derivative of the Lyapunov candidate function must be negative for σ different than 0, and the limit of the Lyapunov candidate function, when the norm of σ goes to infinity, must also go

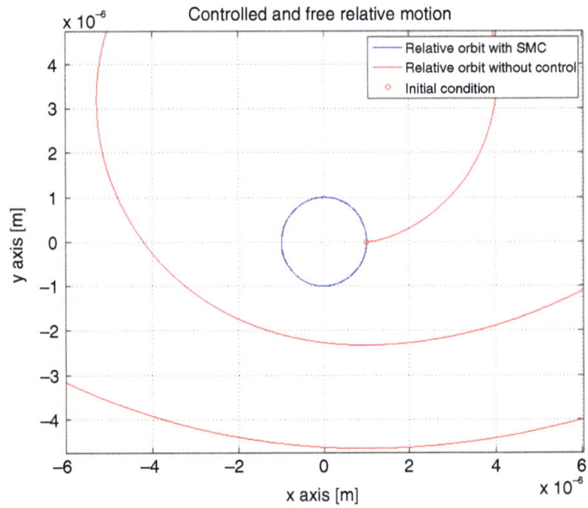

Fig. 14 Relative motion control with the sliding mode controller compared with the free motion

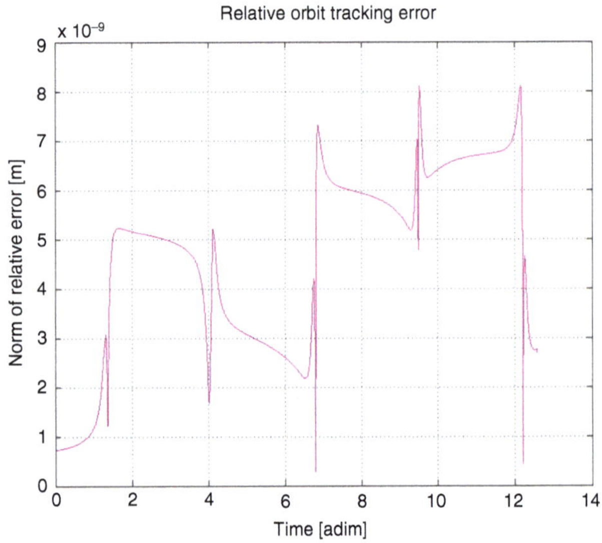

Fig. 15 Relative orbit tracking error using the SMC controller

to infinity. The resulted control function u is:

$$u = -\dot{\sigma} - \rho sign(\sigma). \tag{11}$$

Thus a control law that drives σ to zero in finite time is determined. In the Fig. 14 the uncontrolled relative motion behavior, plotted with the red line, is compared with the sliding mode controller relative motion. From the Fig. 15 it can be seen that

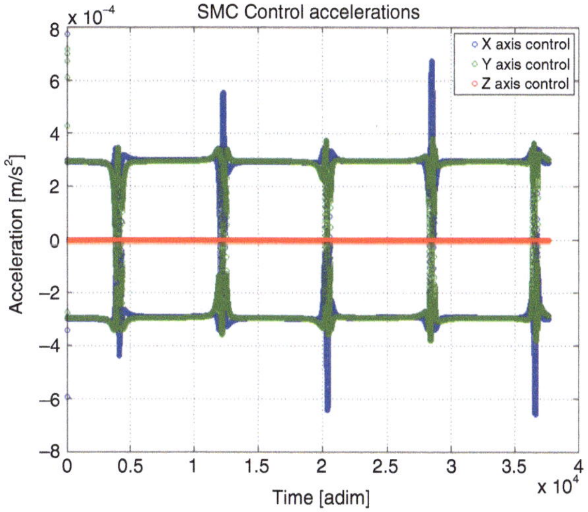

Fig. 16 SMC control accelerations for each axis. As the relative orbit is in the X-Y plane, the control on Z axis is zero

the SMC stabilizes and maintains the desired relative motion of the satellites very accurate, the relative orbit error is in the order of cm.

The plot from Fig. 16 depicts the control acceleration and it can be seen that they does not exceed the thrust levels that the SEP can deliver, that is about 300 mN for a 500 kg spacecraft.

8 Conclusions

Two families of artificial orbits around L_1 and L_2 Lagrange points are found. This artificial orbit families have the same period and have been found using a continuation method from known initial solutions. It was shown that the small and medium orbits in the halo families are unstable. Active control is used to accurately make the spacecraft converge to the reference orbit and nonlinear control is used to control the relative motion of the formation. The reference orbits are constructed using a technique called time-delay feedback mechanism. The reference orbits are tracked with a PD controller. Extremum seeking method is was used to determine the gains of the PD controller. It was shown that stable formation flying around the Lagrange points can be done with low thrust.

Acknowledgements Claudiu Prioroc's research has been funded by the European Commission through the Astrodynamics Network AstroNet-II, under Marie Curie contract PITN-GA-2011-289240.

References

Biggs, J.D., McInnes, C.R.: An optimal gains matrix for time-delay feedback cntrol. In: 2nd IFAC Conference on the Analysis and Control of Chaotic Systems (2009)

Biggs, J.D., McInnes, C.R.: Solar sail formation flying for deep space remote sensing. J. Spacecr. Rocket. **46**, 670–678 (2009)

Biggs, J.D., McInnes, C.R., Waters, T.: Stabilizing periodic orbits above the Ecliptic plane in the solar sail 3-body problem. In: 59th International Astronautical Congress, vol. 8 (2008)

Biggs, J.D., Waters, T.J., McInnes, C.R.: New periodic orbits in the solar sail three-body problem. In: Nonlinear Science and Complexity. Springer, Dordrecht (2009)

Blackman, P.F.: Extremum-Seeking Regulators. The Macmillan Company, New York (1962)

Chen, G., Yu, X.: On time-delayed feedback control of chaotic systems. IEEE Trans. Circuits Syst. I Fundam. Theory Appl. **46**(6), 767–772 (1999)

Chepfer, H., et al.: Estimation of cirrus cloud effective ice crystal shapes using visible reflectances from dual-satellite measurements. J. Geophys. Res. **107**(D23), AAC 21-1–AAC 21-16 (2002)

Dhooge, A., Govaerts, W., Kuznetsov, Yu.A.: MATCONT: a Matlab package for numerical bifurcation analysis of ODEs. ACM SIGSAM Bull. **38**(1), 21–22 (2004)

Doedel, E.J., et al.: Elemental periodic orbits associated with the libration points in the circular restricted 3-body problem. Int. J. Bifurc. Chaos Appl. Sci. Eng. **17**(8), 2625 (2005)

Doedel, E.J., Krauskopf, B., Osinga, H.M.: Global bifurcations of the Lorenz manifold. Nonlinearity **19**(12), 2947 (2006)

Gómez, G., Llibre, J., Martínez, R., Simó, C.: Dynamics and Mission Design Near Libration Points Vol. 1. Fundamentals: The Case of Collinear Libration Points. World Scientific, Singapore (2001)

Huber, M.C.E., Malinovsky-Arduini, M.: The SOHO concept and its realisation. Space Sci. Rev. **61**, 301–334 (1992)

King, M., et al.: Remote sensing of tropospheric aerosols from space: past, present and future. Bull. Am. Meteorol. Soc. **80**(11), 2229–2259 (1999)

Poincaré, H.: Les Methodes Nouvelles de la Mecanique Celeste. Gauthier-Villars, Paris (1892)

Prioroc, C.L., Biggs, J.D.: Low-thrust propulsion formation flying around libration points. Math. Eng. Sci. Aerosp. **6**(1), 87–100 (2015)

Prioroc, C.L., Mikkola, S.: Simple algorithm for relative motion of satellites. N. Astron. Sci. Aerosp. **34**(1), 41–46 (2015)

Pyragas, K.: Continuous control of chaos by self-controlling feed-back. Phys. Lett. A **170**(1), 421–428 (1992)

Slotine, J.J., Sastry, S.S.: Tracking control of nonlinear systems using sliding surfaces with applications to robot manipulator. Int. J. Control **38**(2), 465–492 (1983)

Szebehely, V.: Theory of Orbits. Academic, New York (1967)

Watzin, J.: The Triana Mission - next generation systems architecture ready for flight. In: Proceedings of the Aerospace Conference, vol. 7, pp. 277–285 (2000)

Young, K.D., Utkin, V.I., Ozguner, U.: A control engineer's guide to sliding mode control. IEEE Trans. Control Syst. Technol. **7**(3), 328–342 (1999)

Efficient Modelling of Small Bodies Gravitational Potential for Autonomous Proximity Operations

Andrea Turconi, Phil Palmer, and Mark Roberts

Abstract Maintaining missions in proximity of small bodies requires extensive orbit determination and ground station time due to a ground-in-the-loop approach. Recent developments in on-board navigation paved the way for autonomous proximity operations. The missing elements for achieving this goal are a gravity model, simple enough to be easily used by the spacecraft to steer itself around the asteroid, and guidance laws that rely on a such an inherently simple model. In this research we identified a class of models that can represent well some characteristics of the dynamical environment around small bodies. In particular we chose to fit the positions and Jacobi energies of the equilibrium points generated by the balance of gravity and centrifugal acceleration in the body fixed frame. In this way these gravity models give also a good estimate of the condition of stability against impact for orbital trajectories. Making use of these approximate models we show autonomous guidance laws for achieving body fixed hovering in proximity of the asteroid while ensuring that no impact with the small body will occur during the approach.

1 Introduction

Asteroids represent the next step for space exploration; their detailed study will address open questions on the formation of the Solar System and will support future projects for in-space resources utilisation. Small bodies are also an ideal intermediate step for extending the capabilities of human space flight beyond the Earth-Moon system. Combining the knowledge of asteroids' characteristics and the ability of performing operations in their proximity is also very important for developing the capability of responding successfully to impact threats by Near Earth Objects (NEO). In the past two decades many milestones have been achieved, from the first flybys (Galileo, NASA/ESA, 1991) to missions featuring stable orbits and eventually landing on the surface (Dunham and Farquhar 2002) (NEAR-Shoemaker, NASA, 2000—Rosetta, ESA, 2014) or even getting dust samples back to the

A. Turconi (✉) • P. Palmer • M. Roberts
Surrey Space Centre, University of Surrey, GU27XH Guildford, UK
e-mail: a.turconi@surrey.ac.uk; p.palmer@surrey.ac.uk; m.roberts@surrey.ac.uk

G. Gómez, J.J. Masdemont (eds.), *Astrodynamics Network AstroNet-II*,
Astrophysics and Space Science Proceedings 44,
DOI 10.1007/978-3-319-23986-6_18

Earth (Yoshikawa et al. 2007) (Hayabusa, JAXA, 2005). However, excluding the very last parts of landing trajectories, operations at asteroids have been always performed with the ground-in-the-loop making use of extensive ground station time for measuring the relative positions between the spacecraft and the small body. Additionally, the spacecraft has never had any knowledge of its orbital dynamics and its manoeuvring relied only on the timely execution on preplanned commands. In recent years the developments of the AutoNav software and in particular of its OBIRON component (*On-board Image Registration and Optical Navigation*) has demonstrated the feasibility of autonomous optical navigation in proximity of asteroids (Bhaskaran et al. 2011). In this work we assume that an optical navigation package like the OBIRON is available on the spacecraft providing relative position and velocity with respect to the asteroid centre of mass. To enable autonomous guidance in the proximity of asteroids it is required a model of the gravitational potential than can be used on-board. The limited computational resources available on the spacecraft suggest that it is impractical to use the state of the art detailed polyhedral models (Werner and Scheeres 1996) for this task as well as high degree and order spherical harmonics. Therefore we have to choose a class of approximate models of the gravitational potential that, although not globally accurate, can still represent well some dynamical characteristics that can be useful for the autonomous guidance laws. The comparison between the current concept of operations for small bodies missions and the proposed autonomous operations scenario is outlined in Fig. 1

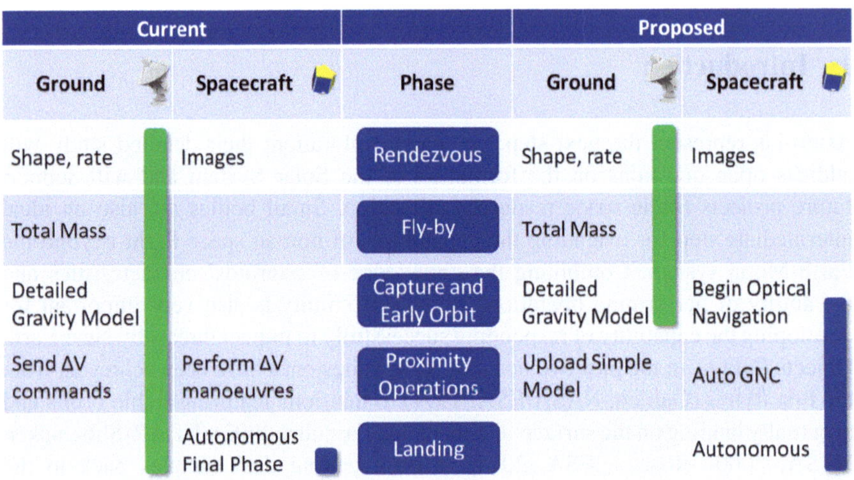

Fig. 1 Current and proposed concept of operations for asteroid missions. By uploading an approximate model based on the state of the art polyhedron, autonomous operations are enabled

2 Background

The irregular shape of asteroids creates an inhomogeneous gravitational field in which the spacecraft generally moves along non-closed and unstable trajectories. In order to guide a spacecraft in the vicinity of an asteroid, elementary characteristics of the small body need to be determined: shape (Torppa and Kaasalainen 2003; Carry et al. 2012; Hudson and Ostro 1994), mass (Hilton 2002) and spin rate (Polishook and Brosch 2009).

The majority of asteroids are in a stable rotation about the axis of their biggest moment of inertia with periods of some hours. Fast rotators and tumbling asteroids exist, but they are not predominant (Pravec et al. 2002) therefore we will consider the small body to be in uniform rotation.

2.1 Equations of Motion in the Body Fixed Frame

A reference frame fixed with the asteroid is used to describe the motion of the spacecraft relative to the rotating body. In such a reference frame the gravitational potential will not vary with time. We will call this reference frame *Body Centred Body Fixed* abbreviated, as *BCBF*.

If we define a principal axis reference frame x, y, z such that the principal moments of inertia will be $I_{zz} > I_{yy} > I_{xx}$, it will be convenient to place the z-axis of the reference frame in the direction of the constant angular velocity vector $\boldsymbol{\omega}$, as shown in Fig. 2.

In such a non-inertial frame the fictitious terms of the accelerations due to the uniform rotation of the reference frame will appear. The total acceleration $\ddot{\mathbf{r}}$ in the BCBF frame is given by the combination of the gravitational, centrifugal and Coriolis accelerations—respectively $\mathbf{a}_{\mathrm{G}}$, $\mathbf{a}_{\mathrm{CF}}$ and $\mathbf{a}_{\mathrm{CO}}$—as shown in Eq. (1). Position, velocity and acceleration in the BCBF frame are represented by $\mathbf{r}$, $\dot{\mathbf{r}}$, $\ddot{\mathbf{r}}$, $\boldsymbol{\omega}$ is the rotation rate vector and U the gravitational potential of the asteroid which

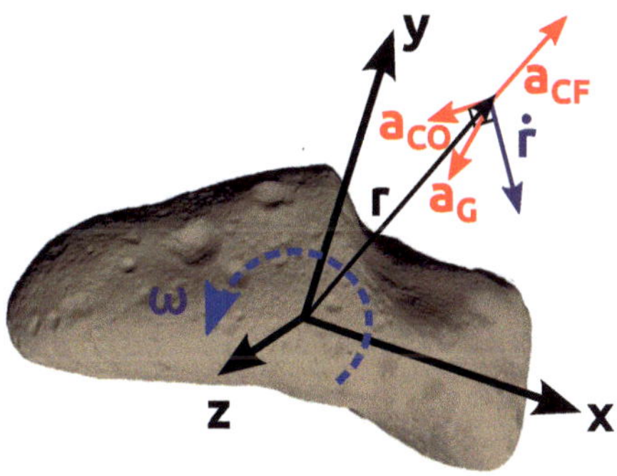

Fig. 2 Body Centred Body Fixed (BCBF) reference frame. x, y and z are principal axis of inertia and $\boldsymbol{\omega}$ is a uniform rotation vector in the same direction of z. $\mathbf{r}$ and $\dot{\mathbf{r}}$ represent position and velocity of the spacecraft relative to the asteroid while $\mathbf{a}_{\mathrm{G}}$, $\mathbf{a}_{\mathrm{CF}}$, $\mathbf{a}_{\mathrm{CO}}$ are the gravitational, centrifugal and Coriolis accelerations

is time invariant in the BCBF rotating frame and therefore depends only on the position $\mathbf{r}$.

$$\begin{cases} \ddot{\mathbf{r}} = \mathbf{a}_{\mathrm{G}} + \mathbf{a}_{\mathrm{CF}} + \mathbf{a}_{\mathrm{CO}} \\ \mathbf{a}_{\mathrm{G}} = \nabla U \\ \mathbf{a}_{\mathrm{CF}} = -\boldsymbol{\omega} \times (\boldsymbol{\omega} \times \mathbf{r}) \\ \mathbf{a}_{\mathrm{CO}} = -2\boldsymbol{\omega} \times \dot{\mathbf{r}} \end{cases} \tag{1}$$

As shown in Fig. 2 the centrifugal acceleration points always in the radial direction. The Coriolis acceleration is always perpendicular to the velocity vector and for a counter-clockwise rotation of the reference frame it imparts a clockwise rotation to the velocity. Finally the gravitational acceleration is in general not simply radial due to the non-uniformity of the gravitational potential.

2.2 *Equilibrium Points*

By setting $\ddot{\mathbf{r}}$ and $\dot{\mathbf{r}}$ to zero in Eq. (1) we find that there are points where the gravitational pull balances the centrifugal force.

$$\mathbf{a}_{\mathrm{G}} = -\mathbf{a}_{\mathrm{CF}} \tag{2}$$

Uniformly rotating asteroids have generally four equilibrium points placed roughly along the short and long axes of the body and lying in the equatorial plane (Scheeres 2012). These equilibrium points are fixed in the BCBF frame in the same way as the Lagrange Points are fixed in the synodic frame of the *Circular Restricted 3-body Problem* (CR3BP) (Vallado 2007). The asteroids' rotational equilibrium points feature various dynamical behaviours depending on the second derivatives of the gravitational potential evaluated there. By writing Eq. (1) in the equatorial plane we find the equations of motion in cartesian and polar components: Eqs. (3) and (4)

$$\begin{cases} \ddot{x} = \dfrac{\partial U}{\partial x} + 2\omega\dot{y} + \omega^2 x \\ \ddot{y} = \dfrac{\partial U}{\partial y} - 2\omega\dot{x} + \omega^2 y \end{cases} \tag{3}$$

$$\begin{cases} \ddot{r} = \dfrac{\partial U}{\partial r} + 2\omega r\dot{\theta} + \omega^2 r \\ r\ddot{\theta} = \dfrac{1}{r}\dfrac{\partial U}{\partial \theta} - 2\omega\dot{r} \end{cases} \tag{4}$$

Setting the velocity and the total acceleration to zero, Eqs. (3) and (4) express the condition to be verified at the equilibrium points.

$$\begin{cases} -\omega^2 x = \dfrac{\partial U}{\partial x} \\ -\omega^2 y = \dfrac{\partial U}{\partial y} \end{cases} \tag{5}$$

$$\begin{cases} -\omega^2 r = \dfrac{\partial U}{\partial r} \\ 0 = \dfrac{\partial U}{\partial \theta} \end{cases} \tag{6}$$

In Eq. (6) is made clear that at the equilibrium points the tangential component of the gradient of the potential shall be zero by itself since it cannot be balanced by the centrifugal acceleration. We computed the positions of the equilibrium points by solving numerically the system of non-linear equations shown in Eq. (5).

2.3 *The Jacobi Energy Integral*

One of the biggest challenges in performing proximity operations is that, in general, low orbits about small bodies are unstable and lead either to impact the asteroid or to escape its sphere of influence.

As it happens in the CR3BP, in presence of a uniformly rotating gravitational potential, a constant of motion exists in the rotating frame where the gravitational potential is time-invariant. Such a conserved quantity is called *Jacobi Integral of Motion* or, more simply, *Jacobi Energy*. It is computed as shown in Eq. (7) by adding the centrifugal potential to the expression of the total mechanical energy:

$$J(\mathbf{r}, \dot{\mathbf{r}}) = \frac{1}{2}|\dot{\mathbf{r}}|^2 - U(\mathbf{r}) - \frac{1}{2}\boldsymbol{\omega}^2|\mathbf{r}|^2 \tag{7}$$

where J is the Jacobi Energy, $\dot{\mathbf{r}}$ is the velocity vector in the body fixed frame, U is the gravitational potential, $\boldsymbol{\omega}$ is angular rate of the small body and $\mathbf{r}$ is the position vector in the body fixed frame.

Natural motion will always keep J constant. If the Kinetic Energy is imposed to be zero, as in Eq. (8), at any given level of Jacobi Energy it will be possible to visualize the so-called Zero-velocity curves. These lines as shown in Fig. 3 will indicate the permitted and forbidden regions of motion at any energy level.

$$J_0(\mathbf{r}) = -U(\mathbf{r}) - \frac{1}{2}\omega^2|\mathbf{r}|^2 \tag{8}$$

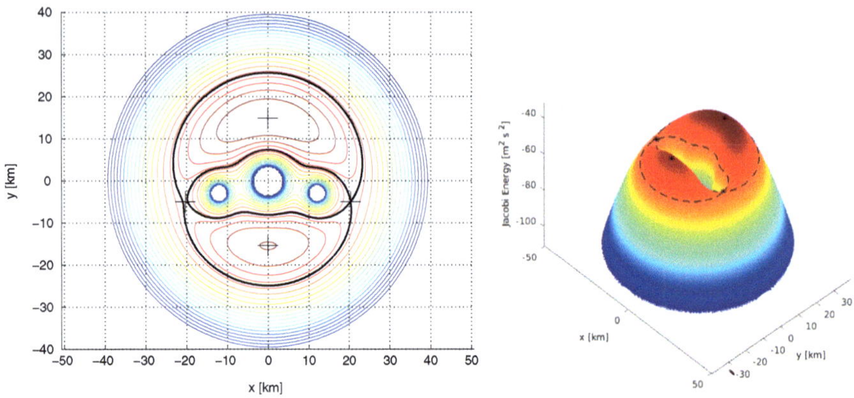

Fig. 3 Zero-velocity curves of the 3-spheres model of 433 Eros. The (+) marks identify the positions of the equilibrium points while the *thick black line* highlights the Zero Velocity Curve at the energy level of the lowest equilibrium point

It is possible to better visualise the evolution of these limiting curves with increasing values of energy, thanks to a 3D plot associating to the x and y positions in the equatorial plane the values of J_0 on the z axis. As it is clear especially from the right hand side of Fig. 3, for levels of the Jacobi energy below the one of the lowest equilibrium point, any trajectory will be prevented from falling onto the asteroid. The 3D plot shows that the collinear equilibrium points can be considered as the energy gateways to the asteroid surface.

In Eq. (9) we set the Jacobi energy of the lowest collinear point as a threshold value, $\min(J_{eq})$: below this energy limit the spacecraft will never impact the asteroid.

$$J(\mathbf{r}, \dot{\mathbf{r}}) \equiv \frac{1}{2}|\dot{\mathbf{r}}|^2 - U(\mathbf{r}) - \frac{1}{2}\omega^2|\mathbf{r}|^2 \leq \min(J_{eq}) \tag{9}$$

3 Generation of the Approximate Model

Given the characteristics of the dynamical environment around asteroids we decided to simplify the representation of their irregular mass distribution with a model consisting of three point masses. This is indeed the lowest number of point masses that can reproduce four non-symmetrical rotational equilibrium points.

Our simple model comes as the generalisation of a 3-spheres model developed at the Surrey Space Centre (Herrera et al. 2013) which is able to represent with good accuracy the position, the dynamical behaviour and the energy level of the equilibrium points of the asteroid (Herrera 2012).

We decided to place the three point masses on the equatorial plane since the equilibrium points either lie on that plane or in its proximity. Therefore the proposed approximate model will be defined by just nine parameters, the value of the masses m_i and their x_i and y_i coordinates on the equatorial plane:

$$m_1, m_2, m_3, x_1, x_2, x_3, y_1, y_2, y_3 \tag{10}$$

In order to relate the approximate models to the higher accuracy ones we need to define an objective function and some constraints on the parameters of the model. Since our aim is to replicate the same characteristics of the four rotational equilibrium points about the asteroid, we would like to satisfy the following equations for each of the four equilibrium points as in Eq. (11)

$$\begin{cases} \left.\dfrac{\partial \tilde{U}}{\partial x}\right|_{(x_e,y_e)} + \omega^2 \cdot x_e = 0 \\ \left.\dfrac{\partial \tilde{U}}{\partial y}\right|_{(x_e,y_e)} + \omega^2 \cdot y_e = 0 \end{cases} \tag{11}$$

where x_e and y_e (with $e = 1 \ldots 4$) are the positions of the equilibrium points computed using the detailed model of the asteroid. The gravitational potential $\tilde{U}$ is the one of the approximate three point mass model as shown in Eq. (12)

$$\tilde{U}(x, y) = \frac{m_1}{|\mathbf{r} - \mathbf{r_1}|} + \frac{m_2}{|\mathbf{r} - \mathbf{r_2}|} + \frac{m_3}{|\mathbf{r} - \mathbf{r_3}|} \tag{12}$$

where the $\mathbf{r}$ is the spacecraft position vector in the BCBF reference frame and the $\mathbf{r_i}$ are the position vectors of the three point masses in the same coordinate system. Moreover, to model correctly the total mass and the first mass moments, we would like the simple model to match the total mass and the position of the centre of mass as well:

$$\tilde{M} = M \tag{13}$$

$$\frac{m_1x_1 + m_2x_2 + m_3x_3}{\tilde{M}} = x_{CM} \tag{14}$$

$$\frac{m_1y_1 + m_2y_2 + m_3y_3}{\tilde{M}} = y_{CM} \tag{15}$$

where $\tilde{M} = \sum m_i$ is the total mass of the approximate model and the coordinates of the centre of mass of the truth model are $(0, 0)$ since we work in the reference frame of the principal axes of inertia. At first we tried to solve a system of nine non-linear equations composed of the eight equations for the equilibrium points, in the form of Eqs. (11), adding Eq. (13) for the total mass. This approach however has shown poor performance and many convergence issues.

We then rewrote the problem as an optimisation one by combining the squares of the residuals of the equations of interest in a single scalar objective function F as shown in Eq. (16)

$$\begin{aligned} g_x(m_i, x_i, y_i) &= \sum_{e=1}^{4} \left[\left. \frac{\partial \tilde{U}}{\partial x} \right|_{(x_e, y_e)} + \omega^2 \cdot x_e \right]^2 \\ g_y(m_i, x_i, y_i) &= \sum_{e=1}^{4} \left[\left. \frac{\partial \tilde{U}}{\partial y} \right|_{(x_e, y_e)} + \omega^2 \cdot y_e \right]^2 \\ g_m(m_i) &= [m_1 + m_2 + m_3 - M]^2 \\ F(m_i, x_i, y_i) &= g_x + g_y + g_m \end{aligned} \tag{16}$$

The optimisation problem was then set as shown in Eq. (17)

$$\min F(m_i, x_i, y_i) : \begin{cases} 0.1\, M \leq m_i \leq M \\ -5\, r_{sync} \leq x_i \leq 5\, r_{sync} \\ -5\, r_{sync} \leq y_i \leq 5\, r_{sync} \end{cases} \tag{17}$$

where (m_i, x_i, y_i) are the parameters of the approximate model, M is the total mass of the truth model and r_{sync} is the synchronous radius for the asteroid when modelled as a point mass:

$$r_{sync} = \sqrt[3]{\frac{GM}{\omega^2}} \tag{18}$$

3.1 Optimisation Test Case

To give an idea of the results obtained from the solution of the optimisation problem we described, we use the following model of ten equal point masses with the associated position non-dimensionalised by r_{sync}:

$$\begin{aligned} \bar{x}_i &= [0.6472, 0.2828, -0.0816, -0.3549, \\ &\quad -0.8104, 0.7099, 0.4366, \\ &\quad -0.1100, -0.3834, -0.3360] \\ \bar{y}_i &= [-0.3153, -0.2389, -0.1625, -0.1052, \\ &\quad -0.0098, 0.2105, 0.2678, \\ &\quad 0.3824, 0.4397, -0.4685] \end{aligned} \tag{19}$$

The results from the optimisation problem described in Eq. (17) using MATLAB's `fmincon` *interior-point* algorithm are the following:

$$x_i = [0.7594, -0.0281, -0.8114]$$
$$y_i = [0.1536, 0.0029, -0.0125] \tag{20}$$
$$m_i = [0.1000, 0.8020, 0.1000]$$

With the new model we computed the new positions of the equilibrium points and the value of the gravitational potential at their locations. Since the potential is defined up to a constant term we determined the correction constant U_{offset} such that the value of the potential of the equilibrium point at the lowest energy is the same between the truth model and the approximation.

$$U_{offset} = U_{eMin} - \tilde{U}_{eMin} \tag{21}$$

Without this last precaution we could not really hope to represent well the level of potential energy with the approximate model since the optimisation is based on the residual forces at the location of the equilibrium points and doesn't take into account the value of the potential energy there.

In Fig. 4 we compare the truth and approximate model described in Eqs. (20) and (21) showing that the collinear equilibrium points are approximated with

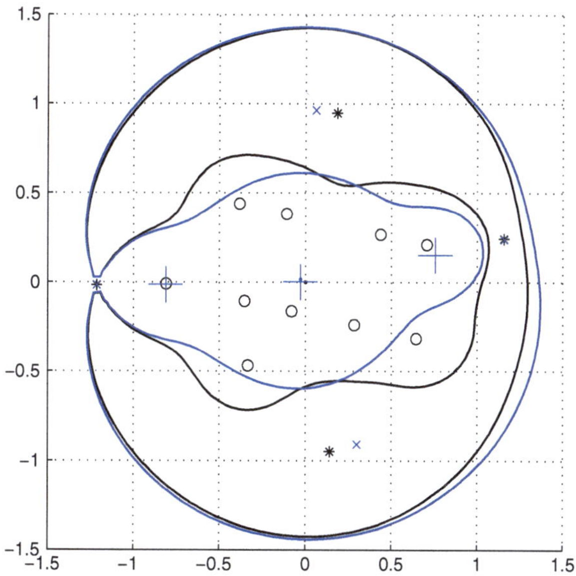

Fig. 4 Comparison between truth and approximate model. The positions of the point masses for both models are highlighted by *circles* (simple) and *plus markers* (truth). The *thicker outer blue line* shows the Zero-Velocity Curve profile of the approximate model while the *inner black line* the same profile computed for the truth model. In *blue* "x" and *black* "*" are also shown the position of the equilibrium points for the two models

remarkable accuracy and that the Zero-Velocity Curves profiles at the energy of the least energetic equilibrium point are reasonably close to each other even globally.

4 Guidance and Control

Thanks to the approximate model of the asteroid which represents correctly the positions of the equilibrium points and their energy level, the spacecraft will be able to use this knowledge to steer itself towards the equilibrium point at the lowest energy while complying with the non-impact condition. Our control strategy is therefore to change the state of the spacecraft in order to increase its energy until it reaches the level of the target equilibrium point. At the same time we need to make sure that we are heading towards the location of the target equilibrium point.

4.1 Reaching the Target Energy Level

Lyapunov functions are scalar functions used to prove the stability of the equilibria of differential equations. Their importance in non-linear control theory relies on the fact that thanks to them it is sometimes possible to devise non-linear controllers with guaranteed stability and that will lead the system to approach the set point asymptotically without any overshoot.

A Lyapunov function shall be positive definite and its derivative shall always be negative for values other than zero. Equation (22) summarizes those conditions.

$$V : \mathbb{R}^n \rightarrow \mathbb{R} \quad ; \quad V(0) = 0 \quad ; \quad V(J) > 0 \quad ; \quad \dot{V}(J \neq J^*) < 0 \tag{22}$$

We have chosen as Lyapunov function $V = \frac{1}{2}(J^* - J)^2$, with J instantaneous Jacobi energy and J^* target Jacobi Energy such that $J^* = \min(J_{\text{eq}})$. The derivative of this Lyapunov function is:

$$\begin{aligned} \dot{V} &= -(J^* - J)\frac{dJ}{dt} \\ \text{with} \quad \frac{dJ}{dt} &= \ddot{x}\dot{x} + \ddot{y}\dot{y} = \mathbf{a}_{CTRL} \cdot \dot{\mathbf{r}} \end{aligned} \tag{23}$$

where r_{dot} is the spacecraft velocity vector and $\mathbf{a}_{CTRL}$ the control acceleration. Equation (23) tells us that the energy will change only because of the component of the control acceleration that modifies the magnitude of the velocity (Turconi et al. 2014).

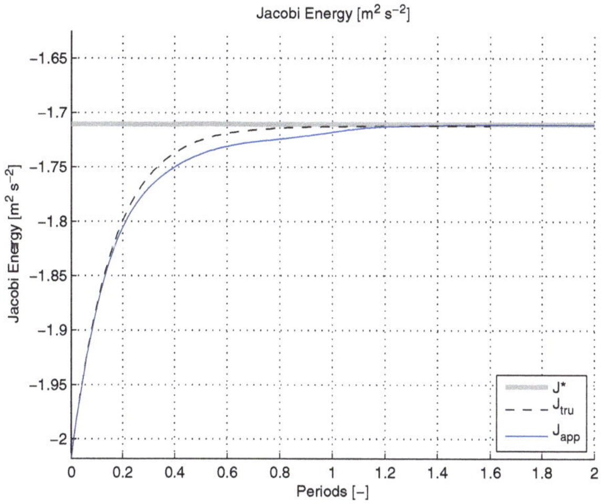

Fig. 5 Comparison of the values of the Jacobi energy while controlling using the truth model (*dashed line*) and the approximate one (*continuous line*). The target energy level is represented by the *thick grey line*

In order to ensure that the derivative of the Lyapunov function will always be negative, it is required an appropriate expression for the control acceleration. Our choice is shown in Eq. (24)

$$\mathbf{a_J} = \frac{1}{|\mathbf{r_2}| + \Delta r_2}(J^* - J)\frac{\dot{\mathbf{r}}}{|\dot{\mathbf{r}}|} \tag{24}$$

where $|\mathbf{r_2}|$ is the distance from the instantaneous position of the spacecraft and the target equilibrium point and Δr_2 is a minimum distance from the target position.

An integrator using the Bulirsch-Stoer method has been developed for the propagation of the equations of motion in the rotating frame. The code has been validated by integrating free motion trajectories and verifying the conservation of the Jacobi Energy. The result of a simulation using only the Lyapunov control law to match the Jacobi energy level is shown in Figure comparing the result of using the truth and the approximate gravity models (Fig. 5).

4.2 Controlling Towards the Equilibrium Point

It has been shown how a Lyapunov control law is able to bring the spacecraft to the correct energy level. However, the final aim of our guidance is in fact to steer the spacecraft towards the location of the equilibrium point. In order to do so we need to add another term to the control acceleration, a term that shall incorporate

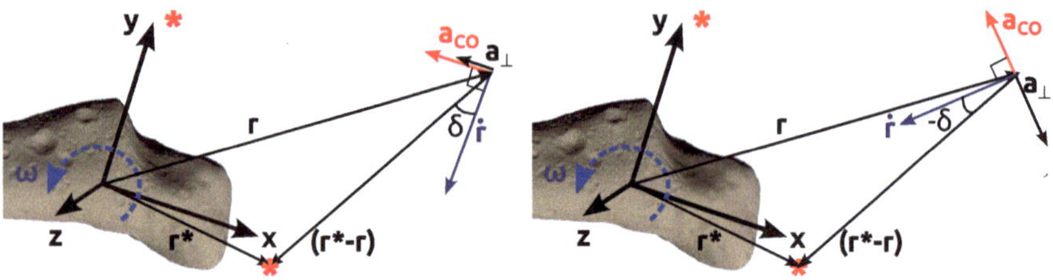

Fig. 6 Geometry of the control acceleration normal to the velocity for $\delta > 0$ and $\delta < 0$

the information of the position of the equilibrium point or of its direction relative to the spacecraft.

With this additional component of the control acceleration we don't want to affect the energy, this implies that the new acceleration has to be always normal to the instantaneous velocity, so that the derivative of the Jacobi Energy is always zero as in Eq. (25)

$$\frac{dJ}{dt} = \mathbf{a}_\perp \cdot \dot{\mathbf{r}} = 0 \tag{25}$$

Since we want to steer the spacecraft towards the target equilibrium point we will use the angle δ as indication of the separation between the direction of the target, as seen by the spacecraft, and the the velocity vector as shown in Fig. 6.

For this component of the acceleration in charge of the rotation of the velocity towards the target, the idea was to design a controller proportional to the angle δ. The intended behaviour of the controller is described by $\mathbf{a}_{\perp \mathrm{DESIRED}}$ in Eq. (26).

$$\mathbf{a}_{\perp \mathrm{DESIRED}} = \frac{|\dot{\mathbf{r}}|^2}{|\mathbf{r_2}| + \Delta r_2} \cdot \delta \cdot \frac{\mathbf{a}_{CO}}{|\mathbf{a}_{CO}|} \tag{26}$$

In the direction normal to the velocity the Coriolis acceleration is always present which, in this case, continuously imparts a clockwise rotation to the velocity. It is sensible to take this into account since, for $\delta > 0$, $\mathbf{a}_{\mathrm{CO}}$ is the same direction of $\mathbf{a}_{\perp \mathrm{DESIRED}}$ while for $\delta < 0$ it happens to be in the opposite one. Thinking of the desired behaviour as the result of the combined effect of the control acceleration $\mathbf{a}_\perp$ and the Coriolis acceleration $\mathbf{a}_{\mathrm{CO}}$, we will write Eq. (27) and subsequently find the expression of the control acceleration including the compensation for the Coriolis force as in Eq. (28).

$$\mathbf{a}_{\perp \mathrm{DESIRED}} = \mathbf{a}_{\mathrm{CO}} + \mathbf{a}_\perp \tag{27}$$

$$\mathbf{a}_\perp = \frac{|\dot{\mathbf{r}}|^2}{|\mathbf{r_2}| + \Delta r_2} \cdot \delta \cdot \frac{\mathbf{a}_{CO}}{|\mathbf{a}_{CO}|} - \mathbf{a}_{\mathrm{CO}} \tag{28}$$

We then extended this steering control law to 3D in order to ensure that the spacecraft is brought on the plane of the equilibrium point. The perpendicular direction to the velocity along which the control acceleration will have to be applied has to take into account also the z component of $\mathbf{r_2}$ and it is defined in Eq. (29)

$$\mathbf{e}_{\perp} = \mathbf{e}_{\dot{r}} \times (\mathbf{e}_{r_2} \times \mathbf{e}_{\dot{r}}) \tag{29}$$

Moreover by compensating the projection of the gravitational and centrifugal accelerations along this newly defined direction we complete the compensation of all the forces already in the direction of the desired acceleration as in Eq. (30)

$$\mathbf{a}_{COMP} = (-\mathbf{a}_{CO} - \mathbf{a}_{G} - \mathbf{a}_{CF}) \cdot \mathbf{e}_{\perp} \tag{30}$$

The final form of the steering component of the control law in 3D is then:

$$\mathbf{a}_{\perp} = \frac{|\dot{\mathbf{r}}|^2}{|\mathbf{r_2}| + \Delta r_2} \cdot \delta \cdot \mathbf{e}_{\perp} + \mathbf{a}_{COMP} \tag{31}$$

4.3 *Guidance and Control Test Case*

Using the full 3D representation of the equations of motion and the 3D formulation of the control law, additional simulations have been then carried out using MATLAB SIMULINK. The typical uncertainties in position and velocity expected from an optical navigation package have been also included: ± 11 m in position and 0.7 mm/s in velocity.

The propagation is carried out around the ten point mass truth model while the control is generated using the approximate three point mass gravity model derived from the optimisation discussed in Sect. 3.

We started the simulation from a prograde circular orbit around the asteroid which is at a lower energy level than the one of the equilibrium point. By never exceeding the Jacobi energy set-point throughout the manoeuvre we ensure that we are always stable against impact with the asteroid.

As shown in Fig. 7 the energy approaches the target value even in presence of uncertainties in relative position and velocity. The magnitude of the control acceleration component parallel to the velocity then vanishes once the energy level is reached. The magnitude of the steering part of the control law is initially up to one order of magnitude larger than the parallel one which takes care of the energy level.

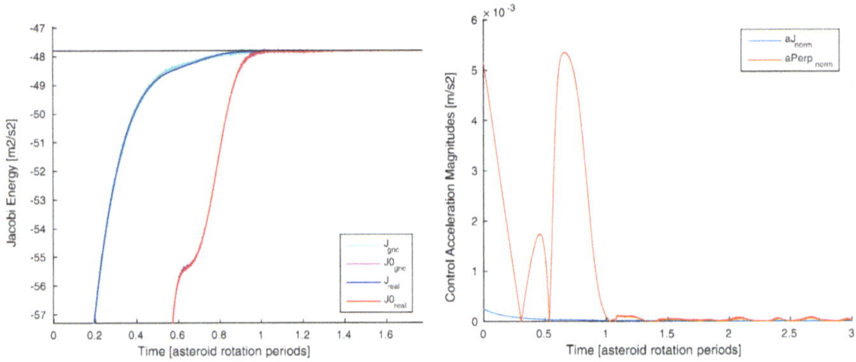

Fig. 7 *Left*: Jacobi energy J and Zero-Velocity Jacobi energy J_0 along the controlled trajectory. *Right*: Magnitude of the two components of the control acceleration: parallel and normal to the instantaneous velocity vector

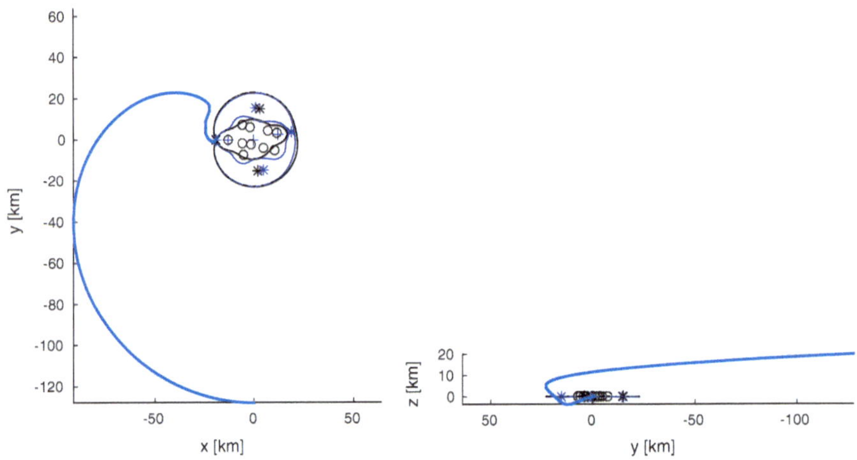

Fig. 8 3D controlled trajectory from prograde circular orbit to the target equilibrium point

In Fig. 8 we show the 3D trajectory of the spacecraft which starts from a prograde circular orbit with a z dispacement above the target plane. As shown by the 3D trajectory and by the plots of the components of relative position and velocity towards the target, the proposed control law is able to guide the spacecraft to the x, y, z coordinates of the target equilibrium point and to reach it with zero velocity (Fig. 9).

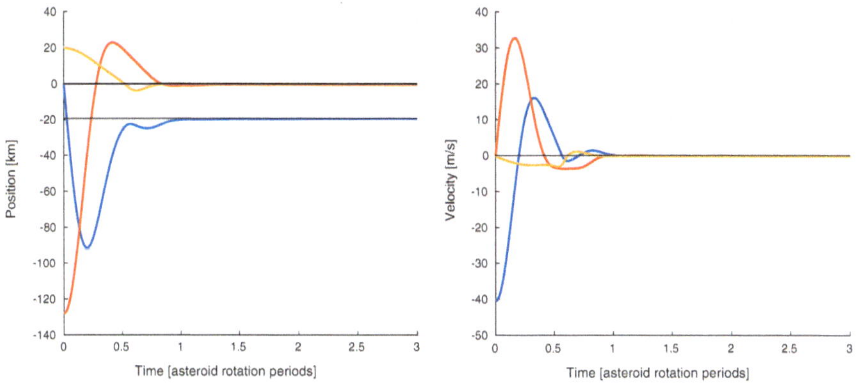

Fig. 9 Components in the body-fixed reference frame for position and velocity relative to target

5 Conclusions and Future Work

From the analysis of the dynamical environment about a rotating asteroid we proposed a simple model of the gravitational potential which gives a good representation of the position and energy level of the equilibrium points. We have shown an optimisation process for the derivation of this class of three point mass models. We also compared the profile of the Zero-Velocity curves of the two models at the level of the equilibrium point at the lowest energy which is needed for ensuring stability against impact.

Thanks to the definition of a control law made of two independent components, parallel and normal to the instantaneous velocity vector, we have shown the use of the proposed approximate model for a transfer from a lower energy prograde orbit around the asteroid to the lowest energy equilibrium point.

Given the large impact on the control law coming from the steering component, further developments are needed in order to improve the guidance. The ideal transfer we can devise by looking at natural trajectories arriving at the equilibrium point has in fact the velocity pointed towards the target only in the final part of the approach. At the moment instead, we are constantly steering the velocity vector towards the target. Even by doing so compensating the natural dynamics that is initially helping we are then forced to counteract the same natural forces to keep the required direction once it is achieved.

Further work is under way for using as truth models the state-of-the-art constant-density polyhedra. While the optimisation strategy and the control approach won't change with the use of these detailed models, higher quality results would be obtained. Approximate models will be generated for real targets performing, at least in simulation, the exact scenario that is outlined in Fig. 1.

References

Bhaskaran, S., Nandi, S., Broschart, S.: Small body landings using autonomous onboard optical navigation. J. Astronaut. Sci. **58**, 409–427 (2011)
Carry, B., Kaasalainen, M., Merline, W.: Shape modeling technique KOALA validated by ESA Rosetta at (21) Lutetia. Planet. and Space Sci. **66**(1), 200–212 (2012)
Dunham, D.W., Farquhar, R.W.: Implementation of the first asteroid landing. Icarus **159**, 433–438 (2002)
Herrera, E.: The Full Problem of Two and Three Bodies: Application to Asteroids and Binaries. University of Surrey, Guildford (2012)
Herrera, E., Palmer, P.L., Roberts, M.: Modeling the gravitational potential of a nonspherical asteroid. J. Guid. Control Dyn. **36**, 790–798 (2013)
Hilton, J.L.: Asteroid masses and densities. Asteroids **III**, 103–112 (2002)
Hudson, R.S., Ostro, S.J.: Shape of asteroid 4769 Castalia (1989 PB) from inversion of radar images. Science **263**, 940–943 (1994)
Polishook, D., Brosch, N.: Photometry and spin rate distribution of small-sized main belt asteroids. Icarus **199**, 319–332 (2009)
Pravec, P., Harris, A., Michalowski, T.: Asteroid rotations. Asteroids **III**, 113–122 (2002)
Scheeres, D.J.: Orbital Motion in Strongly Perturbed Environments: Applications to Asteroids, Comet and Planetary Satellite Orbiters. Springer, London (2012)
Torppa, J., Kaasalainen, M.: Shapes and rotational properties of thirty asteroids from photometric data. Icarus **164**, 346–383 (2003)
Turconi, A., Palmer, P.L., Roberts, M.: Autonomous guidance and control in the proximity of asteroids using a simple model of the gravitational potential. In: 2nd IAA-AAS DyCoSS Conference, Rome (2014)
Vallado, D.: Fundamentals of Astrodynamics and Applications. Microcosm Press, Hawthorne; Springer, New York (2007)
Werner, R.A., Scheeres, D.J.: Exterior gravitation of a polyhedron derived and compared with harmonic and mascon gravitation representations of asteroid 4769 Castalia. Celest. Mech. Dyn. Astron. **65**, 313–344 (1996)
Yoshikawa, M., Fujiwara, A., Kawaguchi, J.: Hayabusa and its adventure around the tiny asteroid Itokawa. In: Proceedings of the International Astronomical Union 2.14 (2007)

Sun-Earth L1 and L2 to Moon Transfers Exploiting Natural Dynamics

Willem J. van der Weg and Massimiliano Vasile

Abstract This work examines the design of transfers from the Sun-Earth libration orbits, at the L_1 and L_2 points, towards the Moon using natural dynamics in order to assess the feasibility of future disposal or lifetime extension operations. With an eye to the probably small quantity of propellant left when its operational life has ended, the spacecraft leaves the libration point orbit on an unstable invariant manifold to bring itself closer to the Earth and Moon. The total trajectory is modeled in the coupled circular restricted three-body problem. The concept of survivability and event maps is introduced to obtain suitable conditions that can be targeted such that the spacecraft impacts, or is weakly captured by, the Moon. Weak capture at the Moon is studied by method of these maps. Some results for planar Lyapunov orbits at L_1 and L_2 are given, as well as some results for the operational orbit of SOHO.

1 Introduction

It has become increasingly accepted by the space community that once a spacecraft has reached an end to its nominal mission lifetime it should be safely disposed of such that future missions are not jeopardized. While this holds especially true for particularly busy orbits around the Earth, such as low earth and geosynchronous orbits, there is also a case to make for safely controlling the disposal of spacecraft in libration point orbits (LPO) at the Sun-Earth L_1 and L_2 libration points. Often these spacecraft will still have propellant left after their mission has been completed, and it is therefore interesting to see what could be done with these spacecraft in terms of disposal or mission extension. In this work we study the disposal options towards the Moon from both Sun-Earth L_1 and L_2 libration points within the framework of the Coupled Circular Restricted Three-Body Problem, or Coupled CR3BP (Koon et al. 2001b). This methodology has been used in the past to study trajectories between

W.J. van der Weg (✉)
University of Zielona Góra, Licealna 9, 65-417 Zielona Góra, Poland
e-mail: wvanderweg@gmail.com

M. Vasile, Ph.D.
University of Strathclyde, Montrose Street 75, G12 1XJ Glasgow, UK
e-mail: massimiliano.vasile@strath.ac.uk

G. Gómez, J.J. Masdemont (eds.), *Astrodynamics Network AstroNet-II*,
Astrophysics and Space Science Proceedings 44,
DOI 10.1007/978-3-319-23986-6_19

the Sun-Earth libration points and the vicinity of the Earth and Moon, where a connection is made using the unstable manifold flowing from the Sun-Earth libration point (within the Sun-Earth CR3BP) and the stable manifold flowing towards the Moon L_2 point (within the Earth-Moon CR3BP) at low Δv cost. Examples include the work done by Canalias and Masdemont (2008); Fantino et al. (2010); Gómez et al. (2001); Koon et al. (2001a).

This work introduces the concept of survivability map and event map to find target conditions, in the vicinity of the Moon, that lead to lunar impact or lunar weak capture. These maps aid in the design of trajectories and effectively replace the use of the stable manifold to design the trajectory arc incoming towards the Moon in the Earth-Moon CR3BP. This approach enables a very simple transfer design where one directly targets a state on the map in order to get the desired capture orbit or impact. In the past, Poincaré maps at the pericentre have been used to find useful target conditions that lead to a capture around a body (Haapala and Howell 2014; Villac and Scheeres 2003). Here we look at finding target states further away from the celestial body, using either variable or constant energy level, that give long permanence within a given region in the configuration space. In doing so we identify also possible capture, impact and escape states. Weak capture (or temporary ballistic capture) is typically defined as a spacecraft moving to within the vicinity of the planet (in this case the Moon) and staying there for some minimum period of time, or by performing at least a single revolution about the planet. There is extensive work in the literature on weak capture in particular to design transfers to the Moon with a reduced propellant cost with respect to a more traditional Hohmann transfer. An algorithmic definition of the weak stability boundary is given by Belbruno (2004), and later expanded upon by García and Gómez (2007). A quite complete overview of the existing literature can be found in the work of Silva PAS and Terra (2012), and a clear definition of weak capture is given by Topputo et al. (2008).

This work begins with a brief overview of the CR3BP, and the method of connecting (often referred to as patching) multiple CR3BPs, in Sect. 2. Then, Sect. 3 introduces the concept of the survival and event maps, which are used to acquire initial conditions (named lunar target states) that lead to lunar impact or capture. Section 4 describes the overall process used to find, and further optimize, transfers. The process of Sect. 4 is used to arrive at some results for a planar Lyapunov orbit at L_2, which are presented in Sect. 5. Finally, a description of future efforts and some concluding remarks are offered in Sect. 6.

2 Properties of the Coupled CR3BP

The process of connecting (or patching together) 3-body problems has been used successfully in the past to obtain suitable results that aid in the creation of transfers in a full ephemeris model. For instance, the methodology has been employed in the study of multi-moon tours (Koon et al. 2001b; Lantoine et al. 2011; Campagnola and Russell 2010) and in the context of the CR3BP (Koon et al. 2001a; Gómez et al.

2001; Marsden and Ross 2006; Canalias and Masdemont 2008; Fantino et al. 2010). The definition and equations of motion of the CR3BP (Sect. 2.1), its equilibrium points (Sect. 2.2), the flow near these equilibrium points (Sect. 2.3), and the method of connecting two CR3BPs (Sect. 2.4) provide the background theory from existing literature on which the subsequent sections (use of the survival maps and the design of the transfers) rely.

The motion of a spacecraft from a Sun-Earth L_1/L_2 libration point orbit towards the Moon is modelled in this work by using two coupled CR3BP models. This effectively divides a trajectory into two separate segments, each using a different gravitational model, where the initial segment is modelled within the framework of the Sun-Earth CR3BP while the second is modelled within the framework of the Earth-Moon CR3BP. The partial trajectories from both CR3BP models are connected at a specified point, via coordinate system conversion, to create a single trajectory that would approximate the trajectory in the actual 4-body dynamics. The Sun-Earth CR3BP has as primary masses the Sun and the Earth-Moon barycentre (the mass of the Earth-Moon barycentre is considered here to be the combined mass of the Earth and the Moon).

2.1 Definition of the CR3BP

The Circular Restricted Three-Body Problem (CR3BP) is a particular case of the three-body problem (being in itself a special case of the more general n-body problem). The restricted problem has been studied extensively in the past, and can be described as two masses (or primaries) of symmetric mass distribution (i.e. they may be considered as point masses) that revolve around their centre of mass in a circular motion. A third massless particle moves within the system of the two revolving primaries without influencing their motion (thus the problem is considered restricted). The CR3BP describes the motion of this third body. The equations of motion of the CR3BP can be derived in several ways, and many reference texts provide a detailed description of the problem formulation. A comprehensive Newtonian approach may be found in the book of Szebehely (1967), and a Lagrangian approach can be found in the book of Meyer et al. (2009). It is convenient to make the system non-dimensional by giving the system a unit of mass (or $m_1 + m_2 = 1$), and by choosing the distance between the primaries to be a unit of length, and by choosing the unit of time such that one full orbital period of both primaries is 2π. As a result of this last choice the angular velocity of the two primaries about the barycentre is $\omega = 1$ (thus making the gravitational constant unity due to this fact and the fact that the total mass is 1). The masses are made dimensionless by dividing each mass by the total system mass. If we assume that $m_1 > m_2$ we may write for the dimensionless masses $\mu_1 = 1 - \mu$ and $\mu_2 = \mu$. The system is now solely defined by the mass ratio of the primaries μ. Due to the rotation of primaries the equations of motion contain the time explicitly in the inertial system. The explicit appearance of time in the equations of motion is commonly eliminated by using a suitable rotating system (non-inertial) where the more massive primary is placed along the x axis at (-μ,0,0) and the less massive primary is placed at (1-μ,0,0). The resulting equations can be written in vectorial form as

$$\frac{d^2\mathbf{r}}{dt^2} + 2\boldsymbol{\omega} \times \frac{d\mathbf{r}}{dt} + \nabla U(\mathbf{r}) = 0, \tag{1}$$

where $\mathbf{r}$ is the position vector of the massless third body. The angular velocity vector $\boldsymbol{\omega}$ of the rotating frame is defined as

$$\boldsymbol{\omega} = \omega \mathbf{e}_z, \tag{2}$$

where $\mathbf{e}_z$ is the positive unit vector along the z axis (as stated above, the magnitude of the angular velocity is $\omega = 1$ in the non-dimensional problem). The 3-body gravitational potential is defined by

$$U(\mathbf{r}) = -\left(\frac{1}{2}|\omega \times \mathbf{r}|^2 + \frac{1-\mu}{r_1} + \frac{\mu}{r_2}\right). \tag{3}$$

In this work, a mass ratio of $\mu_{EM} = 1.2150587 \cdot 10^{-2}$ is used for the Earth-Moon set of primaries, and a mass ratio of $\mu_{SE} = 3.0404234 \cdot 10^{-6}$ is used for the Sun-Earth set (here the smaller primary is considered to be the summed mass of the Earth and Moon). The positions of the third body (i.e. the spacecraft) w.r.t. the primary $\mathbf{r}_1$ and primary $\mathbf{r}_2$ are

$$\begin{aligned} \mathbf{r}_1 &= [x + \mu, y, z]\,, \\ \mathbf{r}_2 &= [x + \mu - 1, y, z]\,. \end{aligned} \tag{4}$$

This system of equations has a first integral, named the Jacobi integral, which relates the value of the Jacobi constant with the gravitational potential and the velocity components of the massless particle. The integral is given by

$$J = -(\dot{x}^2 + \dot{y}^2 + \dot{z}^2) + 2U(x, y, z). \tag{5}$$

2.2 Equilibrium Points and Hill's Region

The CR3BP is known to have five equilibrium points; three unstable collinear points are located along the x axis (named L_1, L_2, L_3) and two equilateral points (named L_4 and L_5), which are stable for the mass ratios considered here. All five equilibrium points lie in the plane of rotation of both primaries (see Fig. 1 for a plot of their locations). These can be found by solving $\nabla U(\mathbf{r}) = 0$ under the assumption of a planar configuration (i.e. all out-of-plane z components are equal to zero). For

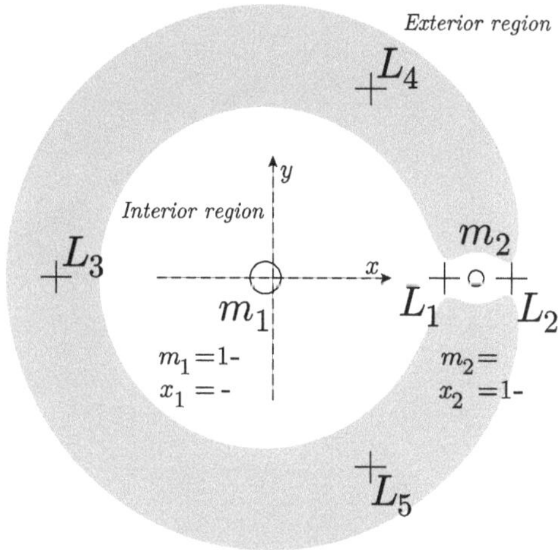

Fig. 1 Diagram showing the equilibrium points in the CR3BP in the rotating frame (with the barycentre being the origin of the axes x and y) and the forbidden region for a particular value of the Jacobi constant

a particular energy level of the system (by setting a constant value for the Jacobi constant) the regions around the primary can be divided into a region where the particle may travel (known as the Hill's region) and a forbidden region (shown for an example energy level as the grey area in Fig. 1) which the particle may not access for the given value of the Jacobi constant.

2.3 Periodic Orbits and Their Flow

As we are studying the departure of spacecraft from periodic orbits at Sun-Earth L_1 and L_2 and their arrival towards the Moon via L_2 from the exterior region in the Earth-Moon system we restrict our discussion to the motion about L_1 and L_2. There are four possible motions (Conley 1968) near each of these two equilibrium points: transit orbits that allow passage between the exterior and the interior regions, non-transit orbits where the particle approaches the equilibrium region but returns back into the region the particle came from, and unstable periodic orbits where the particle remains in the vicinity of the equilibrium point. The fourth type is the particle asymptotically joining or leaving the periodic orbit. These asymptotic orbits are part of a larger structure of invariant manifold 'tubes' (McGehee 1969; Gómez et al. 2000a). The borders of these tubes form the boundary between the transit (inside the tube) and non-transit orbits (outside the tube). There are four manifold 'tubes'; two stable manifolds where the particle flows towards the equilibrium region and two unstable manifolds where the particle flows away from the equilibrium region. These are shown in Fig. 2 for an example periodic orbit at the L_1 libration point in the Earth-Moon system. There exist a number of periodic orbits near the collinear libration points; for instance horizontal Lyapunov orbits (in the plane of the primaries), vertical Lyapunov orbits (figure eight shape where the

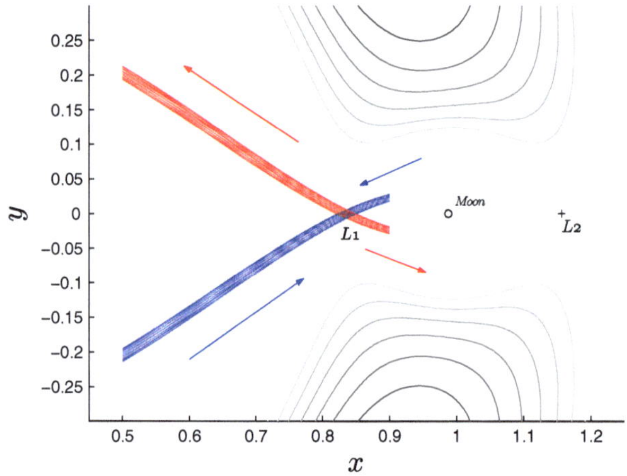

Fig. 2 Illustration of the unstable (*red*) and stable (*blue*) invariant manifolds associated to a periodic orbit at L_1 in the Earth-Moon system

orbit intersects the plane of the primaries in a single location in the rotating reference frame), and three-dimensional halo orbits. The existence of quasi-periodic orbits has also been shown: the Lissajous family of orbits that are around the vertical Lyapunov orbits, as well as the quasi-halo orbits that are around the halo orbits (Gómez et al. 2000a,b).

2.4 Connecting the CR3BPs

Connection between the two CR3BPs is accomplished by converting coordinates from one system to the other. This conversion occurs when the spacecraft, on its way from the Sun-Earth L_1/L_2 equilibrium region, crosses the x location of the second primary (the combined mass of the Earth and Moon) in the Sun-Earth synodical system. Here it is assumed that both systems are coplanar and that both pairs of primaries are in circular orbits around another. To convert the position from Earth-Moon to Sun-Earth reference frame the relation

$$\eta_{SE} = \frac{l_{EM}}{l_{SE}} e^{i\alpha} \eta_{EM} + 1 - \mu_{SE}, \tag{6}$$

where complex notation is used, where the x and y components are given by

$$\eta_{SE} = x_{SE} + iy_{SE}, \eta_{EM} = x_{EM} + iy_{EM}. \tag{7}$$

The distances between the Sun and Earth and Earth and Moon are given by $l_{SE} = 1.495979 \cdot 10^8$ km and $l_{EM} = 384,400$ km. The mass parameter μ_{SE} that defines the

Sun-Earth system is computed by

$$\mu_{SE} = \frac{m_E + m_M}{m_E + m_M + m_S}. \tag{8}$$

The masses are given for the Earth as $m_E = 5.973699 \cdot 10^2 4\,\mathrm{kg}$, for the Moon as $m_M = 7.347673 \cdot 10^2 2\,\mathrm{kg}$, and for the Sun $m_S = 5.973699 \cdot 10^3 0\,\mathrm{kg}$. The angle α representing the relative geometry of both systems (i.e. the angle between the axes spanned along both sets of primaries) is computed using

$$\alpha = \alpha_0 + (\omega_{EM} - \omega_{SE})\frac{t_{EM}}{\omega_{EM}}. \tag{9}$$

where α_0 is the initial relative geometry of the system. The angular velocities for both systems are given by Kepler's third law as

$$\omega_{EM} = \sqrt{\frac{G(m_E + m_M + m_S)}{l_{EM}^3}}, \tag{10}$$

and for the Earth-Moon system, and for the Sun-Earth system as

$$\omega_{SE} = \sqrt{\frac{G(m_E + m_M + m_S)}{l_{SE}^3}}. \tag{11}$$

This gives an angular velocity of $\omega_{EM} = 2.66531437 \cdot 10^{-6}\,\mathrm{rad/s}$ and $\omega_{SE} = 1.99098670 \cdot 10{-}7\,\mathrm{rad/s}$. The velocity from Earth-Moon rotating frame can be converted to Sun-Earth rotating frame by

$$\frac{d\eta_{SE}}{dt_{SE}} = \frac{l_{EM}}{l_{SE}}\frac{\omega_{SE}}{\omega_{EM}}e^{i\alpha}\left(i\left(1 - \frac{\omega_{SE}}{\omega_{EM}}\right)\eta_{EM} + \frac{d\eta_{EM}}{dt_{EM}}\right). \tag{12}$$

Because is it assumed that both rotating frames lie in the same plane (both systems are coplanar) the conversion for any out-of-plane conversion is straightforward. The position is converted using the relation

$$z_{SE} = z_{EM}\frac{l_{EM}}{l_{SE}}, \tag{13}$$

and the velocity is converted using the relation

$$\dot{z}_{SE} = \dot{z}_{EM}\frac{l_{EM}}{l_{SE}}\frac{\omega_{EM}}{\omega_{SE}}. \tag{14}$$

For a comprehensive description of the conversion process (including details on the conversion to and from the inertial reference frame, and from the Earth-Moon to Sun-Earth synodical system) the reader is referred to the work of Castelli (2011).

3 Survival and Event Maps

Regardless of the application, one can be interested in what kind of conditions near the L_2 libration point would be beneficial for establishing a long duration quasi-periodic orbit about the Moon, and what conditions would lead to an impact on the lunar surface. To this end, one can analyse the case of a family of virtual spacecraft placed at $x = x_{L_2}$ and at interspaced points along $-0.25 < y < 0.25$ within the Earth-Moon CR3BP. These spacecraft can then be assigned a velocity, for which two methods are provided in this paper. The first assumes a parallel flow along the x axis, and the second derives the velocity for each point on the basis of a specified value of the Jacobi constant. In the first method, the spacecraft are then given initial velocity components $\dot{y} = 0$ and $\dot{x}$ sampled uniformly from the domain $-0.2 < \dot{x} < 0.2$ (all values in the non-dimensional system) such that the initial flow at $x = x_{L_2}$ is always parallel to the x axis within the Earth-Moon rotating frame. Without loss of generality one can start by restricting the analysis to the planar case so that the position and velocity along the z axis are neglected. This leads to a group of initial states where x and $\dot{y}$ are constant, and y and $\dot{x}$ are varied. The state of the spacecraft can be generally written as

$$\mathbf{x}_{SC} = \left[x_{L_2}, y, 0, \dot{x}, 0, 0\right]^{\mathrm{T}}. \tag{15}$$

This group of states (henceforth referred to as lunar arrival states in this work) is then individually propagated forward in time until the orbit is no longer deemed stable or until the maximum propagation time of 3 months is met. In the framework of this discussion an orbit is considered stable when the spacecraft remains between the locations of the L_1 and L_2 points along the x axis, and does not impact upon the Moon, i.e. when a set of coordinates (x, y) fulfils

$$x_{L_1} < x < x_{L_2} \tag{16}$$

and

$$\sqrt{x^2 + y^2} > R_{Moon}. \tag{17}$$

The result of this propagation can be seen in the survival map shown in Fig. 3.

It can be seen in Fig. 3a that large swaths of the map are accompanied by a low lifetime. Naturally those areas where the value of $\dot{x}$ are positive correspond to a low orbit lifetime as the initial condition will tend to cause the spacecraft to immediately

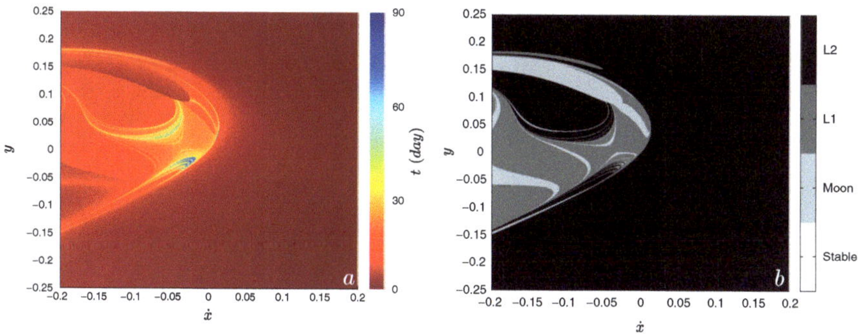

Fig. 3 (**a**) Lunar survival map and (**b**) corresponding lunar propagation event map with constant x and $\dot{y}$ and y and $\dot{x}$ varied along axes

exit the Earth-Moon system past x_{L_2}. The central area in Fig. 3a, however, shows promising areas where the orbit duration is higher. Note that the reason why the areas with positive $\dot{x}$ as initial condition have a non-zero lifetime is because the limit at which the propagation is halted is slightly further out from the Moon than x_{L_2}. This is to allow for a degree of flexibility where the spacecraft may initially move in the opposite direction before moving towards the Moon. Additionally, there is the practical consideration of preventing the propagation from already ceasing at the initial point. To understand which areas of initial conditions are suitable for lunar impact, and which are suitable for lunar capture, the cause of propagation termination is also recorded. This is shown in the event map in Fig. 3b. The possible outcomes are stability (shown in white) as defined previously in Eqs. (16) and (17) for the propagation duration of 90 days, impact on the lunar surface (shown in light grey), passing outside the lunar region via x_{L_1} (shown in dark grey), and passing outside the lunar region via x_{L_2} (shown in black).

In addition to the first method of specifying velocity one may also create a survival map by setting an energy level of the system, or in other words choosing a value of the Jacobi constant of motion

$$J = -(\dot{x}^2 + \dot{y}^2 + \dot{z}^2) + x^2 + y^2 + 2\left(\frac{1-\mu}{r_1} + \frac{\mu}{r_2}\right), \tag{18}$$

obtained from the Jacobi integral of the three-body problem (Szebehely 1967) where r_1 and r_2 are the scalar lengths of the vectors given by Eq. (4). Since this work focuses only on the use of two dimensional maps, the z components will be disregarded ($z = \dot{z} = 0$). By choosing a value of the Jacobi constant, assuming a value of $x = x_{L_2}$, and given a mesh of values of y and $\dot{x}$, the corresponding value of $\dot{y}$ (and $-\dot{y}$) can be computed. Then, as for the previous map the entire set of initial conditions can be propagated forwards in time to study the behaviour. The resulting maps for the set of Jacobi constants $J = [3.00, 3.05, 3.10, 3.15]$ is given in Fig. 4a, along with the corresponding event map in Fig. 4b. The plots contain empty regions, due to no valid real value of $\dot{y}$ existing for particular combinations of the

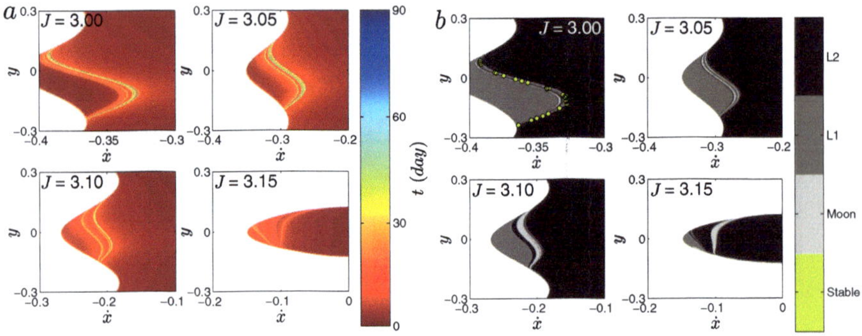

Fig. 4 (**a**) Set of four lunar survival maps and (**b**) corresponding set of four lunar propagation event maps with constant J and y and $\dot{x}$ varied along axes

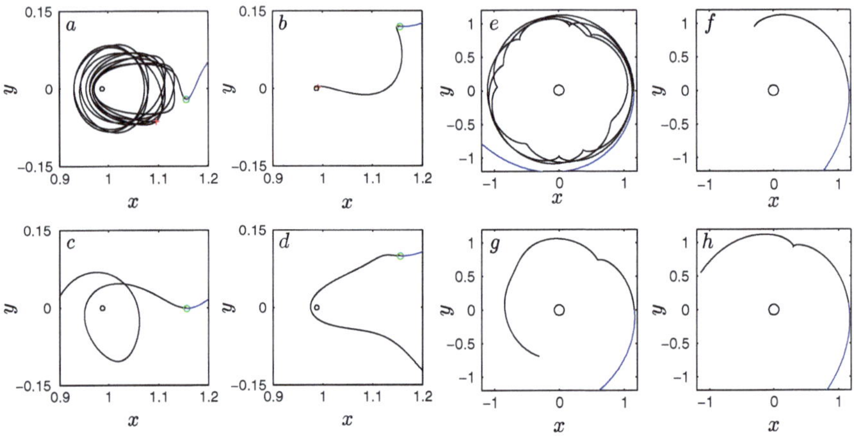

Fig. 5 Example lunar target states (*left four subplots*) in the Earth-Moon rotating reference frame leading to (**a**) weak capture, (**b**) impact, (**c**) leaving the vicinity of the Moon towards the interior region, and (**d**) leaving the vicinity of the Moon towards the exterior region. *Right Plots* (**e**) through (**f**) are the corresponding plots in the inertial reference frame centred at the Earth

Jacobi constant and the other state parameters. The states that are stable for at least 90 days are only found for $J = 3.00$ and for clarity's sake are marked in the event map in Fig. 4b as green dots on the event map. As the Jacobi constant increases the forbidden zone of the Hill's regions increases, and thus the region of interest on the maps becomes smaller and smaller. As a result, increased resolution is generally needed to reveal the structures on the map. This increased resolution comes at an additional computational cost, which is offset by the fact that the region of interest on the maps has also shrunk.

An example lunar target state from each category of event is taken from the survival map shown in Fig. 3 and propagated both forwards and backwards in time. The results are shown in Fig. 5 for a lunar target state leading to weak capture, impact, and exit from the vicinity of the Moon via L_1 and L_2.

Note that the methodology used to create these maps can also be used for non-planar problems by extending the maps with the addition of the z components for position and velocity. Keeping in mind that the maps define a set of target states with associated survivability, one possible extension features the definition of a plane normal to the x axis vector. Points are sampled on this plane, providing a set of positions (x, y, z) where x is fixed. By assuming a value of the Jacobi constant the velocity components $(\dot{x}, \dot{y}, \dot{z})$ can be computed using two free angles. For every pair of coordinates in the $y - z$ plane one can select the optimal velocity components that maximise survivability. This extension merely requires a more involved initial computation to generate the map. Other extensions are possible but require additional assumptions on the velocity components. These extensions, however, are not required to complete the analyses in this work and are left for future work. Finally, it should also be noted that these maps can be constructed for transfers entirely within one CR3BP, for example interior transfers between the Earth and Moon (van der Weg and Vasile 2012). The resulting set of initial conditions, their corresponding orbit lifetime, and their category of decay (impact or exit via libration points) can now serve as the basis for the design of transfers from Sun-Earth libration point orbits towards the Moon.

4 Transfer Design Using the Maps

As described briefly in Sect. 2, the transfer between Sun-Earth libration point orbit and Moon is modelled in two parts: the initial leg in the Sun-Earth CR3BP and the leg describing the motion nearer to the Earth and Moon in the Earth-Moon CR3BP. The transfer from Sun-Earth L_1/L_2 libration point orbit (which defining parameters are given) to the Moon consists initially of following the branch of the unstable manifold, generated from the periodic orbit, towards the Earth-Moon barycentre in the Sun-Earth CR3BP. Instead of utilizing the stable manifold branch (originating from a libration point orbit at L_2 in the Earth-Moon system) in the Earth-Moon CR3BP to bring the spacecraft towards the Moon (as would be typical for a WSB transfer, see Koon et al. 2001a), use is made of the lunar arrival states on the survival map to directly target desired conditions near the Moon (such as weak capture or impact). The procedure outlined in this section is usable for both planar as well as non-planar cases. However, the results generated in the following section assume the two connected CR3BPs to be coplanar and make use of planar survival maps. The procedure remains unchanged; merely z and $\dot{z}$ are always equal to zero for this case. Both individual transfer legs are described here by their position (x, y, z) and their velocity $(\dot{x}, \dot{y}, \dot{z})$ along a discretized period of time, effectively giving two 6×N matrices (where N differs for both legs due to numerical integration and the period of time thereof). The initial leg modelled in the Sun-Earth CR3BP is denoted by $\mathbf{s}_{lpo}$, and the second leg modelled in the Earth-Moon CR3BP is denoted by $\boldsymbol{\sigma}_m$. An example of a stable branch of an invariant manifold in the Earth-Moon CR3BP, as well as a subset of arcs leading to lunar capture and impact, is shown in Fig. 6. The

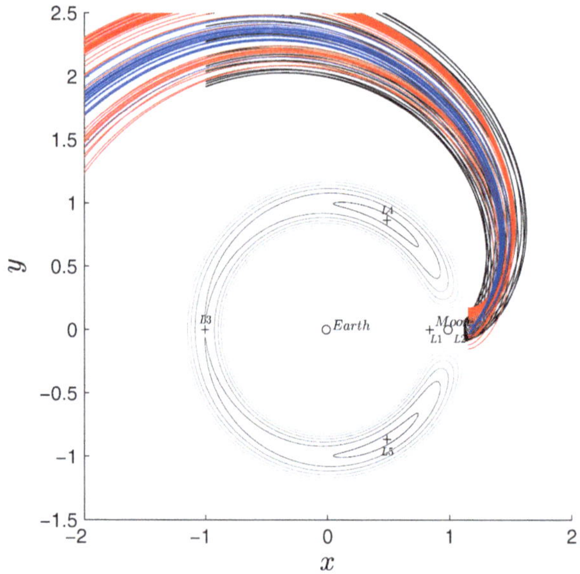

Fig. 6 Stable manifold branch flowing towards the Moon from the exterior of the Earth-Moon system (shown in *black*), the flow towards the Moon based on a representative selection of lunar arrival states targeting weak capture selected from Fig. 3 (shown in *blue*), and the flow towards the Moon based on a representative selection of lunar arrival states targeting lunar impact selected from Fig. 3 (shown in *red*)

stable branch denoting the flow towards the Moon from the exterior regions is shown in black, whereas the weak capture (shown in blue) and impact (shown in red) arcs are obtained from a representative sampling of the survival map in Fig. 3. Figure 6 illustrates that the collection of capture and impact arcs show the same behaviour as the manifold structure flowing towards its associated libration point orbit.

A connection between the arcs $\mathbf{s}_{lpo}$ and $\boldsymbol{\sigma}_m$ from both Sun-Earth and Earth-Moon CR3BPs can be made by transforming one set of states into the reference frame of the other $\boldsymbol{\sigma}_m(\alpha_0) \rightarrow \mathbf{s}_m$, and subsequently searching for intersections on a given Poincaré section. The initial orbital phases α_0^{SE} and α_0^{EM} of both CR3BPs control the geometry of the connection, but this is reduced to a single parameter α_0 ($= \alpha_0^{EM} - \alpha_0^{SE}$) as only the relative phasing between Sun-Earth and Earth-Moon systems is necessary (Fantino et al. 2010). The concept is illustrated in Fig. 7a, where a segment of arcs in the Earth-Moon system (shown in blue) has been converted into the Sun-Earth barycentric synodical reference frame.

A wide selection of lunar arrival states from the lunar survival map that lead to successful capture and to lunar impact are propagated backwards in time and the obtained arcs are translated into the Sun-Earth CR3BP. The resulting plot of lunar arrival states resulting in impact are shown in Fig. 7b, and for capture in Fig. 7c, for an initial orbit phasing of $\alpha_0 = 0$. In both figures, these arcs (a group of arcs $\mathbf{s}_m$)

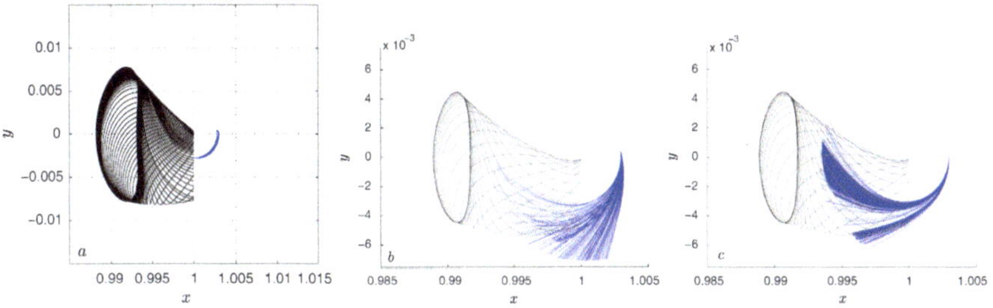

Fig. 7 Both unstable manifold from Sun-Earth L_1 (*black*) and stable manifold from Earth-Moon L_2 (*blue*) shown in Sun-Earth synodical barycentric reference frame, (**b**) unstable manifold from Sun-Earth L_1 LPO (*black*) and initial states leading to lunar impact (*blue*) shown in Sun-Earth synodical barycentric reference frame, and (**c**) unstable manifold from Sun-Earth L_1 LPO (*black*) and initial states leading to lunar quasi-capture (*blue*) shown in Sun-Earth synodical barycentric reference frame

are shown in blue while a segment (a group of arcs $\mathbf{s}_{lpo}$) of an unstable invariant manifold is plotted in black for the sake of comparison.

The connection between the trajectory arcs from both CR3BPs is made on a plane **P** at $x = 1 - \mu$ in the Sun-Earth CR3BP (the barycentre of the Earth-Moon system) whose normal vector is $\mathbf{e}_x = [1, 0, 0]$. An arc $\boldsymbol{\sigma}_m$ flowing towards the Moon—after having its states converted $\boldsymbol{\sigma}_m(\alpha_0) \rightarrow \mathbf{s}_m$ from Earth-Moon to Sun-Earth reference frame—thus has a certain position and velocity $\mathbf{s}_m^{1-\mu}$ when it intersects the plane **P**. This arc must then be connected to an arc $\mathbf{s}_{lpo}$ on the unstable manifold leading away from the Sun-Earth system libration point. This second arc also intersects plane **P**, but at $\mathbf{s}_{lpo}^{1-\mu}$. For the matching of the arcs to be correct the y and z (if the problem is entirely planar z components can be disregarded) position components of $\mathbf{s}_m^{1-\mu}$ should be equal to those of $\mathbf{s}_{lpo}^{1-\mu}$. The two connecting arcs will have a certain disparity in velocity, which is corrected for by manoeuvre.

If the arcs $\mathbf{s}_m$ leading towards the Moon are numerically integrated for a sufficiently long period of time, they will cross the intersection plane multiple times. For each of these intersections a connection can be attempted with the unstable manifold. Naturally, transfer duration will increase when the connection is made at a later intersection (the increase in transfer time is dependent on the specific arc). Another consideration for a trajectory where the connection is delayed until a later intersection is the gravitational influence of the Sun. As the arc leading from the intersection plane towards the Moon takes more and more time (and also generally starts further out on plane **P**) the ability of the Earth-Moon CR3BP to approximate the full body dynamics degrades.

The general solution space for a set of lunar arrival states and a particular libration point orbit can be effectively and quickly mapped by computing and storing the unstable manifold trajectory arcs $\mathbf{s}_{lpo}$ from the Sun-Earth libration point orbit and the trajectory arcs $\boldsymbol{\sigma}_m$ flowing towards the lunar target states. Once this is computed, the transformation of the lunar target state arcs from Earth-Moon to Sun-Earth synodical barycentric reference frame ($\boldsymbol{\sigma}_m(\alpha_0) \rightarrow s_m$) can be performed for a range of values of the orbital phasing angle α_0. For each lunar target state trajectory arc and value of orbital phasing angle α_0 the best matching arc flowing from the Sun-Earth libration point orbit can be found. The criterion is the lowest Δv to connect both arcs, which at the same time satisfies the positional difference on the Poincaré section (on plane **P**) to within set tolerance. Promising pairs of intersections can then be refined further by way of an optimization process. A number of matching pairs can be found based on ranking, which then serve as initial guesses for an optimization process using an SQP gradient solver (Wright and Nocedal 1999). The optimization initially only accounts for two design parameters α_0 and β. This can be expressed as the design variable vector

$$\mathbf{x} = \left[\alpha_0, \beta\right], \tag{19}$$

where α_0 is the initial orbit phasing and β is the position along the Sun-Earth libration point orbit expressed as a curvilinear coordinate within the domain of $[0, 2\pi]$ where 0 is chosen as the position on the libration point orbit at $y = 0$ and with the smallest value for x. The same position along the circuit of the libration point orbit is reached at 2π after clockwise rotation. Note that for initial optimization both parameters are assumed to be independent of each other. When translating this problem to a full ephemeris model an initial time will both proscribe the geometry of the planets α_0 as well as the position of the spacecraft on the LPO β, reducing the number of variables to 1. The state of the arc $\mathbf{s}_{lpo}$ flowing from the libration point orbit at the intersection with plane **P** at $x = 1 - \mu$ is denoted as $\mathbf{s}_{lpo}^{1-\mu} = [\mathbf{p}_{lpo}, \dot{\mathbf{p}}_{lpo}]$, where $\mathbf{p}_{lpo}$ and $\dot{\mathbf{p}}_{lpo}$ are the three element position and velocity vectors at plane **P** in the Cartesian coordinate system in the Sun-Earth synodical reference frame, respectively. In a similar fashion, the state from the arc $\mathbf{s}_m$ flowing towards the lunar vicinity at the intersection with the plane **P** is denoted as $\mathbf{s}_m^{1-\mu} = [\mathbf{p}_m, \dot{\mathbf{p}}_m]$. The objective of the optimization is to minimize the velocity change necessary to change the velocity at the intersection such that the velocity is matched between $\mathbf{s}_{lpo}^{1-\mu}$ and $\mathbf{s}_m^{1-\mu}$. This can be expressed as

$$f(\mathbf{x}) = \Delta v = ||\dot{\mathbf{p}}_{lpo} - \dot{\mathbf{p}}_m||. \tag{20}$$

The positional difference between the two arcs as they meet at plane **P** is added as an equality constraint

$$c(\mathbf{x}) = ||\dot{\mathbf{p}}_{lpo} - \dot{\mathbf{p}}_m|| \tag{21}$$

to the optimization process. This ensures any remaining gap between the arcs meeting at plane **P** is closed. Once a single optimization pass has been completed (after having either satisfied constraint tolerances or having reached the maximum number of evaluations) the design variable vector is expanded to

$$\mathbf{x} = \left[\alpha_0, \beta, \Delta v_{lpo}, \gamma_{lpo}, \delta_{lpo}, \Delta_m, \gamma_m, \delta_m\right], \tag{22}$$

where two manoeuvres are introduced at departure from the libration point orbit and at arrival near the Moon (at the position of the chosen lunar target state). Δv_{lpo} and Δv_m are the magnitudes of the manoeuvres, γ_{lpo} and γ_m are the respective in-plane right ascensions of the manoeuvres (counted from the tangential direction of the velocity change vector to its projection on the orbital plane), and δ_{lpo} and δ_m are the respective out-of-plane declinations of the manoeuvres (the angle between projection of the velocity change vector on the orbital plane and the velocity change vector itself). In the case of a planar transfer from a planar Lyapunov orbit the out-of-plane declinations for both manoeuvres are zero. The optimization process is now repeated with the same objective and constraints.

5 Disposal Results for an L_2 Lyapunov Orbit

As a case study, the prior described algorithm, lunar survival map, and event map are now used to generate Δv maps for both capture and impact transfers from an initial libration orbit (in this case a planar Lyapunov orbit) at L_2 that shares the in-plane amplitude characteristics of the Herschel spacecraft (European Space Agency 1995). Both of these orbits are shown in Fig. 8. The orbit is defined in the Sun-Earth CR3BP by a Jacobi constant of $J = 3.00080469$, an x amplitude of $3.2816 \cdot 10^{-3}$ and a y amplitude of $1.03808 \cdot 10^{-2}$ (non-dimensional units).

Given the libration orbit defined above, a subset of lunar arrival states is selected for the generation of the results. In the case of capture, states with an excellent survival time of at least 65 days are selected from the map (regardless of whether the orbit deteriorates by impacting the Moon, or escaping past L_1 or L_2). For the case of impact, only those states that impact the Moon, and with a not too long survival time (less than 30 days) are selected.

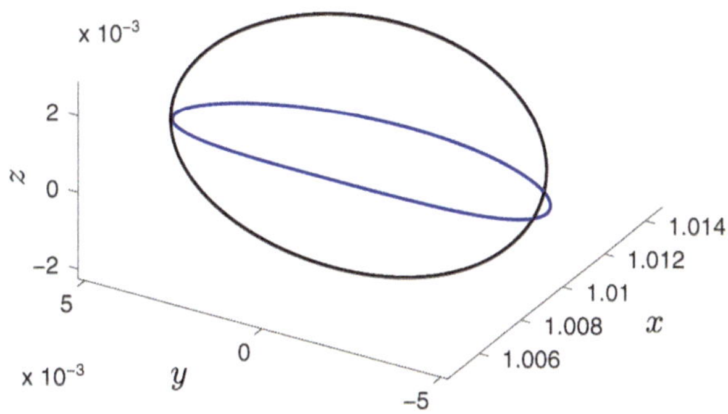

Fig. 8 Representation of Herschel orbit in the CR3BP (*back*) and a planar Lyapunov orbit (*blue*) sharing the same amplitude along x and y axes

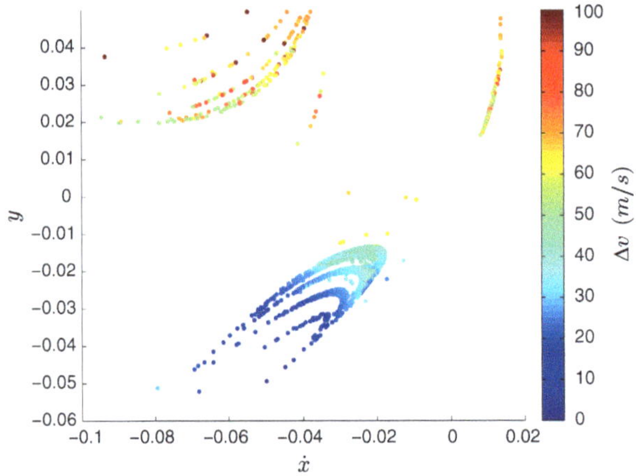

Fig. 9 Δv map of the first intersection for lunar capture from Lyapunov orbit at L_2

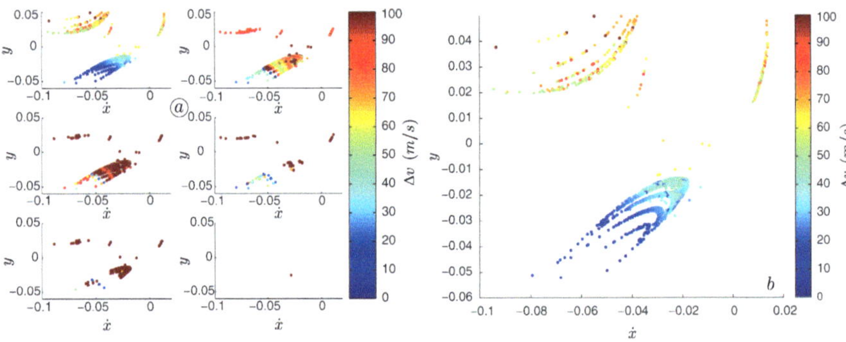

Fig. 10 Δv maps of the first six intersections for lunar capture from Lyapunov orbit at L_2 and (**b**) Δv map of the best results from the first six intersection for lunar capture from Lyapunov orbit at L_2

The results for lunar capture are provided in Figs. 9 and 10. These results were created by sampling the initial orbital phasing angle values α_0 at $1°$ intervals. The results presented in the figures are not fully optimized but fulfil relatively strict constraints on the distance between the meeting points $\mathbf{s}_{lpo}^{1-\mu}$ and $\mathbf{s}_{m}^{1-\mu}$ of the arcs at plane P. In the worst case the constraint violation at plane **P** may be up to 1500 km, but most transfers have a difference of a few 100 km. These constraint violations can be reduced by using the optimization process in Sect. 4. Figure 9 shows the Δv cost in m/s (ranging from 0 to 100 m/s) for each selected lunar arrival state for the very first intersection that occurs at the intersection plane **P**. Figure 10a shows the Δv cost in m/s for the first six intersections of each arc $\mathbf{s}_m$ with the intersection plane. Multiple crossings are achieved by increasing the numerical integration time for each arc; instead of halting propagation after the first intersection it is halted after a number of successive intersections with plane **P**. A plot showing the best Δv

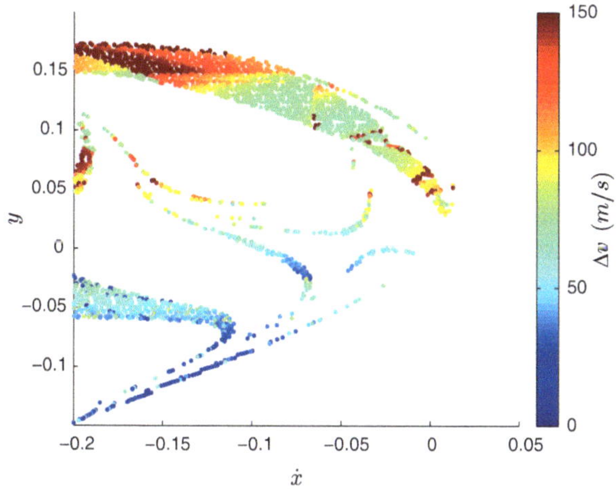

Fig. 11 Δv map of the first intersection for lunar impact from Lyapunov orbit at L_2

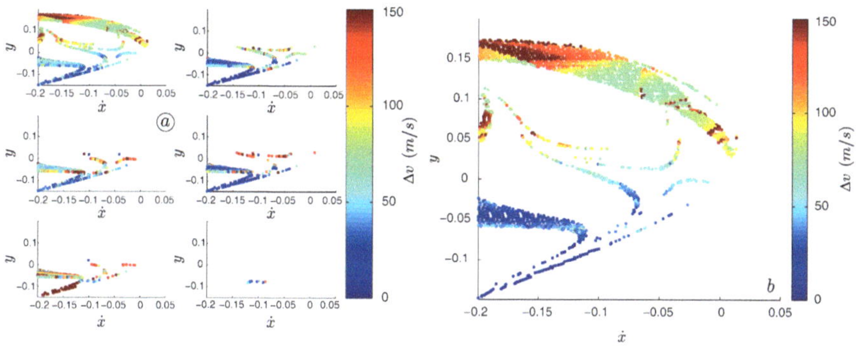

Fig. 12 Δv maps of the first six intersections for lunar impact from Lyapunov orbit at L_2 and (**b**) Δv map of the best results from the first six intersection for lunar impact from Lyapunov orbit at L_2

value found from among all first six intersections per lunar arrival state is given in Fig. 10b.

The lower area of Fig. 9 for the L_2 ranges from near-zero to ca. 30 m/s Δv cost. The lowest value found in the first intersection is 1.526 m/s (before optimization). Δv cost is not substantially improved in the second intersection (Fig. 10a) with the lowest value being 1.424 m/s. Sampling further sections provides no performance benefit in this case.

The results for lunar impact are given in Figs. 11 and 12. As was the case for lunar capture, the results were created by sampling the initial orbital phasing angle values α_0 at 1° intervals. The results presented in the figures are not fully optimized but fulfil the same constraints on the distance between the meeting points $\mathbf{s}_{lpo}^{1-\mu}$ and $\mathbf{s}_m^{1-\mu}$ of the arcs at plane $\mathbf{P}$ as was the case for lunar capture. Figure 11 shows the Δv cost in m/s (ranging from 0 to 150 m/s) for each selected lunar arrival state for the very first intersection that occurs at the intersection plane **P**. Figure 12a shows the Δv cost in m/s for the first six intersections of each arc $\mathbf{s}_m$ with the intersection plane. Multiple intersections are achieved by increasing the numerical integration time for each arc; instead of halting propagation after the first intersection it is halted after a number of successive crossings with plane **P**. A plot showing the best Δv value found from among all first six intersections per lunar arrival state is given in Fig. 12b.

As can be seen from the figures the selected lunar arrival states that lead to impact cover a much wider portion of the generated survival map than those states that lead to capture. Connections between the libration point orbit and lunar impact can be achieved for a number of lunar arrival states at near-zero Δv cost within the first intersection, before optimization. The lowest value found in the first intersection is 2.19 m/s. The Δv cost remains between 1 and 3 m/s for succeeding intersections (Fig. 12a).

Four example trajectories, after optimization, are plotted in Fig. 13, where the libration orbits are shown in red, the segments after the transfer has reached its lunar arrival state are shown in blue, and the connection manoeuvres for the trajectories

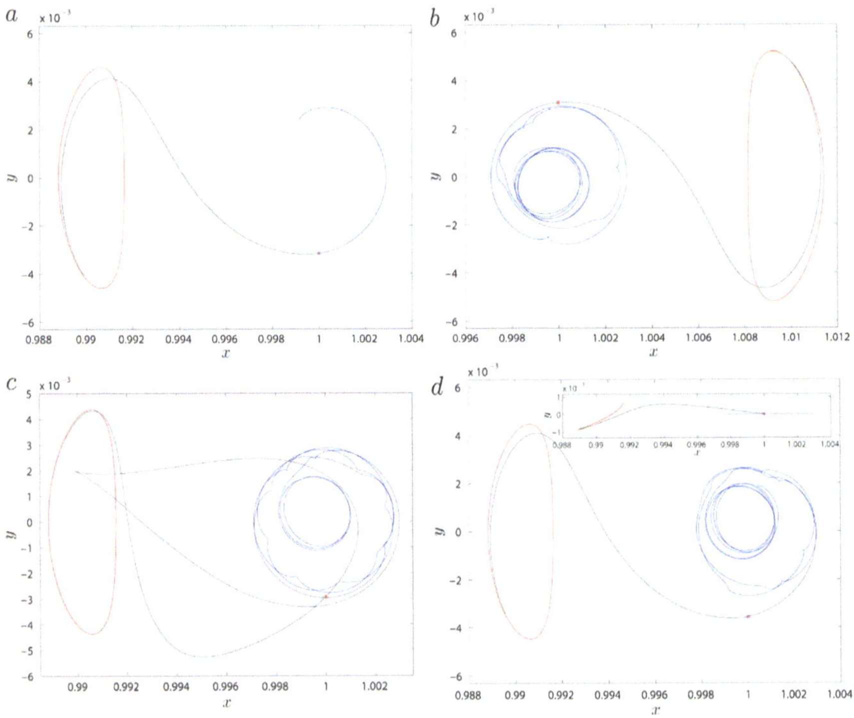

Fig. 13 Plots of example trajectories: (**a**) L_1 lunar impact, (**b**) L_2 temporary lunar capture, (**c**) L_1 temporary lunar capture with two intersections, and (**d**) non-planar L_1 capture in the Sun-Earth synodic reference frame

are shown as stars. The first trajectory (a) is a planar trajectory from the libration point orbit at L_1 that leads to impact upon the lunar surface, costing slightly less than 1 m/s to connect the two legs. The second trajectory (b) is a planar trajectory from the libration point orbit at L_2 that is captured by the Moon for at least 3 months before the spacecraft exits the lunar vicinity via L_1 in the Earth-Moon system. This connection manoeuvre cost 1.6 m/s. The third trajectory (c) shows a capture trajectory from L_1 where two intersections occur before the Sun-Earth and Earth-Moon legs are connected, costing 12 m/s. The fourth trajectory (d) shows a non-planar example (costing 142 m/s to connect) of a lunar capture, including a side view of the trajectory.

6 Conclusions

An algorithm has been presented that efficiently generates transfers from Sun-Earth libration point orbit to the Moon. These transfers can then serve as the basis for further optimization or as the starting point for a transfer in a full ephemeris model. It has been shown that by using the presented survival and event maps lunar impact or weak capture can be directly targeted at low cost in the planar problem. The computational intensive parts of the algorithm have to be computed once; the maps are not linked to the particular problem and thus can be stored for future use. Numerical propagation for the arcs from a particular LPO have to be performed only once and then stored. Due to these facts, the entire search space (across the range of orbital phasing) can be quickly scanned in order to locate where promising initial guesses to generate trajectories lie. Future work will include the generation and study of survival maps with differing Jacobi constant (for instance matching the energy of the map and the LPO) and extending the maps to include a non-planar (z) component.

Acknowledgements This work was done as part of a study for the European Space Agency named "End-Of-Life Disposal Concepts for Lagrange-Point and Highly Elliptical Orbit Missions" (Contract No. 4000107624/13/F/MOS). The authors would like to acknowledge the following team members that contributed to the ESA study: Camilla Colombo, Hugh Lewis, Francesca Letizia, Stefania Soldini, Elisa Maria Alessi, Alessandro Rossi, Linda Dimare, Massimo Vetrisano, and Markus Landgraf. The authors would like to thank Elisa Maria Alessi in particular for her provision of the orbital characteristics of SOHO's operational orbit. This work was supported by the EC Marie Curie Network for Initial Training Astronet-II, Grant Agreement No. 289240.

References

Belbruno, E.: Capture Dynamics and Chaotic Motions in Celestial Mechanics. Princeton University Press, Princeton (2004)

Campagnola, S., Russell, R.P.: Endgame problem part 2: multibody technique and the Tisserand-Poincare graph. J. Guid. Control Dyn. **33**, 476–486 (2010)

Canalias, E., Masdemont, J.J.: Computing natural transfers between Sun–Earth and Earth–Moon Lissajous libration point orbits. Acta Astronaut. **63**, 238–248 (2008)

Castelli, R.: On the relation between the bicircular model and the coupled circular restricted three-body problem approximation. In: Nonlinear and Complex Dynamics, pp. 53–68. Springer, New York (2011)

Conley, C.C.: Low energy transit orbits in the restricted three-body problems. SIAM J. Appl. Math. **16**, 732–746 (1968)

European Space Agency: Herschel portal (1995). http://www.esa.int/Our_Activities/Space_Science/Herschel/. Cited 15 Oct 2015

Fantino, E., Gómez, G., Masdemont, J.J., Ren, Y.: A note on libration point orbits, temporary capture and low-energy transfers. Acta Astronaut. **67**, 1038–1052 (2010)

García, F., Gómez, G.: A note on weak stability boundaries. Celest. Mech. Dyn. Astron. **97**(2), 87–100 (2007)

Gómez, G., Llibre, J., Martínez, R., Simó, C.: Dynamics and Mission Design Near Libration Point Orbits—Volume 1: The Case of Collinear Libration Points. World Scientific, Singapore (2000a)

Gómez, G., Jorba, À., Masdemont, J.J., Simó, C.: Dynamics and Mission Design Near Libration Point Orbits—Volume 3: Advanced Methods for Collinear Points. World Scientific, Singapore (2000b)

Gómez G, Koon WS, Lo MW, Marsden JE, Masdemont J and Ross SD (2001) Invariant manifolds, the spatial three-body problem and space mission design. In: Astrodynamics 2001. Advances in Astronautical Sciences No.109, pp. 3–22. American Astronautical Society , San Diego

Haapala, A.F., Howell, K.C.: Representations of higher-dimensional Poincaré maps with applications to spacecraft trajectory design. Acta Astronaut. **96**, 23–41 (2014)

Koon, W.S., Lo, M.W., Marsden, J.E., Ross, S.D.: Low energy transfer to the Moon. Celest. Mech. Dyn. Astron. **81**, 63–73 (2001a)

Koon, W.S., Lo, M.W., Marsden, J.E., Ross, S.D.: Constructing a low energy transfer between Jovian moons. Contemp. Math. **292**, 129–146 (2001b)

Lantoine, G., Russell, R.P., Campagnola, S.: Optimization of low-energy resonant hopping transfers between planetary moons. Acta Astronaut. **68**, 1361–1378 (2011)

Marsden, J., Ross, S.: New methods in celestial mechanics and mission design. Bull. Am. Math. Soc. **43**, 43–73 (2006)

McGehee, R.P.: Some homoclinic orbits for the restricted three-body problem. Ph.D. thesis, University of Wisconsin (1969)

Meyer, K., Hall, G., Offin, D.: Bifurcations of periodic orbits. In: Introduction to Hamiltonian Dynamical Systems and the N-Body Problem, pp. 271–299. Springer, New York (2009)

Silva, P.A.S., Terra, M.O.: Applicability and dynamical characterization of the associated sets of the algorithmic weak stability boundary in the lunar sphere of influence. Celest. Mech. Dyn. Astron. **113**, 141–168 (2012)

Szebehely, V.: Theory of Orbits: The Restricted Problem of Three Bodies. Academic, New York (1967)

Topputo, F., Belbruno, E., Gidea, M.: Resonant motion, ballistic escape, and their applications in astrodynamics. Adv. Space Res. **42**, 1318–1329 (2008)

van der Weg, W.J., Vasile, M.: IAC-12-C1.4.8 high area-to-mass ratio hybrid propulsion earth to moon transfers in the CR3BP. In: 63rd International Astronautical Congress, Naples (2012)

Villac, B.F., Scheeres, D.J.: Escaping trajectories in the Hill three-body problem and applications. J. Guid. Control Dyn. **26**, 224–232 (2003)

Wright, S.J., Nocedal, J.: Numerical Optimization, vol 2. Springer, New York (1999)

An Introduction to Differential Algebra and the Differential Algebra Manifold Representation

Alexander Wittig

Abstract Differential Algebra techniques have been used extensively in the past decade to treat various problems in astrodynamics. In this paper we review the Differential Algebra technique and present four different views of the method. We begin with the introduction of the mathematical definition of the technique as a particular algebra of polynomials. We then give an interpretation of the computer implementation of the method as a way to represent function spaces on a computer, which naturally leads to a view of the method as an automatic differentiation technique. We then proceed to the set theoretical view of Differential Algebra for representing sets of points efficiently on a computer, which is of particular value in astrodynamics. After this introduction to the well known classical DA techniques, we introduce the concept of a DA manifold and show how they naturally arise as an extension of classical DA set propagation. A manifold propagator that allows the accurate propagation of large sets of initial conditions by means of automatic domain splitting (ADS) is described. Its function is illustrated by applying it to the propagation of a set of initial conditions in the two-body problem.

1 Introduction

Differential Algebra (DA) methods have long been used in the field of accelerator physics to simulate the important nonlinear effects of the repetitive motion of particles in accelerators such as the LHC at CERN or the Tevatron at Fermilab (Berz 1987; Makino and Berz 2005; Berz 1999). The group of Martin Berz and Kyoko Makino at Michigan State University has championed these techniques since 1986 through the development of the COSY INFINITY software package (Makino and Berz 2005) that implements an efficient and easy to use way to perform DA computation on the computer.

A. Wittig (✉)
ESA Advanced Concepts Team, 2200AG Noordwijk, The Netherlands
e-mail: alexander.wittig@esa.int

G. Gómez, J.J. Masdemont (eds.), *Astrodynamics Network AstroNet-II*,
Astrophysics and Space Science Proceedings 44,
DOI 10.1007/978-3-319-23986-6_20

Introduced in the field of astrodynamics starting in 2006 by works of Pierluigi di Lizia and Roberto Armellin (Armellin et al. 2010; Valli et al. 2013; Di Lizia et al. 2008, 2009), Differential Algebra techniques have become quite popular in the field and have spawned many interesting works in research (Alessi et al. 2009) as well as in industry (Bignon et al. 2014).

The purpose of this work is two-fold: in the first part, we provide an introduction to DA techniques and their classical applications in an abstract way so that readers not familiar with the technique can understand the method at an intuitive level. In the second part, we present novel research results obtained by the author during the AstroNet II network. We present an extension of the classical set theoretical view of DA to the concept of a DA manifold, and then describe the Automatic Domain Splitting (ADS) technique which naturally leads to the generation of DA manifolds during the propagation of sets of initial conditions in ODEs. We illustrate the technique by applying it to the propagation of an initial condition set in the two body problem.

2 Differential Algebra

In this section we introduce the mathematical concepts of the DA vector space and present several ways to intuitively grasp what a DA vector represents in several classical applications of DA. In this overview, we largely follow the much more complete definition of Differential Algebra given in Berz (1999).

Similar to other well known mathematical concepts, also Differential Algebra can be used to represent different objects in reality, leading to different conceptual views. A matrix as an element of the algebra of matrices, for example, can represent many different things such as a linear map, a rotation, a data array, or a covariance of a statistical distribution. All of these matrices follow the same mathematical definition, yet represent very different things in the real world.

2.1 *Differential Algebra as a Polynomial Algebra*

The mathematical definition of the DA vector space is most easily introduced by viewing it as an algebra of polynomials. Each DA vector represents a particular polynomial over $\mathbb{R}$

$$P(\boldsymbol{x}) = \sum_{\alpha} a_{\alpha} \boldsymbol{x}^{\alpha}$$

where we use multi-index notation for α. It is uniquely determined by its finitely many coefficients $a_{\alpha} \in \mathbb{R}$. We call the space containing all such polynomials of a fixed maximal order n and v independent variables ${}_nD_v$.

In order to turn this space into a vector space, we introduce the natural addition, subtraction and scalar multiplication as the corresponding operation on the polynomials. In the following, let $P, Q \in_n D_v$ with coefficients a_α and b_α, respectively, be given. Then for any $z \in \mathbb{R}$ we have

$$P(\boldsymbol{x}) + Q(\boldsymbol{x}) = \sum_{\alpha} (a_\alpha + b_\alpha)\, \boldsymbol{x}^\alpha$$

$$P(\boldsymbol{x}) - Q(\boldsymbol{x}) = \sum_{\alpha} (a_\alpha - b_\alpha)\, \boldsymbol{x}^\alpha$$

$$z \cdot P(\boldsymbol{x}) = \sum_{\alpha} z \cdot a_\alpha \boldsymbol{x}^\alpha.$$

It is easy to show that with these definitions, ${}_nD_v$ becomes a complete vector space.

In order to give this vector space a more useful algebraic structure and turn it into what we call the Differential Algebra, we introduce further operations on it. First, we introduce the multiplication between P and Q by simply performing the usual multiplication of the two polynomials and truncating the resulting polynomial to order n:

$$P(\boldsymbol{x}) \cdot Q(\boldsymbol{x}) = \sum_{|\alpha| \leqslant n} \left(\sum_{\beta_1+\beta_2=\alpha} a_{\beta_1} \cdot b_{\beta_2} \right) \boldsymbol{x}^\alpha$$

where the inner sum runs over all pairs of multiindices (β_1, β_2) such that $\beta_1 + \beta_2 = \alpha$ and the outer sum over all multiindices α of order less than or equal to n.

Furthermore, it is possible to introduce a multiplicative inverse P^{-1} for any P with non-zero constant part (i.e. $a_0 \neq 0$) satisfying

$$P^{-1} \cdot P = 1.$$

This relationship uniquely determines $P^{-1} \in_n D_v$ and allows the straight forward introduction of division of two polynomials via $Q/P = Q \cdot P^{-1}$.

Similarly, well defined square roots $\sqrt{P}$ can be introduced for any P with positive constant part (i.e. $a_0 > 0$) through the identity

$$\sqrt{P} \cdot \sqrt{P} = P.$$

Finally, the algebraic structure can be turned into the namesake differential algebra by introducing the derivation operator ∂_i and anti-derivation operator ∂_i^{-1}.

Both operators simply acts on the polynomial P in the usual sense:

$$\partial_i P = \frac{\partial}{\partial x_i} P(\boldsymbol{x}),$$

$$\partial_i^{-1} P = \int_0^{x_i} P(\widetilde{\boldsymbol{x}})\, d\tilde{x}_i$$

where the indefinite integral of $P(\boldsymbol{x})$ with respect to x_i is truncated at order n such that the result lies again in ${}_nD_v$. Since $P(\boldsymbol{x})$ is a polynomial, the derivative and integral are trivial to compute and again of polynomial form. It can be shown that the derivation operator ∂_i is linear and satisfies the Leibnitz product rule with respect to the multiplication introduced on ${}_nD_v$:

$$\partial_{x_i}\left(p(\boldsymbol{x})q(\boldsymbol{x})\right) = \left(\partial_{x_i} p(\boldsymbol{x})\right) p(\boldsymbol{x}) + p(\boldsymbol{x})\left(\partial_{x_i} p(\boldsymbol{x})\right).$$

It is hence a derivation and turns ${}_nD_v$ into a differential algebra.

2.2 *Differential Algebra as Computational Functional Analysis*

As the DA vector space ${}_nD_v$ is high dimensional but finite, it is possible to implement it and the operations as defined above on a computer. In a slight abuse of language, we also refer to such computer implementations as DA.

Similar to the finite algebra of floating point numbers $\mathbb{F}$, which approximates the infinite field of real numbers $\mathbb{R}$ on a computer, the algebra on ${}_nD_v$ can be interpreted as a computer representation of the infinite function space $C^n(0)$ of n times continuously differentiable functions (of v variables) at the origin. In the case of floating point numbers, each real number $x \in \mathbb{R}$ is approximately represented by a floating point number $\bar{x} \in \mathbb{F}$ by essentially storing the first k digits of its binary expansion. We denote this operation by $\bar{x} = \mathscr{F}_x$. Any binary operation $\times$ on the real numbers is then approximated by a corresponding operation $\otimes$ on the floating point numbers $\mathbb{F}$ by requiring that

$$\mathscr{F}_{x \times y} = \mathscr{F}_x \otimes \mathscr{F}_y.$$

For example, the result of adding two real numbers is approximated by first converting them to floating point numbers and then performing floating point arithmetic.

The same relationship can be established for functions in $C^n(0)$ and elements of ${}_nD_v$. For any function $f \in C^n(0)$, denote by $\mathscr{T}_f$ the truncated multivariate Taylor expansion of f up to order n around the origin. Then for any binary operation $\times$ on

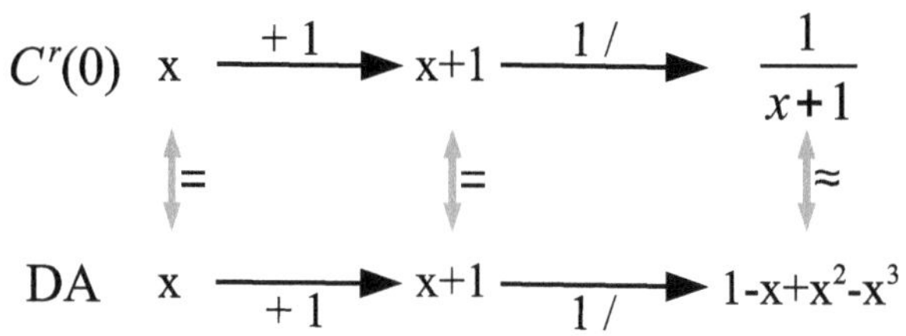

Fig. 1 Step by step evaluation of the expression $\frac{1}{x+1}$. Operations on $C^3(0)$ are approximated by the corresponding operations on ${}_3D_1$

$C^n(0)$ we can define a corresponding operation $\otimes$ on ${}_nD_v$ that satisfies

$$\mathscr{T}_{x \times y} = \mathscr{T}_x \otimes \mathscr{T}_y. \tag{1}$$

This means that for the multiplication of two function $f, g \in C^n(0)$, for example, the result of Taylor expanding $f \cdot g$ is the same as first Taylor expanding f and g separately and then performing a DA multiplication. It can be verified easily that the operations as introduced in the previous section do satisfy this relation.

As a result we can thus perform computations on the function space $C^n(0)$ efficiently on the computer. The result obtained is a Taylor approximation of the final expression evaluated in DA arithmetic. This is illustrated in Fig. 1, where the first two operations yield an exact representation of the result on $C^3(0)$, while the third operations (division) yields and approximation in the form of a Taylor expansion.

Furthermore, this definition allows us to introduce common functions such as trigonometric, exponential, and logarithmic functions as well as their inverses on ${}_nD_v$. We refer to these functions as intrinsic functions. For any $P \in_n D_v$ we define the result of $f(P)$ to satisfy

$$f(P) = \mathscr{T}_{f \circ P}. \tag{2}$$

This way, also for intrinsic functions we have that evaluating the function in $C^n(0)$ and then Taylor expanding the resulting function yields the same as first Taylor expanding the argument and then performing a DA evaluation of the intrinsic function.

As all operators and intrinsic functions satisfy relations (1) and (2), respectively, it follows that the evaluation of any mathematical expression built from those operators and intrinsic functions yields the Taylor expansion of the full expression. This can easily be shown by induction over each operation or intrinsic function.

2.3 Differential Algebra as Automatic Differentiation

An immediate application of the DA computer implementation is to view DA as an automatic differentiation method. For a given function provided as a sequence of basic operators and intrinsic functions, it is straight-forward to obtain the derivatives up to any given order at a given point automatically by evaluating the expression using DA arithmetic. More specifically, let $f(\boldsymbol{x})$ be a function that is n times continuously differentiable at the point $\boldsymbol{x}_0 \in \mathbb{R}^v$. Then we can compute all derivatives with respect to any component of $\boldsymbol{x}$ up to a given order by evaluating the expression

$$P(\boldsymbol{x}) = f(\boldsymbol{x}_0 + \boldsymbol{x})$$

in DA arithmetic. The resulting DA representation of $P \in_n D_v$ by definition of the DA operations is the Taylor expansion of $f(\boldsymbol{x}_0 + \boldsymbol{x})$ evaluated at $\boldsymbol{x} = 0$ and hence it contains the derivatives of f at $\boldsymbol{x}_0$.

Since the algebra on ${}_nD_v$ is a differential algebra, computing the exact derivative at the expansion point $\boldsymbol{x}_0$ can be performed particularly elegantly by applying the derivation operators ∂_i repeatedly to P and evaluating the resulting polynomial at the origin, i.e. taking the constant part. That is, in DA arithmetic we can evaluate the expression

$$\left.\frac{\partial^k}{\partial \boldsymbol{x}^k} f(\boldsymbol{x}_0 + \boldsymbol{x})\right|_{x=0}$$

completely algebraically.

Note that this method for computing the derivatives up to arbitrary order does not at any point use divided differences or other approximations of the derivative. Each derivative is computed accurately up to floating point error. Especially for higher order derivatives this accuracy is much higher than what can be achieved using divided differences.

Furthermore, this method does not require the choice or adaptive determination of some small deviation used in approximating the derivative by divided differences. A single evaluation in DA arithmetic on ${}_nD_v$ automatically yields all derivatives at once. This also sets apart the method from other automatic differentiation techniques which require additional passes for each additional order of the derivative to be computed.

Of great interest in many applications are the low order expansions in linear and quadratic terms, i.e. the first and second derivatives. For example, linearizing the result of an arbitrary algorithm is as simple as evaluating the same algorithm in ${}_1D_v$. The result is the result at the expansion point and for free the transition matrix for that algorithm with respect to its variables.

In particular, this can be applied to algorithms for numerical integration of ordinary differential equations. Those algorithms, such as Runge-Kutta, are formed by a sequence of operations to perform a single time step. By repeating this single time step many times, eventually an initial condition $\boldsymbol{x}_0$ can be propagated to a final condition $\boldsymbol{x}_1$. This sequence of algebraic operations can of course also be performed in ${}_1D_v$, setting the initial condition explicitly to $\boldsymbol{x}_0 + \boldsymbol{x}$. The result is the first order expansion of the flow, more commonly known as the transition matrix of the system at $\boldsymbol{x}_0$. Note that there is no need to derive and integrate any variational equations to obtain this result. The transition matrix is automatically obtained for any dynamics representable as a sequence of basic operations and intrinsic functions on ${}_nD_v$.

2.4 *Differential Algebra as Set Representation*

The final classical way to view Differential Algebra vectors we present here is the set theoretical view. For this view, instead of a single element of ${}_nD_v$, we consider a vector $\boldsymbol{P}$ made of k elements of ${}_nD_v$. We then interpret this vector as a polynomial function $\boldsymbol{P} : \mathbb{R}^v \to \mathbb{R}^k$ with each component of the function given by the corresponding element of ${}_nD_v$ in $\boldsymbol{P}$. We also refer to $\boldsymbol{P}$ as a DA map or a DA set.

Each such vector $\boldsymbol{P} \in_n D_v{}^k$ can be viewed as representing a set of points in $\mathbb{R}^k$. This is done by arbitrarily fixing a domain $D \subset \mathbb{R}^v$ and associating the image of D under $\boldsymbol{P}$, that is

$$P = Im(\boldsymbol{P}) = \{\boldsymbol{P}(x) | \, x \in D\}$$

with $\boldsymbol{P}$. Now the set P is parametrized over the initial domain D, and each element of P has a pre-image in D. The great advantage of this representation is that such a DA set $\boldsymbol{P}$ is capable of representing very accurately even complicated non-linear sets of points, as long as they are compact and simply connected.

While in general arbitrary, often without loss of generality the domain D is chosen as $D = [-1, 1]^v$, as this choice is numerically advantageous. By introducing an appropriate rescaling function, the set $[-1, 1]^v$ can always be scaled to an arbitrary interval product $[a_1, b_1] \times [a_2, b_2] \times \cdots \times [a_v, b_v]$.

With this definition of a DA set in place, it is now trivial to perform algebraic operations on all points in a DA set at once. For example, let us evaluate the result of applying a given function $f : \mathbb{R}^k \to \mathbb{R}^k$ to each element of a set $P \subset \mathbb{R}^k$, i.e. to compute the image $Q = Im(f(P))$. Using a DA set representation $\boldsymbol{P}$ of the set P, one can simply evaluate the function f in DA arithmetic on the DA representation $\boldsymbol{P}$ of the set P. The result is another element $\boldsymbol{Q} \in_n D_v{}^k$ approximating $f(\boldsymbol{P}(x))$, which now over the domain $[-1, 1]^v$ represents the desired set $Q = Im(f(P))$. This relationship is shown in Fig. 2.

Similarly, of course, it is also possible to evaluate any function $f : \mathbb{R}^k \to \mathbb{R}^{k^*}$ in the same way on $\boldsymbol{P}$. In particular, scalar functions $f : \mathbb{R}^k \to \mathbb{R}$ can be evaluated on a given set $\boldsymbol{P}$ by DA evaluation. The result $F = f(\boldsymbol{P})$ in ${}_nD_v$ represents a function over the initial domain D of $\boldsymbol{P}$. With each $x \in D$, there is an associated point $\boldsymbol{P}(x) \in \mathbb{R}^k$ and the value of $f(\boldsymbol{P}(x))$ is given by $F(x)$.

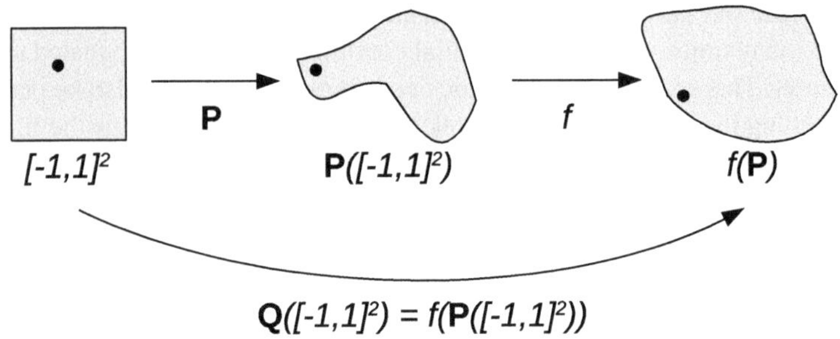

Fig. 2 DA set representation and function evaluation on DA sets. The *highlighted point* indicates the mapping of a single point in the initial domain

A particularly useful application of this set theoretical view in practical applications is for the interpretation of the high order expansions of the flow of an ODE with respect to initial conditions and parameters. As mentioned in the previous subsection, it is possible to expand the flow of an ODE by evaluating common numerical ODE integration schemes in DA arithmetic with the initial conditions expressed as DA variables.When performed to first order, the result obtained can be interpreted as the state transition matrix. However, when higher order terms are computed, it is more convenient to switch to the set theoretical view.

Consider an ODE of the form

$$\dot{x}(t) = f(x, t)$$

with $x \in \mathbb{R}^k$ and $f : \mathbb{R}^k \times \mathbb{R} \to \mathbb{R}^k$. Letting the initial condition of the ODE be a DA set of the form $\boldsymbol{X}_0 = \begin{pmatrix} x_{0,1} + w_1\delta x_1 \\ x_{0,2} + w_2\delta x_2 \\ \vdots \\ x_{0,k} + w_k\delta x_k \end{pmatrix}$ where the w_i are positive and $x_0 \in \mathbb{R}^k$ we have that $\boldsymbol{X}_0$ on the domain $D = [-1, 1]^k$ is an interval box centered around x_0 and of width $2w_i$ along the i-th coordinate axis. Performing the ODE integration in DA arithmetic with this initial condition results in some final state $\boldsymbol{X}_1$. When viewed in the set theoretical view, this final state now represents the final set of points after propagation through the ODE. Each point in the initial condition is uniquely identified by some $\delta x \in D$, and its final state after propagation is simply given by $\boldsymbol{X}_1(\delta x)$.

Thus the DA set theoretical view helps make sense of the high order derivatives of the flow. The higher the expansion order of the flow, the more accurate is the final set represented.

To illustrate this view of DA, consider the following example. The dynamics of an object moving in the solar system is integrated in the framework of the two body problem:

$$\begin{cases} \dot{\boldsymbol{r}} = \boldsymbol{v} \\ \dot{\boldsymbol{v}} = -\frac{\mu}{r^3}\,\boldsymbol{r}, \end{cases} \tag{3}$$

where $\boldsymbol{r}$ and $\boldsymbol{v}$ are the object position and velocity vectors respectively, and μ is the Sun gravitational parameter. The nominal initial conditions are set such that the object starts moving from the pericenter of an elliptic orbit, lying on the ecliptic plane. The pericenter radius is 1 AU, whereas the magnitude of the initial velocity is selected to have a resulting orbit of eccentricity $e = 0.5$. The x and y components of the initial position at time $t_0 = 0$ are chosen in an uncertainty box of size 0.008 AU and 0.08 AU in the x and y direction respectively. The evolution of the boundary of this initial box is then investigated by evaluating the boundary of the DA representation of the propagated set at various times. The evolved box is shown in Fig. 3 corresponding to six integration times uniformly distributed up to time $t = 16$. At each of these times, the entire propagated set of points is accurately represented by a fifth order DA map. For the final time $t = 16$, an analysis of the accuracy at different expansion orders is shown, illustrating how the exact final set (as determined by classical pointwise integration) is approximated better and better by higher order flow expansions.

We conclude the set theoretical view of DA with the following remark. While the DA sets introduced here only approximate the real sets, it is possible to extend the Differential Algebra in such a way to rigorously bound the truncation error in each step and obtain mathematically rigorous enclosures of the correct results (Makino 1998). Such extended DA vectors are referred to as Taylor Models and have been

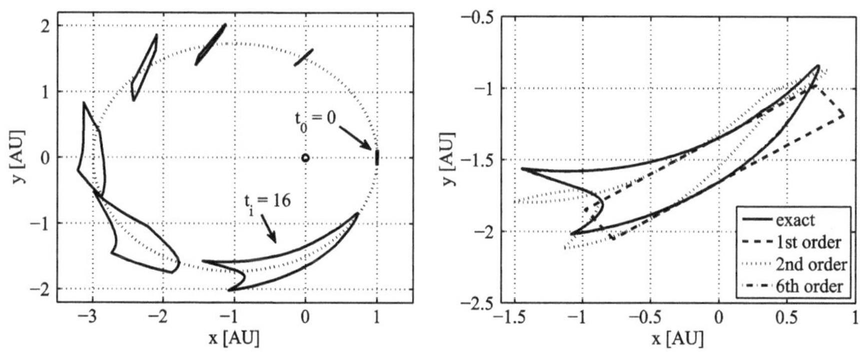

Fig. 3 Propagation of a set of initial positions in the two-body problem using sixth order DA arithmetic (*left*) and accuracy analysis of the final set at time $t = 16$ at different orders (from Wittig et al. 2015)

implemented in COSY INFINITY (Makino and Berz 2005) by Martin Berz and Kyoko Makino for use in verified computing and computer assisted proofs.

3 Differential Algebra Manifold Representation

Within the AstroNet II initial training network work was performed on developing further the numerical techniques used in astrodynamics problems. In particular, the set representation view presented above is of great use when propagating uncertain sets of initial conditions in astrodynamical systems. However, the set representation view of DA is limited by the convergence radius of the resulting polynomials. In strongly non-linear long term propagation of large initial sets, it often happens that the polynomial representation of the set does not converge any more and the approximation becomes very poor. To address this issue, we have developed the simple set representation view into the more advanced concept of a *DA manifold representation*.

To introduce this concept, we begin by reminding the reader briefly of the traditional definition of the atlas of charts of a k-dimensional manifold M (Lee 2009).

Definition 1 We call a pair (U, φ) a chart where $U \subset M$ is an open subset of the manifold and $\varphi : U \to \mathbb{R}^k$ is a homeomorphism of U into Cartesian space.

Definition 2 An atlas $\mathscr{A}$ is a collection of charts $\{(U_\alpha, \varphi_\alpha)\}$ such that the U_α form an open cover of M.

Every topological manifold admits such an atlas, and if all φ_α of the atlas are at least n times differentiable, we refer to the atlas as a C^n-atlas. The atlas is not unique, as many different atlases can describe the same manifold. While not required of an atlas, for many manifolds in practice it is possible to obtain a finite atlas containing only a finite number of charts.

The intuition behind this definition is that the globally complex and non-Cartesian structure of a (differentiable) manifold is represented by covering it with local patches, or charts. On each chart, the manifold is homeomorphic to a subset of Cartesian space (see Fig. 4). Note that in general it is not possible to represent the entire manifold by a single chart, as the topology of the manifold does not have to be compatible with the topology of Cartesian space. In fact, the naming of the charts and the atlas is a reminder of cartography: while it is possible to make local rectangular maps that form a 1:1 mapping between some point on the globe and the Cartesian space of the map, it is not possible to put a map of the entire globe on a rectangular poster without introducing degenerate points.

This definition of a manifold is very convenient for the theoretical study of many dynamical systems. The existence of the charts at each point of the manifold allows the pull-back of dynamics in the manifold to dynamics in Cartesian space and the subsequent push-back onto the manifold. This automatically provides a differential

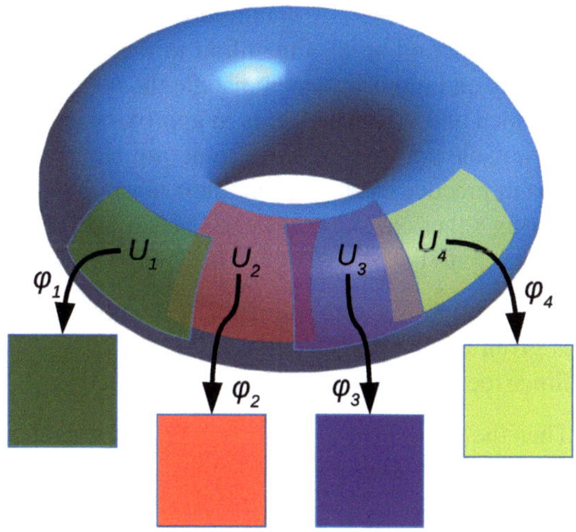

Fig. 4 Illustration of a 2-dimensional manifold partially covered by some charts of its atlas

structure on the manifold in an abstract way. It is thus possible to reduce the treatment of any local property on the manifold to be done in familiar Cartesian space, and then simply pushing back into the manifold.

3.1 Manifold Computer Representation

While mathematically nice, representing and manipulating such manifolds in the computer is not straight forward. Typically manifolds on the computer are globally represented as some form of grid of points in the manifold (e.g. Topputo et al. 2013). The obvious advantage of this method is its simplicity in the implementation.

However, obtaining an efficient grid can be quite difficult especially in higher dimensions, and for accurate interpolation between the grid points a dense grid is required over the entire manifold.

Another problem with this representation is that it does not allow for easy manipulation of the manifold. A typical task, for example, is the evaluation of a function on a manifold. While it is possible to evaluate such a function on each stored grid point on the manifold, the quality of the result varies with the function being evaluated. A coarse grid may approximate a simple initial manifold well, but a function that varies quickly over the manifold may not be represented well using the same grid. To obtain a good approximation of the function on the manifold, it is in general necessary to resample the manifold over a grid chosen to not only approximate the manifold well, but also adapted to the function being evaluated.

Lastly, when compared to the elegant mathematical definition of a manifold, the pointwise computer representation loses the key feature of the underlying analytical structure of the manifold setting it apart from a mere set of points.

The goal of the DA manifold representation is to retain as much of the mathematical definition of a manifold in the computer representation, while providing an easy and intuitive way to handle the resulting representation of a manifold. The key idea of a DA manifold is to approximate the charts in an atlas of the manifold by suitable polynomial representations. However, instead of storing the φ_α directly, it is in practice easier to store their inverses φ_α^{-1}. As the φ_α are homeomorphisms, storing either one is equivalent. Furthermore, for simplicity we restrict the domain of all φ_α^{-1} to the set $(-1,1)^k$. This yields the following definition of a DA manifold.

Definition 3 A v-dimensional DA manifold M embedded in w-dimensional space is given by its finite DA atlas $\mathscr{A}$ of DA charts $((-1,1)^v, \varphi_\alpha^{-1})$ with $\varphi_\alpha^{-1} \in_n {D_v}^w$. Since the domain for each DA chart is fixed to $(-1,1)^v$, it is usually omitted.

Then the open sets $U_\alpha = \varphi_\alpha^{-1}((-1,1)^k)$ form an open cover of the manifold M, and the charts are given by $(U_\alpha, \varphi_\alpha)$. The concept of the DA-manifold is illustrated in Fig. 5.

The reason for storing the inverse function in the DA charts of a DA manifold is simply that it is difficult to store the sets U_α on which the functions φ_α are defined on a computer. While it is a useful theoretical concept, in a practical implementation it is much easier to store the Cartesian domain of the inverse maps. A further practical simplification in our definition of a DA manifold is the fixing of the domain of each DA chart to the common domain $(-1,1)^v$, which is then implicitly understood and does not need to be stored any more, further reducing the memory requirements of a DA manifold.

Apart from its close relationship with the elegant mathematical definition of a manifold, another advantage of the DA manifold is closely related to the fact that the DA representations of the φ_α^{-1} are given by Taylor expansions around some expansion point. As is common with Taylor expansions, the expansion of the φ_α^{-1} functions has a limited convergence radius within which the truncated

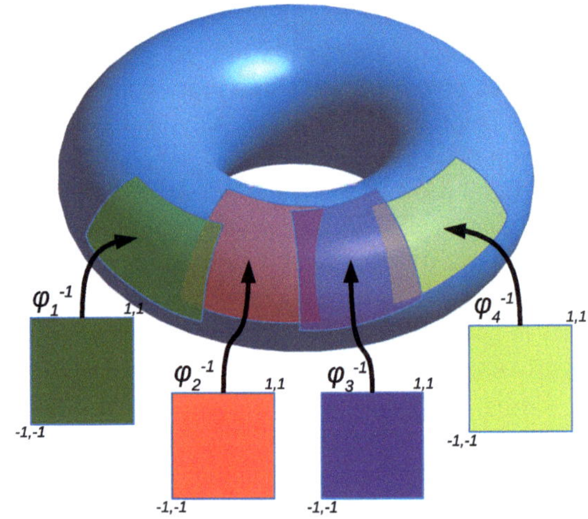

Fig. 5 Illustration of a 2-dimensional DA manifold partially covered by some DA charts of its atlas

Taylor expansion represents the real function sufficiently well. Especially in high-dimensional cases, it is not practical to increase the expansion order arbitrarily to meet a given accuracy requirement due to the factorial growth of the number coefficients in the polynomials. Instead, it is more efficient to subdivide the domain into smaller pieces, and represent the original function by several local Taylor expansions. This observation gives rise to the concept of Domain Splitting, which we will introduce in the next Section.

4 Manifold Propagation with Automatic Domain Splitting

Of great practical importance is the use of the DA maps for the representation of the evolution of sets of initial conditions under the effect of strongly non-linear dynamical systems. In particular, the evolution of a set of initial conditions under the effect of a non-linear ordinary differential equation (ODE) is of great practical interest as was shown in Sect. 2.4. Unfortunately, a combination of various factors can cause a single polynomial to not represent this relation between initial and final conditions sufficiently accurately any more. This effect is typically caused by some combination of the strong non-linearity of the ODE, long integration times, and large initial sets.

To illustrate this problem, we return to the example of the two body problem in Sect. 2.4. For a single period of the system, it was possible to represent the propagated set by a single DA map. However, integrating further to about 1.96 revolutions, a fifth order Taylor expansion is no more sufficient to represent the final set any more, as is shown in Fig. 6. The black set represented by the polynomial

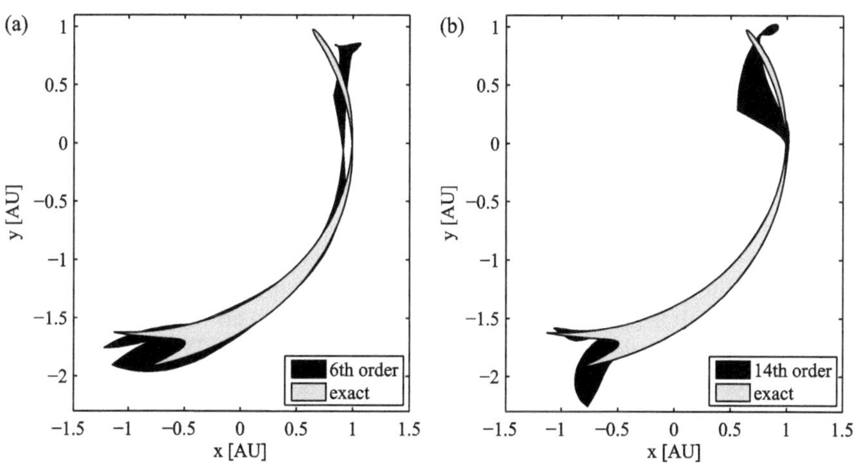

Fig. 6 A propagated set of initial conditions in the two body problem after 1.96 revolutions (*left*: order 6, *right*: order 14)

expression is clearly not congruent with the exact gray set determined by pointwise integration.

To solve this problem, we have developed a technique for Manifold Propagation with Automatic Domain Splitting (ADS) (Wittig et al. 2015). At its core, this method considers the initial condition of the initial value problem as a trivial DA manifold containing only a single DA chart. After each integration step, ADS estimates the truncation error of the polynomial set representation, and automatically subdivides the domain of the DA chart being propagated if the error is above a specified threshold. This creates two new DA charts, which are added to the manifold instead of the old chart. Then the propagation continues with one of the new charts until it either needs to be split again or the final propagation time has been reached. The result of this propagation is a DA manifold consisting of a finite number of charts that, taken together, represent the entire final set at the specified final time.

For the estimation of the truncation error, several methods can be employed. The simplest is to look at the terms of highest order and to use their size as an estimate of the truncation error. This is similar to the methods used in many other areas of numerical analysis such as step-size control for numerical integration of ODEs. However, since there is more information available in the polynomials than just the highest order terms, we use a different technique. On some sufficiently small domain, the terms of any convergent power series converge at least exponentially. This is why we first compute the size S_n of the coefficients of order n by taking S_n to be the largest absolute value of all coefficients of order n. We then perform an exponential fit of the parameters A and B to match the S_n to the function

$$S(n) = A\exp(Bn).$$

Finally we can use this exponential fit to estimate the size of the $n+1$ order terms by simply evaluating $S_{n+1} = S(n+1)$. In our experience in practice this yields a quite accurate estimate of the truncation error that is more precise than just the highest order terms because it uses all information available in the polynomial.

A similar method is employed to determine the direction in which to split the domain when a split becomes necessary. In this case, however, the polynomial is factored by powers of one of the independent variables x_i in the polynomial, and the exponential fit is performed over all coefficients containing a given power of x_i. This allows an estimate of the contribution of the x_i direction to the total truncation error. Performing this estimate for all independent variables in the domain, we can pick the direction with the largest contribution to the truncation error as the direction to split. For a more detailed description of the ADS algorithm see Wittig et al. (2015).

In the setting of the two body problem above, Fig. 7 illustrates this procedure. In subfigure (a), after almost one complete period the entire manifold is still represented by a single chart. However, in the next time step in subfigure (b), the automatic domain splitting has identified the need to split the chart into three charts representing the propagated manifold accurately. Proceeding with the manifold propagation until close to two periods, the manifold is now represented by four

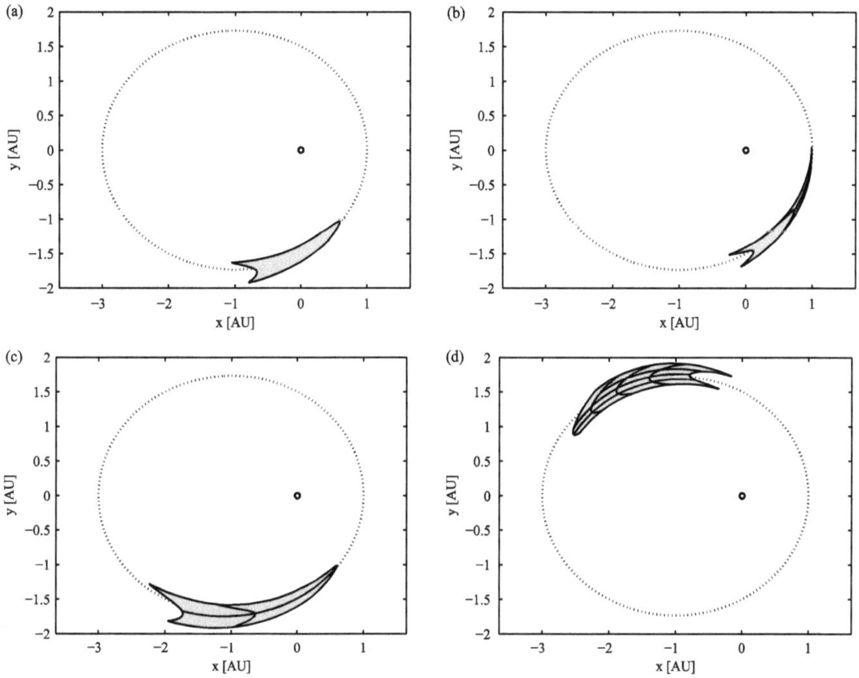

Fig. 7 Propagation of the initial uncertainty set in the two-body dynamics using 14th order Taylor expansions of the flow and automatic domain splitting (clockwise from *top left*): $t_a = 930.1$ day (0.90 revolutions); $t_b = 988.3$ day (0.96 revolutions); $t_c = 1918.4$ day (1.86 revolutions); $t_d = 2325.3$ day (2.25 revolutions)

charts. After another perigee passage, the ADS algorithm has split the manifold even further, yielding 15 charts which still accurately represent the whole set.

Focusing again on $t = 1.96$ revolutions, when the single DA map was visibly wrong, we can compare the result of the manifold propagation with ADS. Figure 8 shows both a single 14th order polynomial and the DA manifold consisting of nine separate charts generated automatically by the ADS algorithm. As can be seen, the DA manifold coincides with the exact solution generated by pointwise integration shown in Fig. 6.

This simple example illustrates how the concept of a DA manifold combined with Automatic Domain Splitting vastly increases the utility of DA based propagation compared to the simple DA set propagation using a single DA map. Of course in a real application the propagation would not be performed in Cartesian coordinates and the problem of domain splitting would be completely avoided by using e.g. Keplerian elements to describe the orbits. In more complicated dynamics such as the full solar system dynamics, however, elements providing an analytical solution are not readily available and even a propagation in a more suitable set of coordinates will suffer from high non-linearities (Di Lizia et al. 2009). It has been shown in

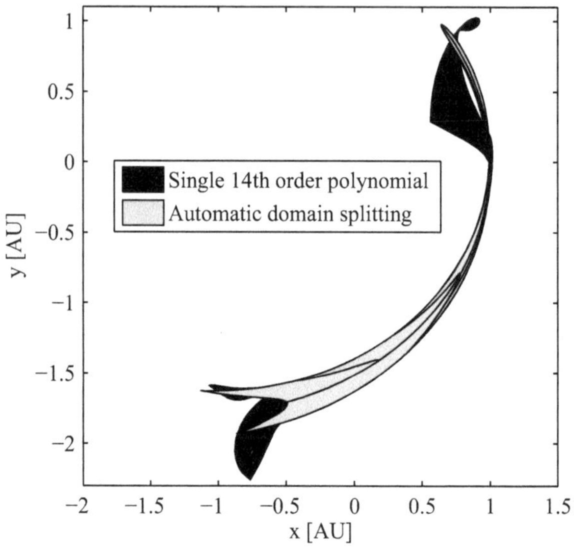

Fig. 8 Comparison of a single 14th order polynomial and the DA manifold of nine DA charts generated by ADS. The ADS DA manifold coincides with the exact solution

Wittig et al. (2015) that manifold propagation with ADS can successfully handle also these cases.

5 Conclusions

In this work, we presented an introduction to the Differential Algebra on ${}_nD_v$ and outlined various ways to view this algebra in applications. An extension of the set theoretical view has been proposed which allows the efficient and elegant representation of manifolds on the computer in the form of DA manifolds. These manifolds are straight forward to manipulate in ${}_nD_v$ while maintaining the elegant structure of the mathematical definition of a manifold. Furthermore, DA manifold propagation was introduced, allowing for the efficient and accurate propagation of a (potentially large) initial manifold in strongly non-linear ordinary differential equations. The technique has been illustrated by applying it to the propagation of an initial set through the Keplerian dynamics of the two body problem.

References

Alessi, E.M., Farres, A., Vieiro, A., Jorba, À., Simó, C.: Jet transport and applications to neos. In: Proceedings of the 1st IAA Planetary Defense Conference, Granada (2009)

Armellin, R., Di Lizia, P., Bernelli-Zazzera, F., Berz, M.: Asteroid close encounters characterization using diffferential algebra: the case of apophis. Celest. Mech. Dyn. Astron. **107**(4), 451–470 (2010)

Berz, M.: The method of power series tracking for the mathematical description of beam dynamics. Nucl. Instrum. Methods **A258**(3), 431–436 (1987)

Berz, M.: Modern Map Methods in Particle Beam Physics. Academic, New York (1999)

Bignon, E., Pinède, R., Azzopardi, V., Mercier, P.: Jack: an accurat numerical orbit propagator using taylor differential algebra. In: Presentation at KePASSA Workshop, Logroño, 23–25 April 2014

Di Lizia, P., Armellin, R., Lavagna, M.: Application of high order expansions of two-point boundary value problems to astrodynamics. Celest. Mech. Dyn. Astron. **102**, 355–375 (2008)

Di Lizia, P., Armellin, R., Zazzera, F.B., Jagasia, R., Makino, K., Berz, M.: Validated integration of solar system dynamics. In: Proceedings of the 1st IAA Planetary Defense Conference, Granada (2009)

Lee, J.: Manifolds and Differential Geometry. Graduate Studies in Mathematics, vol. 107. American Mathematical Society, Providence (2009)

Makino, K.: Rigorous analysis of nonlinear motion in particle accelerators. PhD thesis, Michigan State University (1998)

Makino, K., Berz, M.: Cosy infinity version 9. Nuclear Instrum. Methods **A558**, 346–350 (2005)

Topputo, F., Zhang, R., Zazzera, F.B.: Numerical approximation of invariant manifolds in the restricted three-body problem. In: Proceedings of the 64th International Astronautical Congress. International Astronautical Federation, Paris (2013). IAC-13,C1,9,11,x18153

Valli, M., Armellin, R., Di Lizia, P., Lavagna, M.R.: Nonlinear mapping of uncertainties in celestial mechanics. J. Guid. Control Dyn. **36**(1), 48–63 (2013)

Wittig, A., Di Lizia, P., Armellin, R., Makino, K., Bernelli-Zazzera, F., Berz, M.: Propagation of large uncertainty sets in orbital dynamics by automatic domain splitting. Celest. Mech. Dyn. Astron. **122**(3), 239–261 (2015)

Identification of New Orbits to Enable Future Missions for the Exploration of the Martian Moon Phobos

Mattia Zamaro and James D. Biggs

Abstract One of the paramount stepping stones towards NASA's long-term goal of undertaking human missions to Mars is the exploration of the Martian moons. In this paper, a showcase of various classes of non-Keplerian orbits are identified and a number of potential mission applications in the Mars-Phobos system are proposed. These applications include: low-thrust hovering around Phobos for close-range observations; Libration Point Orbits in enhanced three-body dynamics to enable unique low-cost operations for space missions in the proximity of Phobos; their manifold structure for high-performance landing/take-off maneuvers to and from Phobos' surface; Quasi-Satellite Orbits for long-period station-keeping and maintenance. In particular, these orbits could exploit Phobos' occulting bulk as a passive radiation shield during future manned flights to Mars to reduce human exposure to radiation. Moreover, the latter orbits can be used as an orbital garage, requiring no orbital maintenance, where a spacecraft could make planned pit-stops during a round-trip mission to Mars.

1 Introduction

Since the discovery of Phobos and Deimos in 1877, the two natural satellites of Mars have become increasingly interesting astronomical objects to investigate. Phobos is closer to Mars than Deimos and almost double its size, but despite this, they are very similar, since they share common physical, orbital and geometrical features. Their origin is still largely unknown (Bell et al. 1993; Rosenblatt 2011), and is currently debated to have been either an asteroid capture by Mars, or coalescence from proto-Mars or Solar System material, or even accretion of material from Mars ejected from its surface after an impact with a previous small body. This uncertainty is enhanced by the mysterious composition of these moons inferred from infrared spectral analysis. Due to their relative low density and high porosity, they could hide

M. Zamaro (✉) • J.D. Biggs
Advanced Space Concepts Laboratory, University of Strathclyde, 75 Montrose Street, G1 1XJ, Glasgow, UK
e-mail: mattia.zamaro@strath.ac.uk; james.biggs@strath.ac.uk

G. Gómez, J.J. Masdemont (eds.), *Astrodynamics Network AstroNet-II*,
Astrophysics and Space Science Proceedings 44,
DOI 10.1007/978-3-319-23986-6_21

a considerable amount of iced water (Rosenblatt 2011), which is an attractive in-situ resource that could be exploited by human missions.

Due to its proximity to Mars, Phobos is currently of great interest for future missions to the Red Planet. During its Ministerial Council Meeting of November 2012, ESA confirmed post-2018 mission concepts: the Mars Robotic Exploration Programme would include a mission (Phootprint) to return back to Earth a sample from Phobos (Koschny 2012). In addition, NASA has identified a mission to Phobos as a key milestone to achieve before bringing humans to Mars (Hopkins and Pratt 2011; Lee 2012), since the absence of atmosphere on Phobos and Deimos makes landing and take-off easier for a manned spacecraft than on Mars. Therefore, the Martian moons could be exploited as outposts for astronauts: Phobos' proximity and fast orbital period can provide a relay for robotic exploration on Mars, and protection from space radiation hazards for manned spacecraft orbiting Mars (Phobos' bulk shielding the spacecraft). At the beginning of 2013, with the development of a new rover platform for the exploration of minor bodies, consisting of robotic hedgehogs, it has been reported that NASA is taking into consideration a mission (Surveyor) to Phobos as a test-bed for this new technology (Pandika 2012).

The purpose of this paper is to present a breakdown of different kinds of orbits that could be exploited in future space missions to Phobos. Section 2 introduces the reader to the physical environment connected to the orbital dynamics and constraints of a spacecraft in the vicinity of Phobos. The following Sects. 3–5 showcase each of the different kinds of orbits around this moon, such as: hovering points using Solar Electric Propulsion (SEP); natural Libration Point Orbits and their Invariant Manifold trajectories, and their artificial equivalent with constant low-thrust; Quasi-Satellite Orbits around Phobos. Section 6 provides a summary of the different solutions focusing on their applications in space missions to Phobos.

2 Preliminary Analysis for a Space Mission Around Phobos

In this section the basic design aspects of the dynamics and physics of a spacecraft in orbit of Phobos are introduced.

2.1 Physical and Astrodynamical Characteristics

The immediately noticeable characteristics of Phobos are its small size and its irregular shape: in particular the surface is marked by a dense texture of grooves and by several big craters. Phobos has an almost circular and equatorial orbit around Mars, and it rotates with synchronous period and almost zero-tilt with respect to its orbital motion. The low altitude of its orbit is lower than that required for a Mars-synchronous rotation. The physical and orbital parameters of Mars and Phobos, used in the analysis of the orbits undertaken in this paper, have been retrieved from NASA JPL online database.

2.2 Relative Dynamics

The general equations of motion of the relative orbital dynamics are stated in Eq.(1),

$$\ddot{\mathbf{q}} = -\mathbf{a}_\mathrm{A} + \mathbf{a}_\mathrm{G} + \mathbf{a}_\mathrm{P} + \mathbf{a}_\mathrm{C} + \mathbf{a}_\mathrm{D} \tag{1}$$

where $\mathbf{q}$ is the position of the spacecraft and $\mathbf{a}_\mathrm{A}$ is the apparent acceleration of the general relative frame of reference. $\mathbf{a}_\mathrm{A}$ is a function of the frame's translational acceleration and angular velocity with respect to an inertial reference. $\mathbf{a}_\mathrm{G}$ is the sum of the gravity accelerations of the celestial bodies of interest, each defined as the gradient of the gravitational potential $u_{G,\oplus} = Gm_\oplus/||\mathbf{q} - \mathbf{q}_\oplus||$, where G is the gravitational constant, $m_\oplus$ and $\mathbf{q}_\oplus$ are the mass and position of the body $\oplus$. $\mathbf{a}_\mathrm{P}$ indicates the thrusting acceleration of the propulsion system of the spacecraft required for artificial orbits, while for natural orbits $\mathbf{a}_\mathrm{P} = \mathbf{0}$. These three terms constitute the model of the dynamics where the reference signal of the orbit over time $\mathbf{q}(t)$ is solved, to be used by the guidance system in the mission. This motion will be perturbed in the real world by the disturbance $\mathbf{a}_\mathrm{D}$, consisting of the forces not considered in the model, and by the perturbations on the initial condition $\mathbf{q}_0$, due to the inaccuracies of the navigation system. To track the guidance law, such perturbations need to be counteracted by the station-keeping action $\mathbf{a}_\mathrm{C}$ of the orbital control system.

The study of the dynamics of a spacecraft about Phobos is conducted in the first instance with the model of the classical circular restricted three-body problem (CR3BP) (Koon et al. 2011), where the two massive bodies are Mars ($\oplus = 1$) and Phobos ($\oplus = 2$). Using non-dimensional units, the only parameter of the CR3BP is the mass factor μ, the normalized mass of the secondary body with respect to the total mass of the primaries, while the semi-major axis of their orbit provides the length unit l. For the case of Mars and Phobos, two peculiarities stand out when comparing them to the other couples of primaries studied in the Solar System with the CR3BP model. The mass parameter is very small ($\mu = 1.66 \cdot 10^{-8}$), and the length unit is significantly small too since the altitude of Phobos' orbit is less than twice the radius R_1 of Mars ($R_1/l = 36\,\%$).

The Hill's sphere of influence (SOI) is the region around each body where the dynamics are dominated by its own gravity field, and its radius for Phobos is 0.17 % of the distance from Mars. The related maximum altitude of the SOI is only 3.5 km, therefore it is impossible to naturally orbit around Phobos with a Keplerian motion, as shown in Fig. 1(left). The reduced SOI highly affects the orbital dynamics of a spacecraft around Phobos. Moreover, Phobos is tidally-locked, which means that Phobos' attitude is approximately fixed in the rotating frame of the CR3BP (Duxbury and Callahan 1981).

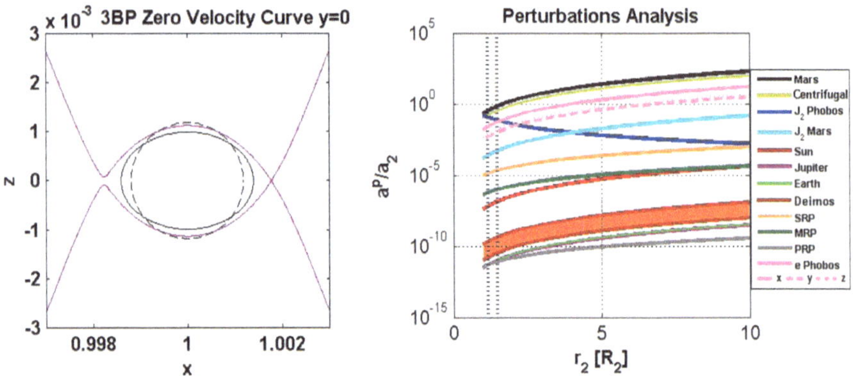

Fig. 1 *Left*: Hill's surface for L_2 energy. *x-z* projection. Phobos mean sphere (*dashed line*) and ellipsoid (*plain line*). *Right*: Differential perturbations analysis. *Vertical dotted lines* indicate Phobos' major size and Hill's SOI radius

2.3 *Orbital Perturbations*

An analysis is undertaken to quantify approximately the errors that occur in the Mars-Phobos system when it is approximated with a CR3BP. The orbital perturbations considered are the following. The dominant gravity harmonic J_2, for both Mars and Phobos' gravity fields, related to the oblateness of the body. The gravitational perturbation of additional bodies, in the closest configuration with Phobos. The radiation pressures of the Sun (SRP), Mars (MRP), and Phobos (PRP). Finally, the maximum effect of the eccentricity of the Mars-Phobos' orbit in the CR3BP dynamics. This analysis is undertaken with the reference of the same relative state with respect to Phobos, considering the resulting differential perturbation. Moreover, it is applied for the case of fixed relative points. The results are presented in Fig. 1(right). The perturbations are shown as a ratio $\mathbf{a}^{\mathbf{P}}/\mathbf{a}_2$ with respect to Phobos' Keplerian gravity term in the point, as a function of the radial distance from Phobos along the Mars-Phobos direction. In conclusion, the CR3BP does not provide an accurate approximation to describe the Mars-Phobos system's dynamics: the gravity harmonics and the orbital eccentricity of Phobos are the main orbital perturbations in proximity of the moon, and outside its Hill's SOI boundary the eccentricity becomes the dominant term, with Mars J_2 being the second largest.

2.4 *Radiation Environment*

The ionizing part of the space radiation in the Solar System, which is not shielded by the atmosphere and magnetic field as it happens here on the Earth surface, is currently considered the most challenging engineering aspect in designing a safe manned mission in deep space (Genta et al. 2014). The Mars' magnetic field is very weak, so no trapped particles constitute the radiation environment for a mission following the orbit of Phobos, which is similar to a deep space environment at the Sun-Mars distance. The two main sources of radiation are the protons from the Solar Energetic Particle Events (SEPEs) and the protons and alpha particles from the Galactic Cosmic Rays (GCRs). An estimation analysis is conducted with the open-source SPENVIS program and its dedicated model for Mars MEREM (Gonçalves et al. 2009). To derive an approximated figure of the gross effect of the radiation environment to human factors, the dosimetry quantity called the Effective Dose (Ef.D.) is considered. The result for a mission in Phobos' orbit from 2010 to 2030 is Ef.D.=1.9 Sv/year, 1.1 Sv/year from SEPEs protons and 0.8 Sv/year from GCRs protons and alpha particles. Considering the case of a 35-year old astronaut, the dose for 1 year, without any structural shielding, falls inside the range 1.75–2.5 Sv that indicates the maximum amount of radiation dose that such human crew could be allowed to absorb throughout the entire mission (Reitz 1998). Thus, the development of a strong shielding strategy for crewed missions is required. An interesting idea that has recently gained attention, is that a manned spacecraft during a Mars orbital mission segment could exploit Phobos as a passive radiation shield. Staying in its shadowing wake would theoretically counteract the gross Ef.D. of the directional part of the SEPEs, while remaining close to the moon will block any incoming isotropic particles (remaining SEPEs and GCRs) as much as its bulk covers the sky.

2.5 *Lighting Conditions*

Phobos' orbital plane is inclined with respect to its ecliptic plane by Mars' rotational tilt $\theta_M = 25.19°$. The Sun in the CR3BP frame rotates clockwise with an angular velocity equal to the difference between Phobos and Mars revolution rates, and a declination in the range $[-\theta_M, \theta_M]$ according to the seasonal phase of Mars. In this section the analysis of the lighting conditions is undertaken using eclipse modeling, which is to derive the zones of light and shadow produced by a shadowing central body when illuminated by a radiating body. This is described by a scalar light function field L, ranging from 0 to 1 to express the ratio of incident light with respect to the complete light case.

The approach used is to analyze the shadowing effect of each couple of bodies. The first case is the Sun-Mars L_1 over time at Phobos, computed with a cylindric eclipse model. During winter and summer Phobos is constantly in light, without Martian eclipses. The maximum eclipse time at the equinoxes is 54 min, corresponding to 12 % of Phobos daytime. Summer's total light period is about 164 days, and in the winter this is about 110 days.

The second case is the Sun-Phobos L_2, which is a time-variant 3D field, computed with a cylindric eclipse model and the mean spherical shape of Phobos. The mean integral value $\bar{L}_2$ along one Phobos' revolution period, for a given distance to

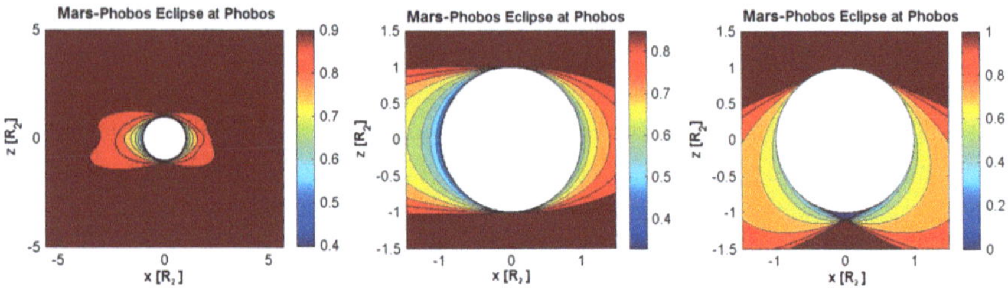

Fig. 2 Lighting conditions around Phobos. Light field around Phobos in the radial-vertical plane of the CR3BP frame, averaged over a year, a spring equinox month, a summer solstice month

Phobos, is minimum on the surface of motion where the conjunction line between the anti-Sun and Phobos revolves, shortly becoming 1 in points out of the surface. Such a minimum value rapidly increases with the distance from Phobos, starting from $\bar{L} = 50\,\%$ at the body's surface.

Since the radiation of Mars (without the albedo) is inside the infrared spectra, the analysis of the Mars-Phobos eclipse is neglected. This is because it would bring little variation to the lighting conditions and radiation hazards.

The conclusion of the analysis is obtained combining the previous single couples into the system of three massive bodies. The results of the Sun-Mars-Phobos mean 3D light field $\bar{L}_{12}$ are shown in Fig. 2 on the x-z plane. During summer and winter, complete light and one cone of complete shadow appear in the Phobos polar regions. During equinoctial or long observation periods, the lighting conditions around Phobos are close to experience continuous light, up to 88 % due to the unavoidable Martian eclipses.

2.6 Sky Occultation

In this section the possible exploitation of Phobos as a natural shield against the isotropic cosmic rays (SEPEs and GCRs) is considered. The idea is that the incoming radiation on a spacecraft is lowered proportionally to the fraction filled in the sky by the apparent size of the body's bulk, as seen by the spacecraft's location. Since the hidden body is the total background sky, in this paper this action is referred to as sky occultation. The occulting bodies are Mars and Phobos. The approach is similar to the one undertaken for the lighting conditions, defining an occultation function field O which represents the bulk/sky fraction of the occulting body. In the CR3BP frame Mars and Phobos are fixed, so their O does not depend on time. For a first analysis, the mean spherical shape for the two bodies is considered. The occultation of Mars, at the Phobos location, is $O_1 = 3.4\,\%$. The occultation of Phobos depends only from the radial distance from the body, starting from $O_2 = 50\,\%$ on the surface (astronauts staying inside of a deep crater would be shielded also laterally by the mountain ridge), and then decreases rapidly. The

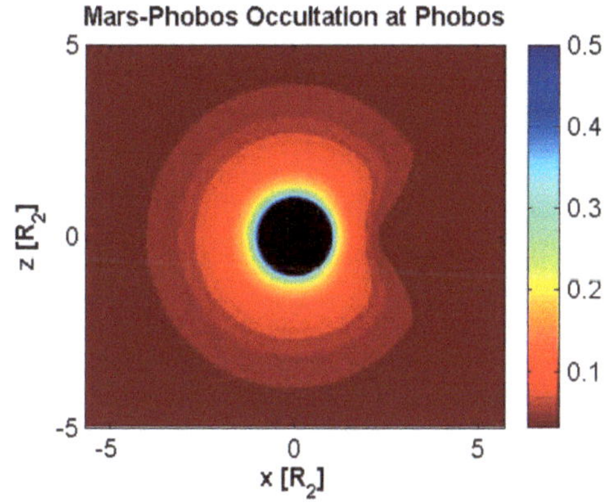

Fig. 3 Sky occultation around Phobos. Occultation field around Phobos in the radial-vertical plane of the CR3BP frame

conclusion of the analysis is obtained combining the previous single effects. This requires to determine if the apparent shapes of the two bodies' bulks intersect, and how much they overlap. This corresponds to the light function of the Mars-Phobos couple $L_{1,2}$, not computed in Sect. 2.5 but required here using the dual-cone eclipse model. The resulting two-body occultation is $O_{12} = O_2 + L_{1,2}O_1$, and it is shown in Fig. 3. This analysis highlighted that significant reduction of the isotropic SEPEs and GCRs by using the bulk of Phobos to occult part of the celestial sphere is obtained inside the SOI of the moon. Orbits that remain inside the Phobos' SOI are suitable to enhance the radiation protection of the spacecraft by exploiting Phobos as a passive shield.

3 Hovering Points

A simple trajectory for a mission around Phobos is provided by maintaining a fixed position with respect to its figurative frame. Due to the small μ, this is similar to a Martian Keplerian orbit close to Phobos, analogous to the trailing/leading configurations used in Formation Flying (Zamaro 2011). The analysis of these trajectories is undertaken adding a constant propulsive acceleration $\mathbf{a_P}$ to the equations of motion of the CR3BP. The aim is to counteract the natural acceleration of the CR3BP

$$\mathbf{a_P} = \mathbf{a_A} - \mathbf{a_G} \tag{2}$$

and leading to an Artificial Equilibrium Point (AEP). Since Phobos is tidally-locked, this represents a body-fixed hovering. Figure 4 presents the iso-surfaces of the thrusting acceleration level required to hover around Phobos. By increasing $\mathbf{a_P}$,

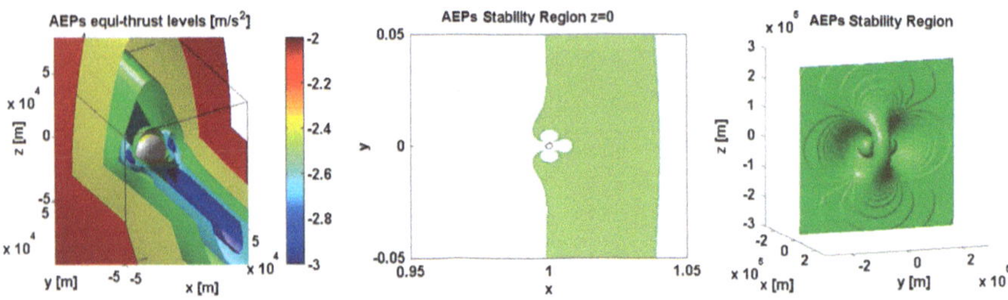

Fig. 4 AEPs of the Mars-Phobos system. On the *left*, iso-surfaces slices of propulsive acceleration magnitude (logarithmic scale) required for body-fixed hovering. On the *right*, planar stability region (in *green*) and inner boundary of the 3D stability region around Phobos

AEPs could be further displaced from the natural equilibrium points of the CR3BP nearby the secondary body, which are the two collinear libration points (LPs) L_1 and L_2, and the two equilateral LPs L_4 and L_5. The cis/trans couple of LPs L_{1-2} is located on the boundary of the Hill's SOI, at an altitude of 3.5 km from Phobos. Despite this proximity, the thrust level required to establish an AEP displacing a collinear LP is very demanding. Instead, the equilateral LPs L_{4-5} can be displaced close to Phobos along the y-axis still with levels of thrust affordable by SEP. Establishing AEPs over polar regions requires higher thrust levels.

The next step in the hovering analysis is to look for the stability of the AEPs. The linear Lyapunov marginal stability for the AEP requires all the eigenvalues of the linearized dynamics to be purely imaginary. For the Mars-Phobos system, the 3D stability region is made up of three realms: one central ring, and two symmetric half hyperbolic coronas placed over it. As presented in Fig. 4, the inner surface of the ring is distorted in proximity to the second massive body leaving outside the body's SOI. The planar stability region in the orbital plane is a thin corona extending along the Mars-Phobos orbital distance, that comprises the equilateral LPs and cuts off the three collinear LPs. In proximity of the secondary body, the inner stability region boundary is distorted to represent a three-leaf clover: this corresponds to the outcome in (Bombardelli 2012).

If we compare the stability region with the equi-thrust curves in the orbital plane, six attractive positions for medium distance observation of Phobos are identified. They are four minimum-distance AEPs at 25 km, and two minimum-control AEPs with 0.4mm/s^2, all obtained by the displacement of the equilateral LPs and affordable by SEP. Further trailing/leading orbits around Mars provide attractive cheap, stable, in-light fixed positions with respect to Phobos at long distances from the moon. All the AEPs available with a low-thrust level are not stable near the collinear LPs, and AEPs closer to Phobos, used for radiation shielding and to perform short dedicated operations, are feasible only with heavy or multiple thrusters.

4 Libration Point Orbits and Their Invariant Manifolds

In the framework of the classical CR3BP (Koon et al. 2011), around each of the collinear LPs L_1 and L_2 there exist a central manifold characterized by families of periodic orbits (POs) (the two branches of planar and vertical Lyapunov orbits, and the two branches of Northern and Southern Halo orbits), and quasi-periodic orbits (QPOs) around them (known as Lissajous orbits). These Libration point orbits (LPOs) are highly unstable, so their natural motion needs to be computed by the guidance system with high precision to provide trajectories tracked with low cost (Koon et al. 2011). Moreover, these LPOs are separatrices of motion between transit and non-transit orbits to enter or escape from the SOI of the second massive body. The boundary of these tubes is given by the Invariant Manifolds (IMs) of the LPOs that provide the energy-efficient trajectories to minimize the fuel consumption of spacecraft for interplanetary transfer phases.

From the preliminary analysis of the dynamics in proximity of Phobos in Sects. 2.2–2.3, it was found that to describe the natural relative motion inside this moon's SOI, its highly inhomogeneous gravity field and its orbital eccentricity must be taken into account. In Zamaro and Biggs (2014, 2015) the dynamical substitutes of the LPOs of the CR3BP were derived in a more realistic model that considers these two major orbital perturbations. The modeling of the complete gravity field is provided by the spherical harmonics series expansion of gravity harmonics (GHs) of Chao and Rubincam (1989). Since Phobos is tidally-locked, the addition of the GHs keeps the dynamics autonomous. The extended elliptic restricted three-body problem is coined ER3BP-GH, where the gravitational potential of Phobos $u_{G,2}$ is yielded by a truncated series expansion of GHs $(J, \lambda)_{m,n}$ through Legendre polynomials $P_{n,m}$ (R is Phobos mean-volume radius).

$$u_G = \frac{Gm}{R}\sum_{n=0}^{\infty}\left(\frac{R}{r}\right)^{n+1}\sum_{m=0}^{n} J_{n,m}\cos m\,(\psi - \lambda_{n,m})P_n^m\,(\cos\vartheta) \tag{3}$$

The potential in Eq.(3) is a function of the spherical coordinates r,ϑ,ψ of the Phobos figurative frame.

The procedure in Zamaro and Biggs (2014, 2015) to compute the LPOs in the ER3BP-GH makes use of the numerical continuation technique, starting from the LPOs in the CR3BP. The dynamical substitutes of the LPOs are families of POs and QPOs no longer symmetric and highly tilted and distorted from the classical case. They are presented in Fig. 5: around each cis/trans-side of Phobos, they are constituted by the oscillating LPs and a 1-parameter family D of iso-periodic POs, three 1-parameter families A, B, and CD (made by two branches C and D) of 2-tori QPOs, and two 2-parameter families AB and C of 3-tori QPOs. Their lighting and occulting conditions, and surface coverage are reported in Fig. 6. It is worth noting that the larger LPOs can provide passive radiation shielding over 20 %.

Since the orbits are close to Phobos, no homoclinic nor heteroclinic connections are available, but the IMs of these LPOs could be exploited as natural landing or

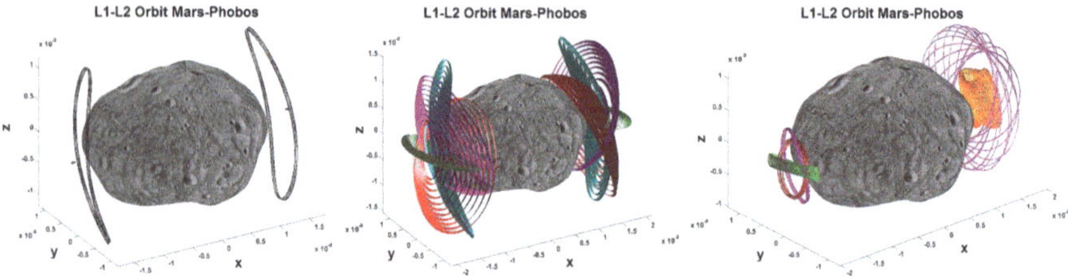

Fig. 5 LPOs in the Mars-Phobos ER3BP-GH. On the *left*, the two iso-periodic families of POs around the oscillating LPs. In the *center*, the families of 2-tori: A (*red*), B (*green*), C (*magenta*), D (*cyan*). On the *right*, example of 3-tori of different size and width around the LPs

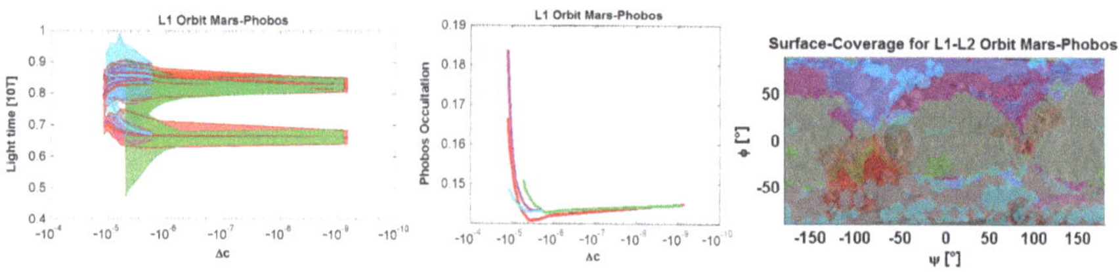

Fig. 6 LPOs performances. On the *left*, light function of the families of LPOs around L_1, averaged over ten periods, at the days of equinoxes (*lower cluster*) and solstices (*upper cluster*). Filled area spans values for different starting phases of the Sun. In the *center*, sky occultation function by the Phobos' bulk of the families of LPOs around L_1, averaged over one period (additional occultation by Mars' bulk will be 3.4 %). On the *right*, surface of full geometrical coverage filled by the largest orbits of the families of LPOs around L_1 and L_2. The LPOs are parameterized by the differential Jacobi integral Δc with respect to L_1, and families colors are coherent with Fig. 5

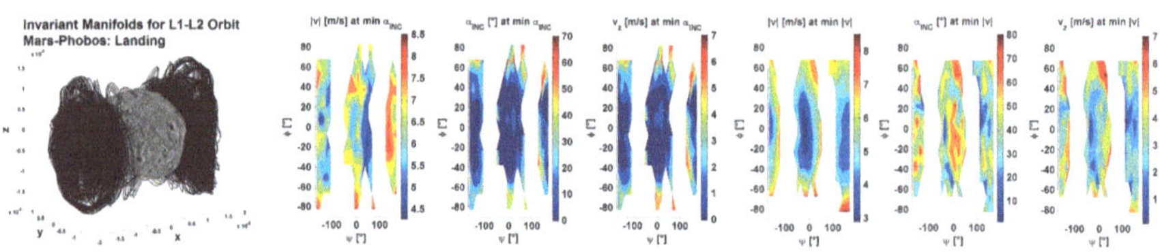

Fig. 7 IMs of the LPOs in the Mars-Phobos ER3BP-GH. On the *left*, inside branch of the tube of unstable IMs from the families of 3-tori LPOs. In the *center*, related performances of the trajectories that provide the minimum incidence at the touch-down as a function of the longitude and latitude of the landing site: landing velocity modulus, angle of incidence, downward vertical velocity. On the *right*, performances of the stable IMs of the same families that provide the minimum velocity magnitude at the launch

take-off gateways to and from the surface of Phobos. As presented in Fig. 7, the inside branch of the IMs provides high-efficient natural tangential landing paths and low escape velocity injections (30 % of the Keplerian Δv value) on the two collinear faces of the moon. These trajectories have the potential to be exploited for sample-and-return missions to this moon, where free-fall is required to avoid contamination of the sample's soil by the exhaust plume of the thrusters or rockets' nozzle.

4.1 Artificial LPOs and Their IMs

The natural LPOs computed in Sect. 4 are investigated in the framework of the addition of a constant acceleration representing SEP. The idea is the same used in Sect. 3 for hovering points, computing the artificial orbits around the displaced L_{1-2}. The numerical continuation is undertaken increasing the constant acceleration magnitude $\|\mathbf{a_P}\|$. The analysis is undertaken six times to consider thrust directions along all coordinated axes $\pm\hat{\mathbf{x}}$, $\pm\hat{\mathbf{y}}$, and $\pm\hat{\mathbf{z}}$.

Figure 8 shows some examples of the effects that the addition of a constant acceleration produces on the natural LPOs and IMs. In particular, thrusting towards Phobos moves the orbits further from the moon, so the effect of the GHs rapidly decreases and the LPOs tend to become similar to the families of the classical CR3BP. Figure 9 shows the effect on the longitudinal period of the LPOs: the addition of constant low thrust allows cheap artificial LPOs to be obtained with period equal to the 2:1 orbital overresonance of Phobos around Mars. This means that they remain periodic also in the ER3BP-GH. Figure 9 shows also the stability properties of the LPOs: despite the LPOs remain unstable, the Floquet instability index could be massively lowered with the thrust required to displace the LP far from

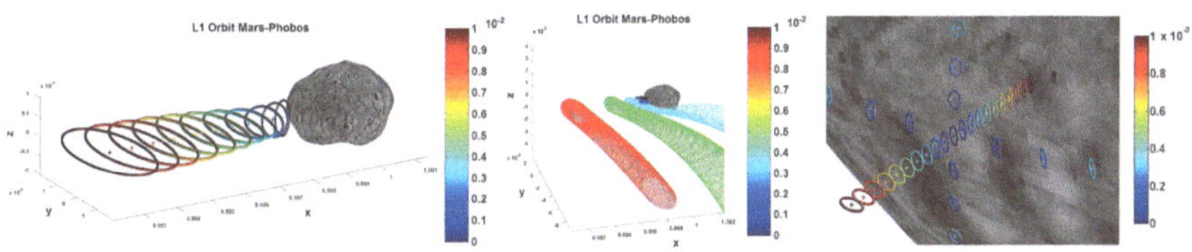

Fig. 8 Artificial LPOs and their IMs with constant acceleration around Phobos. First two graphs show the example of one natural LPO of the family A around L_1 and the trajectories of the inner branch of its unstable IM, both modified by different levels of constant acceleration magnitude (m/s^2) along the direction $+\hat{\mathbf{x}}$ (*crosses* represent the current LP). On the *right*, example of a small LPO of the family A around L_1 with different levels of constant acceleration magnitude (m/s^2), along all coordinate axes directions

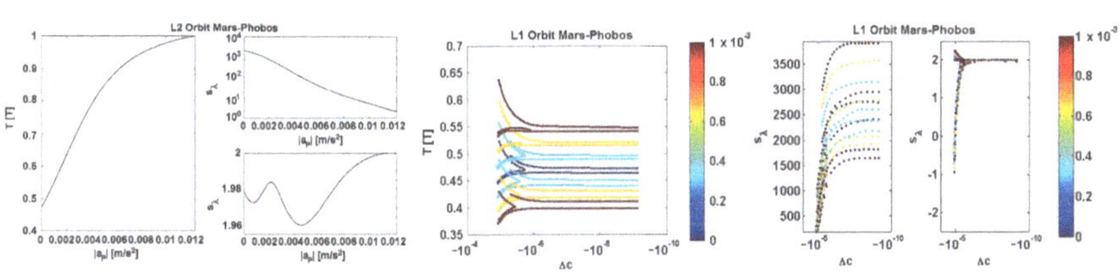

Fig. 9 Artificial LPOs around Phobos. On the *left*, example of one natural LPO of the family A around L_2 modified by different levels of constant acceleration magnitude along the direction $-\hat{\mathbf{x}}$: longitudinal period and stability properties (stability indexes of the two non-unit couples of eigenvalues of the monodromy matrix). Following graphs show the characteristic curves of the same properties for all the four families of LPOs around L_1 with different constant acceleration magnitude (m/s^2) along directions $\pm\hat{\mathbf{x}}$

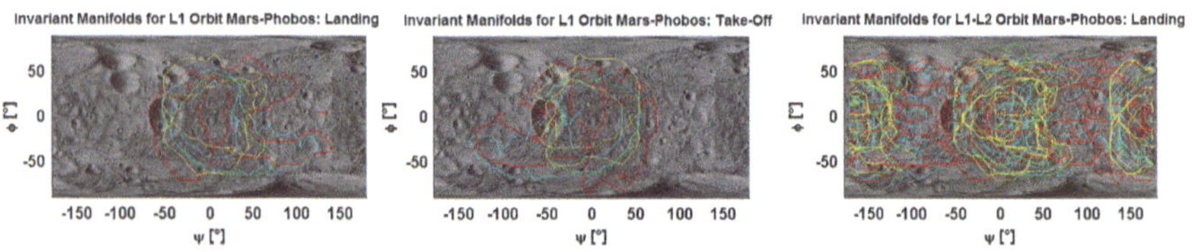

Fig. 10 IMs of the artificial LPOs around Phobos. First two figures show the possible landing/take-off sites through the IM of the family A of LPOs around L_1, third figure shows the region of landing sites for all the families of artificial LPOs around L_{1-2}. Constant acceleration magnitude of 1 mm/s^2 along all coordinate axes directions (*green line* for $\mathbf{a_P} = \mathbf{0}$, *cyan* for directions $\pm\hat{\mathbf{x}}$, *red* for directions $\pm\hat{\mathbf{y}}$, *yellow* for directions $\pm\hat{\mathbf{z}}$)

Phobos. This has a great impact on the frequency demand for the control subsystem, reducing the duty cycle up to the 25 % for artificial LPOs displaced at an altitude over 60 km from Phobos along the Mars-Phobos radial. Stable artificial LPOs along the y-axis are obtained displacing the equilateral LPs. Finally, displacing LPOs away from the natural SOI, in addition to reducing instability, it has other important advantages. It provides simplification of mission operations constraints, enhancement of light conditions, and in particular reduction of the computational load of tracking these orbits. The effect of Phobos' gravity field quickly lowers with the distance, so the convergence of the solution of the LPOs (and so its reliability) will be obtained with a far lower order of the truncated GHs model to be used by the guidance subsystem.

The IMs of the artificial LPOs have been computed in Fig. 10, and they have revealed that using constant thrust, along an appropriate direction, allows to enlarge the region of landing and take-off sites.

5 Quasi-Satellite Orbits

Other orbits that can be used in a mission to Phobos lie outside its SOI. They are a family of QPOs which are called by different names: Quasi-Satellite (QSOs), Quasi-Synchronous, Distant Satellite, Distant Retrograde orbits. In this section, the dynamics of the QSOs are studied as the relative motion between two close synchronous Keplerian orbits around the primary body of the 3BP, like in Formation Flying. The solution of the Keplerian synchronous Formation Flying in the Hill's rotating frame is a relative retrograde elliptical orbit, called epicycle (Zamaro 2011). This results in an artificial satellite of the secondary body, but due only to the attraction of the primary, and since Phobos is tidally-locked the spacecraft rotates in its figurative frame in 1:1 resonance. In particular, for slightly eccentric Keplerian orbits (as in the case of Phobos) the epicycle resulting from a difference in eccentricity Δe between the spacecraft and Phobos is an ellipse centered on Phobos. A difference in inclination Δi inclines the epicycle, and the relative motion is 3D.

The QSO is the dynamical substitute of the epicycle in the three-body dynamics, and is characterized by an oscillation along the y-axis of the 3BP frame, and a secular precession of the relative line-of-nodes for 3D orbits. The longitudinal period increases with the size of the epicycle.

The analysis of the QSOs around Phobos in this section is conducted with the dynamics of a Martian Formation Flying, using the Gauss' Planetary Equations (Battin 1999). The state variables are the orbital elements (OEs) of the spacecraft around Mars, and the ER3BP is retained using as perturbation the gravity of Phobos. In Cabral (2011), a linearized stability analysis of the QSOs is undertaken in the ER3BP around their original epicycle in the Keplerian dynamics, and the results are applied to the Mars-Phobos case. The minimum distance condition from Phobos is 29.4 km ($\Delta e = 0.00315$), above which the planar QSO is stable. The 3D stability condition that bounds the admissible difference in inclination to the difference in eccentricity below which the inclined QSO is stable is $\Delta i / \Delta e < 96\,\%$.

The stability of this class of orbits, combined with the reduced SOI and the synchronous rotation, makes the QSOs attractive solutions to orbit Phobos. In this section the nonlinear stability of the QSOs around Phobos is probed using the linear stability region of Cabral (2011) (a trapezoid in the plane Δe-Δi as shown in Fig. 11) as a first guess in order to limit the boundary of the state-space in terms of osculating OEs where conducting the nonlinear simulations. We further limit the region by setting a range of minimum altitudes of the epicycle from Phobos between 20 and 60 km. The QSOs are simulated up to 1 year: a QSO is considered stable when it does not drift away by the end of the simulation. The resulting nonlinear stability region boundary is presented in Fig. 11. The planar QSO at minimum stable Δe has now a minimum altitude of 25 km.

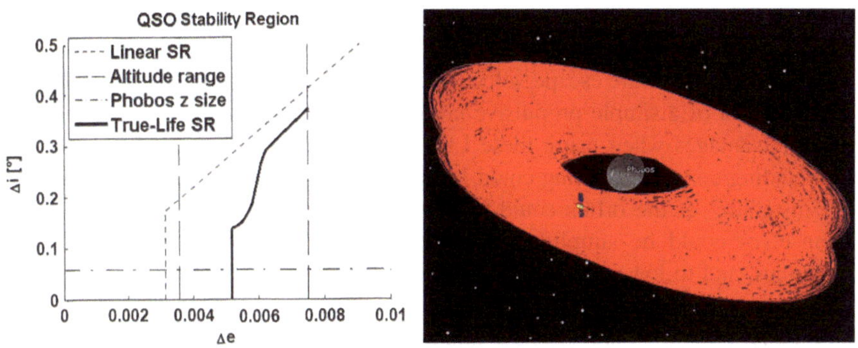

Fig. 11 QSOs around Phobos. On the *left*, stability region of QSOs around Phobos tested by high-fidelity 1 year simulations, defined by initial conditions on osculating OEs around Mars for the spacecraft and Phobos: positive differences in eccentricity and inclination, starting at perimars epoch. On the *right*, example of a 3D stable QSO in the Phobos Hill's frame for 1 year propagation

6 Conclusions

In this paper several kinds of orbits around the Martian moon Phobos are analyzed. Each orbit has a number of potential applications and their performance can be assessed against the requirements of each mission segment.

Trailing/Leading orbits around Mars, starting from 25 km distance from Phobos, are attractive configurations, because they are cheap and affordable by SEP even for heavy human modules, they are stable to perturbations, and they are mostly in full light. They are the best orbits to start to approach Phobos SOI, however their ground-track on the moon is stationary and limited. Other distant configurations or close-range AEPs requires either high thrust or high station-keeping cost for hovering over long-time: they can be used only for short and dedicated operations of small unmanned spacecraft.

We then analyzed the LPOs computed in an improved system of the relative dynamics in proximity of Phobos, upgrading the Mars-Phobos ER3BP with the real gravity field of the moon, modeled with a gravity harmonics series expansion. These orbits are very close to the moon surface, therefore they are similar to close-range points but with an extended ground-track and range of lighting conditions, and the Phobos' bulk occultation of the sky could provide relevant passive shielding from the cosmic rays radiation. Despite their instability, the LPOs are natural motion and so will require no propulsion and low station-keeping cost to provide observation on Phobos and communication bridges to manage robotic scouts on Mars and Phobos. However, they require the high accuracy of an optical navigation subsystem, and high-load on the guidance subsystem, whose reference signal must be computed with a high-fidelity gravity field of the moon. Another useful application is to exploit their IMs as landing/taking-off gateways to and from the moon. In this paper it is proved that there exist natural trajectories for a specific range of longitude-latitude sites able to land tangentially, facilitating a soft controlled touch-down, and depart with a very little escape velocity, less than 30 % of the Keplerian Δv value. The addition of a simple propulsive law, to obtain a constant acceleration, offers some advantages when using these LPOs for short-phases: the surface coverage and landing/take-off targeting could be extended to the whole surface of Phobos, the instability of the orbits could be lowered, and the computation of the orbits themselves could be simplified to maintain them periodic also in the elliptic dynamics, and to lower the accuracy of the required model of the gravity field of the moon.

Finally, the QSOs are the best solution for a precursor unmanned mission to Phobos. They are both natural orbits with no need of propulsion, and self-stable up to very long time with no need of station-keeping, and so they can be used as parking orbits with distance starting from 25 km from Phobos. Closer 3D QSOs provide a fast complete coverage to map the surface of Phobos and identify the landing site, and they are mostly in light. In particular, the long-time stability of the QSOs around Phobos could be exploited as an orbital repository to send, in advance, unmanned propulsion modules, fuel stockpiles, and provisions, to remain parked in a secure Martian orbit without orbital maintenance costs and with short-period phasing maneuvers to dock the modules. To allow the first human expeditions to visit Mars and return to the Earth, the spacecraft could make scheduled pit-stops at this orbital garage on Phobos' orbit.

References

Battin, R.H.: An Introduction to the Mathematics and Methods of Astrodynamics. American Institute of Aeronautics and Astronautics Inc., Renston (1999)

Bell, J.F., Fanale, F., Cruikshank, D.P.: Chemical and physical properties of the Martian satellites. In: Resources of Near-Earth Space, Space Science Series, pp. 887–901. The University of Arizona Press, Tucson (1993)

Bombardelli, C.: Stable artificial equilibrium points in the Mars-Phobos system. In: 1st IAA-AAS Conference on Dynamics and Control of Space Systems, Porto, 19–21 March 2012

Cabral, F.S.P.: On the stability of quasi-satellite orbits in the elliptic restricted three-body problem. Master's thesis, Universidade Técnica de Lisboa (2011)

Chao, B.F., Rubincam, D.P.: The gravitational field of Phobos. Geophys. Res. Lett. **16**(8), 859–862 (1989)

Duxbury, T.C., Callahan, J.D.: Pole and prime meridian expressions for Phobos and Deimos. Astron. J. **86**(11), 1722–1727 (1981)

Genta, G., Dupas, A., Salotti, J.M.: IAA study group on global human Mars system missions exploration. In: 65th International Astronautical Congress, Toronto, 29 September–3 October 2014

Gonçalves, P., Keating, A., Nieminen, P.: MarsREM: the Mars energetic radiation environment models. In: 31st International Cosmic Ray Conference, Łódź, 7–15 July 2009

Hopkins, J.B., Pratt, W.D.: Comparison of Deimos and Phobos as destinations for human exploration, and identification of preferred landing sites. In: AIAA Space 2011 Conference & Exposition, Long Beach, 27–29 September 2011

Koon, W.S., Lo, M.W., Marsden, J.E., Ross, S.D.: Dynamical Systems, the Three-Body Problem and Space Mission Design. Springer, New York City (2011)

Koschny, D.: Phootprint: an ESA mission study. In: Mars Concepts and Approaches Workshop, Houston, 12–14 June 2012

Lee, P.: Phobos and Deimos update. In: 7th Small Bodies Assessment Group Meeting, Pasadena, 10–11 July 2012

Pandika, M.: Stanford researchers develop acrobatic space rovers to explore moons and asteroids (2012). news.stanford.edu/news/2012/december/rover-mars-phobos-122812.html

Reitz, G.: Biological effects of space radiation. In: ESA Space Weather Workshop (1998)

Rosenblatt, P.: The origin of the Martian moons revisited. Astron. Astrophys. Rev. **19**, 44 (2011)

Zamaro, M.: Coupled orbital and attitude SDRE control system for spacecraft formation flying. Master's thesis, Politecnico di Milano (2011)

Zamaro, M., Biggs, J.D.: Dynamical systems techniques for designing libration point orbits in proximity of highly-inhomogeneous planetary satellites: application to the Mars-Phobos elliptic three-body problem with additional gravity harmonics. In: 10th International Conference on Mathematical Problems in Engineering, Aerospace and Sciences, International Congress on Nonlinear Problems in Aviation and Aeronautics, Narvik, 15–18 July 2014; AIP Conf. Proc. **1637**, 1228–1240 (2014)

Zamaro, M., Biggs, J.D.: Natural motion around the Martian moon Phobos: the dynamical substitutes of the libration point orbits in an elliptic three-body problem with gravity harmonics. Celest. Mech. Dyn. Astr. **122**(3), 263–302 (2015)

MIX
Papier aus verantwortungsvollen Quellen
Paper from responsible sources
FSC® C105338

If you have any concerns about our products,
you can contact us on
ProductSafety@springernature.com

In case Publisher is established outside the EU,
the EU authorized representative is:
Springer Nature Customer Service Center GmbH
Europaplatz 3, 69115 Heidelberg, Germany

Printed by Libri Plureos GmbH
in Hamburg, Germany